Lecture Notes in Networks and Systems

Volume 878

The series "Lecture Notes in Networks and Systems" publishes the latest developments in Networks and Systems—quickly, informally and with high quality. Original research reported in proceedings and post-proceedings represents the core of LNNS.

Volumes published in LNNS embrace all aspects and subfields of, as well as new challenges in, Networks and Systems.

The series contains proceedings and edited volumes in systems and networks, spanning the areas of Cyber-Physical Systems, Autonomous Systems, Sensor Networks, Control Systems, Energy Systems, Automotive Systems, Biological Systems, Vehicular Networking and Connected Vehicles, Aerospace Systems, Automation, Manufacturing, Smart Grids, Nonlinear Systems, Power Systems, Robotics, Social Systems, Economic Systems and other. Of particular value to both the contributors and the readership are the short publication timeframe and the world-wide distribution and exposure which enable both a wide and rapid dissemination of research output.

The series covers the theory, applications, and perspectives on the state of the art and future developments relevant to systems and networks, decision making, control, complex processes and related areas, as embedded in the fields of interdisciplinary and applied sciences, engineering, computer science, physics, economics, social, and life sciences, as well as the paradigms and methodologies behind them.

Indexed by SCOPUS, INSPEC, WTI Frankfurt eG, zbMATH, SCImago.

All books published in the series are submitted for consideration in Web of Science.

For proposals from Asia please contact Aninda Bose (aninda.bose@springer.com).

M. Shamim Kaiser · Juanying Xie ·
Vijay Singh Rathore
Editors

ICT: Smart Systems and Technologies

Proceedings of ICTCS 2023, Volume 4

Springer

Editors
M. Shamim Kaiser
Jahangirnagar University
Dhaka, Bangladesh

Vijay Singh Rathore
Department of Computer Science
and Engineering
Jaipur Engineering College and Research
Centre
Jaipur, Rajasthan, India

Juanying Xie
School of Computer Science
Shaanxi Normal University
Xi'an, China

ISSN 2367-3370 ISSN 2367-3389 (electronic)
Lecture Notes in Networks and Systems
ISBN 978-981-99-9488-5 ISBN 978-981-99-9489-2 (eBook)
https://doi.org/10.1007/978-981-99-9489-2

This Springer imprint is published by the registered company Springer Nature Singapore Pte Ltd.
The registered company address is: 152 Beach Road, #21-01/04 Gateway East, Singapore 189721, Singapore

Paper in this product is recyclable.

Preface

Eighth International Conference on Information and Communication Technology for Competitive Strategies (ICTCS-2023) targets state-of-the-art as well as emerging topics pertaining to information and communication technologies (ICTs) and effective strategies for its implementation for engineering and intelligent applications.

The conference is anticipated to attract a large number of high-quality submissions, stimulate the cutting-edge research discussions among many academic pioneering researchers, scientists, industrial engineers and students from all around the world and provide a forum to researcher; propose new technologies, share their experiences and discuss future solutions for design infrastructure for ICT; provide a common platform for academic pioneering researchers, scientists, engineers and students to share their views and achievements; enrich technocrats and academicians by presenting their innovative and constructive ideas; and focus on innovative issues at international level by bringing together the experts from different countries.

The conference was held on December 8 and 9, 2023, physically at Hotel—Four Points by Sheraton Jaipur, India, and Digitally on Zoom organized by Global Knowledge Research Foundation and Managed by G R Scholastic LLP.

Research submissions in various advanced technology areas were received, and after a rigorous peer review process with the help of program committee members and external reviewers, 200 papers were accepted with an acceptance rate of 17%. All 200 papers of the conference are accommodated in five volumes; also, papers in the book comprise authors from 22 countries.

This event success was possible only with the help and support of our team and organizations. With immense pleasure and honor, we would like to express our sincere thanks to the authors for their remarkable contributions, all the technical program committee members for their time and expertise in reviewing the papers within a very tight schedule, and the publisher Springer for their professional help.

We are overwhelmed by our distinguished scholars and appreciate them for accepting our invitation to join us through the virtual platform and deliver keynote speeches and technical session chairs for analyzing the research work presented by the researchers. Most importantly, we are also grateful to our local support team for their hard work for the conference. This series has already been made a continuous series which will be hosted at different locations every year.

Dhaka, Bangladesh — M. Shamim Kaiser
Xi'an, China — Juanying Xie
Jaipur, India — Vijay Singh Rathore

Contents

Editors and Contributors

About the Editors

Dr. M. Shamim Kaiser is currently working as Professor at the Institute of Information Technology of Jahangirnagar University, Savar, Dhaka-1342, \Bangladesh. He received his Bachelor's and Master's degrees in Applied Physics Electronics and Communication Engineering from the University of Dhaka, Bangladesh, in 2002 and 2004, respectively, and the Ph.D. degree in Telecommunication Engineering from the Asian Institute of Technology (AIT) Pathumthani, Thailand, in 2010. His current research interests include data analytics, machine learning, wireless network and signal processing, cognitive radio network, big data and cyber security, renewable energy. He has authored more than 100 papers in different peer-reviewed journals and conferences. He is Associate Editor of the *IEEE Access Journal*, Guest Editor of *Brain Informatics Journal*, and *Cognitive Computation Journal*. Dr. Kaiser is Life Member of Bangladesh Electronic Society; Bangladesh Physical Society. He is also a senior member of IEEE, USA and IEICE, Japan, and an active volunteer of the IEEE Bangladesh Section. He is the founding Chapter Chair of the IEEE Bangladesh Section Computer Society Chapter.

Dr. Juanying Xie is currently Professor at Shaanxi Normal University, Xi'an, China. I have been a full professor at the School of Computer Science of Shaanxi Normal University in PR China. My research interests include machine learning, data mining, and biomedical data analysis. I have published around 50 research papers and published two monograph books. I have been an associate editor of *Health Information Science and Systems*. I have been a program committee member of several international conferences, such as the International Conference on Health Information Science. I have been a senior member of China Computer Federation (CCF), a member of Chinese Association for Artificial Intelligence (CAAI), a member of Artificial Intelligence and Pattern Recognition Committee of CCF, and a member of Machine Learning Committee of CAAI, etc. I have been a peer reviewer for many journals, such as "Information Sciences", and "IEEE Transactions on Cybernetics."

I was awarded my Ph.D. in signal and information processing from Xidian University. I cooperated with Prof. Xiaohui Liu at Brunel University in the UK from 2010 to 2011 in machine learning and gene selection research. I received an engineering master degree in the application technology of computers at Xidian University and a bachelor degree of science in computer science at Shanxi Normal University.

Dr. Vijay Singh Rathore is presently working as Professor in the Department of Computer Science and Information Technology, JECRC University, Jaipur. He was previously with IIS (Deemed to be) University, Jaipur (India). He received Ph.D. from the University of Rajasthan and has teaching experience of 20 years. He is Secretary, ACM Jaipur Chapter, Past Chairman, CSI Jaipur Chapter, got two patents published, Ph.D. Supervised (Awarded: 16, Under Supervision: 07), 80+ research papers and 10+ books got published. He is handling international affairs at the IIS University, Jaipur. His research areas are internet security, cloud computing, big data, and IoT.

Contributors

Padma Adane Shri Ramdeobaba College of Engineering and Management, Nagpur, India

R. Adline Freeda KCG College of Technology, Karapakkam, Chennai, India

Parul Agarwal Jamia Hamdard, New Delhi, India

Jawed Ahmed Jamia Hamdard, New Delhi, India

Basam Akshitha B V Raju Institute of Technology, Hyderabad, India

Nisha Amin Department of Computer Science, Karnataka State Akkamahadevi Women's University, Vijayapura, India

A. Anju KCG College of Technology, Karapakkam, Chennai, India

P. J. Antony A. J. Institute of Engineering and Technology, Mangalore, VTU, Belagavi, India

Arifuzzaman Department of Visual Communication, School of Arts Humanities and Commerce, Amrita Vishwa Vidyapeetham, Mysuru, India

S. Arpitha Mangalore University, Mangalore, Karnataka, India

S. Arunkumar Department of ECE, Sri Ramakrishna Engineering College, Coimbatore, India

Vibhor Bansal Jindal Global Business School, O.P. Jindal Global University, Sonipat, Haryana, India

Hari Krishna Bhagavatham OUCCBM, Osmania University, Hyderabad, India

Sakshi P. Bhavsar MCA Department, Charotar University of Science and Technology, Anand, Gujarat, India

Harsh Namdev Bhor Department of IT, K J Somaiya Institute of Technology, Mumbai, Maharashtra, India

Ram Babu Buri Computer Science and Engineering, Arya College of Engineering & I.T., Jaipur, India

Venkamaraju Chakravaram Jindal Global Business School, O.P. Jindal Global University, Sonipat, Haryana, India

Vaishnav Chaudhari Dr. Vishwanath Karad MIT World Peace University, Pune, India

Harsh N. Chavda MCA Department, Charotar University of Science and Technology, Anand, Gujarat, India

R. K. Chethan Department of Commerce and Management, School of Arts Humanities and Commerce, Amrita Vishwa Vidyapeetham, Mysuru, India

Amit Chhabra Department of Computer Engineering and Technology, Guru Nanak Dev University, Amritsar, India

Sonali Chopra Jamia Hamdard, New Delhi, India

Tanupriya Choudhury Department of Computer Science and Engineering, Graphic Era Deemed to Be University, Dehradun, India;
CSE Department, Symbiosis Institute of Technology, Symbiosis International University, Pune, India

Souresh Cornet Amrita School for Sustainable Futures, Amrita Vishwa Vidyapeetham, Amritapuri, India

Rohit Dardige Department of Computer Engineering, Faculty Science and Technology, Vishwakarma University, Pune, India

K. Darshan Department of Commerce and Management, School of Arts Humanities and Commerce, Amrita Vishwa Vidyapeetham, Mysuru, India

Swapnil Deshmukh School of Engineering, Ajeenkya DY Patil University Pune, Pune, India

Avanti Dhiran Shri Ramdeobaba College of Engineering and Management, Nagpur, India

S. Duraisamy Chikkanna Government Arts College, Tirupur, India

Amit K. Gaikwad G H Raisoni University Amravati, Amravati, India

Jyotsna S. Gaikwad Deogiri College, Aurangabad, Maharashtra, India

Abhishekh Gangwar Department of Biometrics, Centre for Development of Advanced Computing, Pune, India

B. Geeta Department of Applied Electronics, Gulbarga University, Kalaburgi, India

M. Gokul Department of ECE, Sri Ramakrishna Engineering College, Coimbatore, India

Chandrashekhar Goswami MIT School of Computing, MIT ADT University, Pune, India

Gugan KCG College of Technology, Karapakkam, Chennai, India

T. Gunanandhini Amrita School of Business, Amrita Vishwa Vidyapeetham, Coimbatore, India

M. Harshavardhan Department of Computer Science and Engineering, Sri Krishna College of Technology, Coimbatore, Tamil Nadu, India

Yash Honrao Dr. Vishwanath Karad MIT World Peace University, Pune, India

S. Inbasudan Department of Computer Science and Engineering, Sri Krishna College of Technology, Coimbatore, Tamil Nadu, India

A. Jagan B V Raju Institute of Technology, Hyderabad, India

Aniketh V. Jambha Department of Computer Science and Engineering, Amrita School of Engineering, Amrita Vishwa Vidyapeetham, Bengaluru, India

S. Jayabal Department of Computer Science and Engineering, Sri Krishna College of Technology, Coimbatore, Tamil Nadu, India

Sajithkumar K. Jayaprakash Amrita School for Sustainable Futures, Amrita Vishwa Vidyapeetham, Amritapuri, India

H. B. Jethva Department of Computer Engineering, GEC Patan, Katpur, Gujarat, India

J. Jeyamani ECE, United Institute of Technology, Coimbatore, India

Amal Megha John Department of ECE, Sri Ramakrishna Engineering College, Coimbatore, India

Jasno Elizabeth John Christ (CHRIST (Deemed to be University), Delhi NCR, India

Shruti Kallurwar Shri Ramdeobaba College of Engineering and Management, Nagpur, India

A. Karnan Department of Computer Science and Engineering, Faculty of Engineering and Technology, Annamalai University, Annamalainagar, Tamil Nadu, India

Arwinder Kaur Department of Computer Engineering and Technology, Guru Nanak Dev University, Amritsar, India

Neelu Khare Vellore Institute of Technology Vellore, Vellore, India

N. Kireeti Sai Bharadwaj Department of Computer Science and Engineering, Amrita School of Computing, Amrita Vishwa Vidyapeetham, Amritapuri, India

Ketan Kotecha Symbiosis Centre for Applied Artificial Intelligence, Symbiosis Institute of Technology, Symbiosis International University, Pune, India

V. S. Krithikaa Venket KCG College of Technology, Karapakkam, Chennai, India

Jayasmita Kuanr KIIT (Deemed to be University), Bhubaneswar, India

Ankit Kumar Department of Computer Engineering and Applications, GLA University, Mathura, Uttar Pradesh, India

Gaurav Kumar Department of Computer Engineering and Applications, GLA University, Mathura, Uttar Pradesh, India

Sushmita Mahapatra Shri Ramdeobaba College of Engineering and Management, Nagpur, India

Nafees Mansoor Department of Computer Science and Engineering, University of Liberal Arts Bangladesh (ULAB), Dhaka, Bangladesh

Shamla Mantri Dr. Vishwanath Karad MIT World Peace University, Pune, India

Rajeev Mathur School of Engineering and Technology, Jaipur National University, Jaipur, India

Jayant Mehare G H Raisoni University Amravati, Amravati, India

Chandrabhan Mishra Computer Science and Engineering, Arya College of Engineering & I.T., Jaipur, India

Deepanjali Mishra KIIT (Deemed to be University), Bhubaneswar, India

Saumya Mishra Computer Science and Engineering, Arya College of Engineering & I.T., Jaipur, India

Tushar Mulwani Vellore Institute of Technology Vellore, Vellore, India

Deep Nandu Shah and Anchor Kutchhi Engineering College, Mumbai, India

Ch. Nikhilesh Krishna Department of Computer Science and Engineering, Amrita School of Computing, Amrita Vishwa Vidyapeetham, Amritapuri, India

Deepti Nikumbh Pimpri Chinchwad College of Engineering, Pune, India

Ahmed J. Obaid University of Kufa, Kufa, Iraq;
Al-Ayen University, Thi-Qar, Iraq

Rajat Pandey Chandubhai S. Patel Institute of Technology, Charotar University of Science & Technology (CHARUSAT), Changa, Gujarat, India

Killol Pandya Chandubhai S. Patel Institute of Technology, Charotar University of Science & Technology (CHARUSAT), Changa, Gujarat, India

Chetana Pareta School of Engineering and Technology, Jaipur National University, Jaipur, India

Asfaq Parkhetiya Shree L. R. Tiwari College of Engineering, Mumbai, India

Shaktikumar V. Patel Gujarat Technological University, Ahmedabad, Gujarat, India

Upesh Patel Chandubhai S. Patel Institute of Technology, Charotar University of Science & Technology (CHARUSAT), Changa, Gujarat, India

Vishal P. Patel Department of Computer Engineering, SPCE Visnagar, Visnagar, Gujarat, India

Vibhakar Pathak Computer Science and Engineering, Arya College of Engineering & I.T., Jaipur, India

Chirag Pathania Department of Electrical and Electronics Engineering, Amrita School of Engineering, Amrita Vishwa Vidyapeetham, Bengaluru, India

Sanjay Patidar Department of Software Engineering, Delhi Technological University, Delhi, India

Deepali Patil Dwarkadas J. Sanghvi College of Engineering, Mumbai, India

Jayamala Kumar Patil Bharati Vidyapeeth's College of Engineering, Kolhapur, Kolhapur, India

Shreeya Patil Dr. Vishwanath Karad MIT World Peace University, Pune, India

Vaibhav Patil Department of Computer Engineering, Faculty Science and Technology, Vishwakarma University, Pune, India

Vikas Dattatray Patil Bharati Vidyapeeth's College of Engineering, Kolhapur, Kolhapur, India

N. Pavitha Department of Computer Engineering, Faculty Science and Technology, Vishwakarma University, Pune, India

M. Pavithra Chikkanna Government Arts College, Tirupur, India

Ameya Pawar Department of Computer Engineering, Faculty Science and Technology, Vishwakarma University, Pune, India

Sanjay Shamrao Pawar Bharati Vidyapeeth's College of Engineering, Kolhapur, Kolhapur, India

M. Praksha Mangalore University, Mangalore, Karnataka, India

S. Pramila Christ (CHRIST (Deemed to be University), Delhi NCR, India

Aman Pushp Symbiosis Institute of Business Management, Pune Symbiosis International (Deemed University), Pune, India

D. S. Radhika Shetty Vivekananda College of Engineering and Technology, Puttur, VTU, Belagavi, India

V. Raghu Ramaiah University of Applied Sciences, Bangalore, Karnataka, India

Ragul KCG College of Technology, Karapakkam, Chennai, India

R. Ragupathy Department of Computer Science and Engineering, Faculty of Engineering and Technology, Annamalai University, Annamalainagar, Tamil Nadu, India

R. L. Raibagkar Department of Computer Science, Karnataka State Akkamahadevi Women's University, Vijayapura, India

Sujay Bharath Raj Department of Computer Science and Engineering, Amrita School of Computing, Amrita Vishwa Vidyapeetham, Amritapuri, India

G. G. Rajput Department of Applied Electronics, Gulbarga University, Kalaburgi, India

Rakesh KCG College of Technology, Karapakkam, Chennai, India

Vidya Sagar Rao OUCCBM, Osmania University, Hyderabad, India

Shailesh Rastogi Symbiosis Institute of Business Management, Pune Symbiosis International (Deemed University), Pune, India

Sunitha Ratnakaram Jindal Global Business School, O.P. Jindal Global University, Sonipat, Haryana, India

Avishek Rauniyar Department of Computer Science and Engineering, Amrita School of Computing, Amrita Vishwa Vidyapeetham, Amritapuri, India

V. Ravikumar Pandi Department of Electrical and Electronics Engineering, Amrita School of Engineering, Amrita Vishwa Vidyapeetham, Amritapuri, India

Subhashree Rout School of Humanities, KIIT Deemed to be University, Bhubaneswar, India;
School of Management, Centurion University of Technology and Management, Bhubaneswar, India

Anatte Rozario Department of Computer Science and Engineering, University of Liberal Arts Bangladesh (ULAB), Dhaka, Bangladesh;
Department of Computer Science and Engineering, Islamic University of Technology, Gazipur, Bangladesh

Sahal Bin Saad Department of Computer Science and Engineering, University of Liberal Arts Bangladesh (ULAB), Dhaka, Bangladesh

Sadi Mahmud Sagar Department of Computer Science and Engineering, University of Liberal Arts Bangladesh (ULAB), Dhaka, Bangladesh

Ankit Saha Vellore Institute of Technology Vellore, Vellore, India

Bhavna Saini Central University of Rajasthan, Jaipur, India

Swati Samantaray School of Humanities, KIIT Deemed to be University, Bhubaneswar, India

G. Sandhya Department of Computer Science and Engineering, Sri Krishna College of Technology, Coimbatore, Tamil Nadu, India

Sanjay Department of Computer Science, Amrita School of Computing, Amrita Vishwa Vidyapeetham, Mysuru, Karnataka, India

Soumya Sathyan Department of Electrical and Electronics Engineering, Amrita School of Engineering, Amrita Vishwa Vidyapeetham, Amritapuri, India

Sumit Sawant Department of IT, K J Somaiya Institute of Technology, Mumbai, Maharashtra, India

K. Sebasthirani EEE, Sri Ramakrishna Engineering College, Coimbatore, India

Bhavesh Shah Department of Computer Engineering, Faculty Science and Technology, Vishwakarma University, Pune, India

R. Shankar Chikkanna Government Arts College, Tirupur, India

A. K. Sharma Department of Creative Technologies and Product Design, National Taipei University of Business, Taipei City, Taiwan

Sangeeta Sharma Computer Science and Engineering, Arya College of Engineering & I.T., Jaipur, India

Ansar Sheikh St.Vincent Palloti College of Engineering & Technology Nagpur, Nagpur, India

Devesh Shetty Shree L. R. Tiwari College of Engineering, Mumbai, India

Vishal Shrivastava Computer Science and Engineering, Arya College of Engineering & I.T., Jaipur, India

Abhilash Shukla MCA Department, Charotar University of Science and Technology, Anand, Gujarat, India

Mutyala Sai Sri Siddhartha Department of Artificial Intelligence, Amrita School of Engineering, Amrita Vishwa Vidyapeetham, Bengaluru, India

Anureet Virk Sidhu Symbiosis Institute of Business Management, Pune Symbiosis International (Deemed University), Pune, India

Ashish Kumar Singh Department of Software Engineering, Delhi Technological University, Delhi, India

Kamred Udham Singh School of Computing, Graphic Era Hill University, Dehradun, India

Teekam Singh CSE Department, Symbiosis Institute of Technology, Symbiosis International University, Pune, India;
Department of Computer Science and Engineering, Graphic Era Deemed to Be University, Dehradun, India

S. Sivakumar Amrita School of Business, Amrita Vishwa Vidyapeetham, Coimbatore, India

Kamini Solanki MCA Department, Charotar University of Science and Technology, Anand, Gujarat, India

Ashutosh Somavanshi Department of IT, K J Somaiya Institute of Technology, Mumbai, Maharashtra, India

Manik Sadashiv Sonawane Bharati Vidyapeeth's College of Engineering, Kolhapur, Kolhapur, India

Pratham Soni Department of IT, K J Somaiya Institute of Technology, Mumbai, Maharashtra, India

Aswathy Sreenivasan Amrita School of Business, Amrita Vishwa Vidyapeetham, Coimbatore, India

Thompson Stephan Graphic Era Deemed to be University, Dehradun, Uttarakhand, India

Kavya Suresh Department of Electrical and Electronics Engineering, Amrita School of Engineering, Amrita Vishwa Vidyapeetham, Amritapuri, India

M. Suresh Amrita School of Business, Amrita Vishwa Vidyapeetham, Coimbatore, India

Anuradha Thakare Pimpri Chinchwad College of Engineering, Pune, India

Poonam Thanki Chandubhai S. Patel Institute of Technology, Charotar University of Science & Technology (CHARUSAT), Changa, Gujarat, India

Shivani Thapar Department of Computer Engineering and Technology, Guru Nanak Dev University, Amritsar, India

Jaimin N. Undavia MCA Department, Charotar University of Science and Technology, Anand, Gujarat, India

Trushit Upadhyaya Chandubhai S. Patel Institute of Technology, Charotar University of Science & Technology (CHARUSAT), Changa, Gujarat, India

Shraddha Utane H V P M's COET, Amravati, India

Vipina Valsan Department of Electrical and Electronics Engineering, Amrita School of Engineering, Amrita Vishwa Vidyapeetham, Amritapuri, India

Jinesh Varma Chandubhai S. Patel Institute of Technology, Charotar University of Science & Technology (CHARUSAT), Changa, Gujarat, India

Vivek Kumar Verma Manipal University Jaipur, Jaipur, India

Vijay Department of ECE, Sri Ramakrishna Engineering College, Coimbatore, India

K. M. Yogesh Ramaiah University of Applied Sciences, Bangalore, Karnataka, India

Fayeq Zaidi Shree L. R. Tiwari College of Engineering, Mumbai, India

Digitalization of the Apparel Industry—The Impact of COVID-19

Sunitha Ratnakaram, Vibhor Bansal, Venkamaraju Chakravaram, Hari Krishna Bhagavatham, and Vidya Sagar Rao

1 Introduction

It would be correct to say that we all are somehow responsible for the development of the apparel industry. The apparel industry is one of the dominant industries in our environment, given the investment we make in clothing; however, being a fragile industry as fashion keeps updating constantly, it is being affected by multiple factors. The apparel industry continually adapts its nature and keeps changing its course to suit the changed circumstances. The world witnessed the same during the outbreak of COVID-19. When the whole world came to a halt, the apparel industry was one of the few industries which became an early adapter of this outbreak. This paper provides a glimpse of how the apparel industry got affected by the pandemic and how it brought people's lives back to normalcy. Now, technological advancement has played a prominent role and helped different people in this industry sustain themselves during the most challenging times. Various companies implemented digitalization in their sectors which helped them in adapting their changing course and sustaining their businesses which became a crucial aspect of survival then. This paper also talks about various challenges people face during the implementation of digitalization in the apparel industry. People were not keen on new changes like changes in working structure, managing day-to-day tasks, working style, and others. Each day was a new challenge for them as it was not possible to predict the radical developments

S. Ratnakaram · V. Bansal · V. Chakravaram (✉)
Jindal Global Business School, O.P. Jindal Global University, Sonipat, Haryana, India
e-mail: vchakravaram@jgu.edu.in

S. Ratnakaram
e-mail: sratnakaram@jgu.edu.in

V. Bansal
e-mail: 17jgbs-vbansal@jgu.edu.in

H. K. Bhagavatham · V. S. Rao
OUCCBM, Osmania University, Hyderabad, India

M. S. Kaiser et al. (eds.), *ICT: Smart Systems and Technologies*, Lecture Notes in Networks and Systems 878, https://doi.org/10.1007/978-981-99-9489-2_1

occurring in the industry given the pandemic situation. Zara did one of the best implementations of digitalization in these testing times; a part of this paper discusses the same. Hence, it would be correct to state that digitalization helped develop the whole apparel industry to a new level by bringing more comfort to its end-users. In this paper, we try to understand the effects of COVID-19 on the textile industry through data collected from the interviews with the aggrieved parties; further, we try to present the effect of digitalization on the industry through the recent literature surveyed.

2 Literature Review

In today's era, digitalization plays a prominent role in our society which is all possible due to the advancements in technology. Due to people's wide access to the internet and online convenience buying, customers prefer buying things online instead of visiting stores. The same has directly impacted their purchasing behavior and pattern. This digitalization played a crucial role during the COVID-19 outbreak. We all are aware of the fact how COVID-19 impacted our lives. Millions of people were infected, and many lost their lives. The whole world economy was hit badly due to this outbreak.

We can see its impact on the industry's growth patterns. Due to the imposition of lockdown, various companies took advantage of the situation and started the application of digitalization in their organizations. Adaption of digitalization has helped various companies survive during these difficult times.

3 The Pandemic Era—Effect on the Apparel Industry

The apparel industry took a tremendous hit due to this outbreak. Since imposition of lockdown affected the industry's supply chain as this industry is majorly dependent on physical workers. Various analysts have stated that this outbreak will have a long-term impact on industry worldwide. The asymmetric distribution of profits and expenses among companies also shocked the foundation of various companies, which ultimately resulted in the shutting down of businesses. Its impact can also be noticeable in the global production system, i.e., "triple hit," demand disturbance, supply demolition, and reduction [5]. COVID-19 impacted global value chains; as seen in the apparel industry, primary and support activities in the value chain took a hit. The biggest reason behind the disturbance in the supply chain is due to illness of workers. It is primarily due to the sickness of these workers, which resulted in lower output from production houses. The vicious triad of fear of job loss, health loss, and low wages that we saw during this pandemic was unprecedented in recent times, i.e., post great depression times.

A significant reason behind the disturbance of the global supply chain was the lack of raw material supply. As most countries were suffering from the outbreak's impact, they could not fulfill their raw material supply orders, which created havoc in fulfilling the received orders in the apparel industry [4]. Although situations were getting in control with time, travel restrictions worsened the situation. Major apparel hubs like Europe and the USA followed strict travel guidelines to reduce the spread of the virus. We can see this effect in the supply of apparel industry orders [4].

4 Change in Working Structure

To cope with the prevailing situation, various companies have to work by applying several strategies that will support their survival during these challenging times. Given the pandemic's long-term impact on the whole economy, companies are trying to apply various working models to cope with this crisis. Few such strategies are: firstly, targeting different geographical markets for the production of the apparel that have ease of doing business restrictions; secondly, making changes in their distribution channels as per new norms set by the government; and finally, exploring various operation models. Companies are also looking to diversify into the manufacturing of face masks and PPE kits promoting self-care due to the advent of ill-effects caused by COVID-19. These products are in high demand. Various major giants are doing these products as it can be part of their social responsibility [14].

In the early stage of the COVID-19, the entire fashion industry reached a complete halt. The pandemic effects are readily noticeable within the companies working as disturbances in working patterns, seizure of stocks, blocked payments, canceled fashion shows, and others. Order fulfillment became a massive challenge as fashion companies faced restrictions shifting on deadlines as companies filed for bankruptcy [10]. Working conditions were being restricted, which directly affected labor in the industry.

Moreover, to discuss the financial crisis, it would not be wrong to state the profound impact of the pandemic on the same. Companies were forced to shut their operations permanently, but companies with deep pockets sustained it gracefully. The governments of various countries also provided financial support through various schemes to avoid the global crash in the industry. However, apparel companies are still looking for a few more boosters; nevertheless, they still retain their optimistic hopes of sustainability and taking advantage of other opportunities of the future [8].

Researchers called this pandemic the "Perfect Storm," as companies lost billions of dollars. Through the study's findings, Brydges et al. knew which industry segment is most affected concerning sustainability. It is primarily due to increased discount rates for clearing the blocked stock from inventory. Due to demand reduction, the supply chain has shifted to a low growth stage which will take time to recover from the phase [4].

The travel industry also plays a prominent role in affecting the apparel industry. The demand reduction can be seen due to the implementation of lockdown in various places. The travel industry came to a halt. Due to this, various fashion shows and mega fashion events were forced to be canceled as their investors from around the globe will not be able to attend these events. People around the globe used to travel to shop in high-end fashion streets. Hence, immobility resultant from travel restrictions profoundly impacted the apparel industry and became a challenge for them to fulfill their demand without being physically present [13].

5 Application of Digitalization in the Apparel Industry

Today, most customers prefer the online mode for purchasing in the current digitalized environment. In a way, this pandemic acted as a booster dose for applying digitalization to business models by shifting its dependency from an offline model to online outlets. As studied, digitalization provides manufacturers and marketers with the flexibility of time and provides safety from the COVID-19 outbreak without compromising on fulfilling their demands. Various scholars have also stated that areas with improved digitalized infrastructure have fulfilled their advantages during the pandemic compared with areas without improved digital infrastructure. Scholars have given the word "breach," highlighting the importance of digitalized infrastructure in the country. They also raised their query about considering access to the internet as a right instead of a privilege as a breach of digital infrastructure has realized its importance [6]. Digitalization comes with immense benefits and equivalent costs.

With the opportunities and threats it brings, digitalization creates positive and negative feedback loops. Hence, creating a balance between digitalization in an organization is essential and developing flexibility equally. Therefore, companies are working together to avoid these negative aspects from their employees' perspectives [2].

Studies prove that when companies started adopting the digitalization model, it helped them comply with customers' demands [1]. However, it has also raised various questions on work from home in the form of trust issues, loss of control over employees, employee surveillance, and more. One of the significant issues noticed was the inaccessibility of the digital crowd. However, people have been using digital platforms for the past few years but not enough to support the industry. It was a great challenge to shift to the digital platform in a short period as making a balance between maintaining health and work-life was quite challenging [6]. Another noticeable challenge was the lack of advanced skillset among employees. There are various companies where most employees are not skilled enough to adapt to new technological changes. These companies faced tough challenges while adapting to digitalization during the pandemic. Hence, companies should develop the skillsets of their employees [7].

Even government can take multiple steps to focus on the development of digital infrastructure as it provides ease of doing work for the apparel industry in doing their business on online platforms. For example, the government can provide subsidiaries in setting internet services to companies or helping internet companies expand their reach to businesses and support them in every way possible. To provide data security, the government can create strict guidelines and regulations, which will help provide confidence in storing confidential information in safe hands as it will help avoid a data breach and addresses privacy issues. In the post-pandemic conditions, government support can play an essential role in boosting small apparel businesses as incentives could be molded to help employees better. Even various analysis companies collect information from people, which helps them know their consumption patterns to plan their sales methodology accordingly [2]. Moreover, digitalization does support them to an extent about the customer base, but it took time for them to adopt the same. Increased customer support could be seen during the end of the year 2020 as people started adapting to the internet in fulfilling their requirements.

6 Application of Digitalization by Zara During Pandemic

Zara is one of the most successful brands in the apparel industry, with multiple stores across the nations globally. These stores' net sales were estimated to be around $450 million. As all the industries were being affected, Zara's sales got dipped. A majority of sales of Zara were from physical stores; the temporary shutdown of these stores led to the piling up of inventory in the stores [3]. The company listed a 43% decline in its sales in its initial quarter of the financial year. At this time, Zara's digital infrastructure helped support the company's sales on the online platform. Zara's website and application for smartphones helped gain back the company's sales during the pandemic. People prefer spending online to avoid crowd exposure in fear of being infected with the virus [11]. Zara was able to clear up its inventory in a short period by providing huge discounts on its website and application compared with its physical stores. Zara could recover its sales by 90% compared with the pre-pandemic times by the end of October. This result was only possible due to its focus on the digitalization organization [9].

Zara also invested in upgrading its mobile application to ease its customers' purchasing. Various companies also use AI to understand their customer's requirements to provide the most suitable products. Various apparel industries survey their customers to know about their preferences and suggest articles accordingly. It is done by analyzing customers' answers through an algorithm. Companies are also using AI chat boxes that act as artificial assistants for the customers to help them with their queries. The most suitable example for this would be Zara. Zara provides this feature which helps its customers clarify doubts related to product care, availability, and substitute product. Furthermore, these interventions through technology helped Zara regain its hold [12].

7 Methodology

In order to study the effect of the pandemic on the fashion industry, we used a qualitative research method. For this research, primary research was most suitable as it helps provide in-depth insights related to the problems in the industry. We conducted in-person, in-depth interviews for the same. We surveyed a sample of 50 respondents for the study. These interviews helped us understand respondents' perspectives on the issue. All of these people were involved in the apparel industry and have been part of the COVID-19 outbreak. The selected respondents belong to the same industry, but they play different roles. The majority of the respondents belong to metropolitan cities. Moreover, the disruption affected these respondents more than others due to high COVID-19 cases areas.

The chosen methodology was appropriate as the research was trying to understand a multitude of problems the apparel industry faces due to the pandemic. It required in-depth exploration from the sampled people, which ultimately helped us glimpse their situation during the pandemic. We successfully analyzed the results thematically from the sample taken, which ultimately helped us conclude our findings.

8 Results

We successfully took the survey and analyzed the results. After a few sample responses, we found that responses were repeating similarly. Therefore, we concluded the survey and tried to recognize patterns from the collected data points. Based on the content, all the responses fell into two major categories: one is financial challenges and concerns, and the second is after-effects of a pandemic. These responses evolved from the challenges people faced during the pandemic.

The majority of the people faced challenges in financial terms due to the pandemic; there was a downfall in people's spending. Generally, people were spending less on apparel compared with earlier times. This reduction in spending led to financial challenges as the income source was receding. People's mind-set was more toward purchasing necessary goods; they said,

> There is no harm in stocking necessities rather than regretting after as our inventory will get empty at the end; we can postpone buying apparel for now.

This statement holds its value as people were spending above their limits to stock their necessities which led to a shortage of goods and avoiding spending on lifestyle products.

Respondents' responses also reflected their concern with labor issues like.

> due to lack of labour, our supply has taken a big hit, although demand is good; as it is to do with health, labour is not ready to come back to work like earlier days.

It was correct to state this fact as some of the apparel products like loungewear were top sellers as the market was hit during the lockdown. There was a massive demand for loungewear on websites. However, distribution channels and production houses took a colossal hit due to a lack of labor, negatively impacting the market supply. These issues ultimately led to the loss of customers, which negatively impacted the company's image.

At the same time, other people faced the scarcity of raw materials supply.

> Due to lack of raw material, we lost major sales target as we could not fulfill as per customer needs. Most raw material suppliers were scared to put money at stake to fulfill any futuristic targets; they were sceptical of any positive happenings shortly.

Respondents say that due to raw material supply scarcity, people faced a supply crisis that led to a loss of customer base, ultimately resulting in incurring losses.

> apparel market faced a huge downfall. All of the inventory of formal clothes, smart wear, and others became dead stocks for months. We even tried for debit help from corporate companies to exchange our work clothes inventory with casual clothing, but what we received was a big No. All we did was sell work clothes at par or a below our purchase price to collect money to buy casual clothing, which was as per market demand.

The above statement given by a renowned multi-brand outlet sales manager was a complete shock. This statement provides a glimpse of the challenges people face in the apparel industry.

When discussing the possible ways to resolve these different problems mentioned, few responses were directed toward taking advantage of credit facilities from the financial institutions. In contrast, others focused on utilizing digitalized solutions for their problems. One of the respondents was taking short-term loans for the businesses and adapting to new trends in the industry. For instance, with a manifold increase in the work-from-home concept, customer demand rose for loungewear clothes. Realizing the trend, manufacturers focused on supplying and delivering these goods through couriers and home delivery. It is the ultimate solution for offline shopping and the most convenient way of providing goods during the pandemic. This sudden shift in consumption patterns led to a rise in the financial needs of people in business. People were leaning on financial institutions to fulfilling their financial needs. Hence, respondents stated,

> it is more feasible to take the loan now than bluffing on market stability as one can lose everything due to high-risk factors. To the possible extent, we are trying to get support from banks in the worst case, its money lenders.

Due to this reason, many respondents prefer taking financial help either from the organized sector or from the unorganized sector. However, the financial world was diverting its focus to the upcoming viable opportunities given changed dynamics across industries. For example, the adoption of digital platforms for ease of their business leads to great advantage for the people.

> Digitalization act as a life-saver during this difficult time which provides us the confidence to sustain our work at this difficult time. Digitalization of my business opened a new path for diversifying my business at a whole new level.

As people get exposed to another market segment, it helps them boost their product demand as per market needs, especially in the trading sector of the apparel industry, which also helps in providing exposure to different types of customer markets.

> Whether it is loungewear or sportswear, we are always one step ahead to supply in the market. Due to low inventory turnover cost, we can take the necessary risk; with financial aid & digitalization, we can easily diversify our business smoothly.

This adoption of digitalization at the right time ultimately leads to positive effects among people in the apparel industry. Apart from that, few people faced various challenges while implementing digitalization as due to lack of knowledge and implementation of the right strategy can bring adverse outcomes. For example, this respondent did not get a great outcome while selling due to poor implementation of the digitalized strategy.

> To sell quickly and empty inventory, we listed a huge discount on our product online, which backfired our sales. We were not aware that people do judge products based on pricing. After putting excess discount, it reflected negatively on our product quality leading to a reduction in sales. We are happy we learned this lesson sooner.

Still, various other people found successful ways to boost their businesses by implementing small initiatives. For example,

> Our company successfully converted our cloth wastage into face masks sold at a premium margin & sold online. Who knew that with digitalization, one could find a way to boost sales in a short period. We implemented the strategy of supplying loungewear with matching face masks and got an amazing response.

Though digitalization has its disadvantages, it comes with its merits also. Hence, it will not be wrong to state that every person who implemented digitalization was not benefited, especially during the pandemic.

9 Discussion and Conclusion

It would be correct to state that the COVID-19 pandemic affected people's lives both in a good and bad manner. It depended upon the people who utilized the opportunity and made the best out of it. We found that people adopted the change in the industry as an indicator for future opportunities and took full advantage of it with the implementation of digitalization. It could be seen by seeing the change in its operational structure. However, some people did fall apart as they could not foresee the market trends' signals, which led to the downfall of their businesses. After analyzing both the aspects of positive and negative, it would be correct to conclude that COVID-19 outbreak effects did outrage the whole industry. However, only a handful of people could make the best of the situation, which acted as life-changing decisions by bringing revolutionized change in their working patterns in the industry, providing a competitive edge for future endeavors.

References

1. Almeida F, Santos JD, Monteiro JA (2020) The challenges and opportunities in the digitalization of companies in a post-COVID-19 World. IEEE Eng Manage Rev 48(3):97–103
2. Amankwah-Amoah J, Khan Z, Wood G, Knight G (2021) COVID-19 and digitalization: the great acceleration. J Bus Res 136:602–611
3. Bloomberg (2021) Zara Owner's Lean Business Model Helps it cope with pandemic (2020). https://www.bloomberg.com/news/articles/2020-09-16/inditex-earnings-beat-analyst-estimates-as-zara-stores-reopen. Accessed 9 Dec 2021
4. Brydges T, Retamal M, Hanlon M (2020) Will COVID-19 support the transition to a more sustainable fashion industry? Sustain Sci Pract Policy 16(1):298–308
5. Castañeda-Navarrete J, Hauge J, López-Gómez C (2021) COVID-19's impacts on global value chains, as seen in the apparel industry. Dev Policy Rev 39(6):953–970
6. Faraj S, Renno W, Bhardwaj A (2021) Unto the breach: what the COVID-19 pandemic exposes about digitalization. Inf Organ 31(1):100337
7. Gangoda A, Krasley S, Cobb K (2023) AI digitalisation and automation of the apparel industry and human workforce skills. Int J Fashion Des Technol Educ 1–11
8. Gjoni A (2021) The impact of Covid-19 on the creations of fashion designers. Arts Des Stud
9. Harbott A (2021) How Zara adapted fast fashion to Covid. Manage Today. https://www.managementtoday.co.uk/zara-adapted-fast-fashion-covid/innovation/article/1709103. Accessed 10 April 2021
10. Kaplanidou A (2018) Digitalization in the apparel manufacturing process (Master's thesis)
11. Shabir S, AlBishri NA (2021) Sustainable retailing performance of Zara during COVID-19 pandemic. Open J Bus Manage 9(03):1013
12. Silvestri B (2020) The future of fashion: how the quest for digitization and the use of artificial intelligence and extended reality will reshape the fashion industry after COVID-19. ZoneModa J 10(2):61–73
13. Uğur NG, Akbıyık A (2020) Impacts of COVID-19 on global tourism industry: a cross-regional comparison. Tourism Manage Perspect 36:100744
14. Zhao L, Kim K (2021) Responding to the COVID-19 pandemic: practices and strategies of the global clothing and textile value chain. Cloth Text Res J 39(2):157–172

Efficient Fire Detection and Automation Using Haar Cascade

G. Sandhya, M. Harshavardhan, S. Inbasudan, and S. Jayabal

1 Introduction

The detection of fire is more important task in order to safeguard people. This also helps to prevent the damages that are occurred. For this type of issue, use sensor for technical solutions. This works only when the condition is satisfied during the fatal flaw. In the very serious case, the sensors could also get damaged or it could not be able to function properly when the cause of fire is high. This type of sensor needs to be ionized by smoke and fire. Moreover the user receives the alert by using digital camera and video processing technique. In the case it may use large and open spaces for video-based system. In the current scenario the fire monitoring system is installed in many indoors [6] and outdoors. This vision-based detecting provides more accuracy and responds quickly; it could also monitor large area.

2 Literature Review

Wang et al. had did their research on fire detecting that mainly conquered with multiexpert system [1]. This is used on only indoor environment. Here the fire is detected using video fire detection technique which provided the best results. As the initial step they used RGB color to segregate the region instead of traditional method. Later than they used centroid motion to flicker the characteristic of the flame [11, 13]. Then they identified the flames. Finally, they used large dataset to acquire the best detection in indoor environment.

G. Sandhya (✉) · M. Harshavardhan · S. Inbasudan · S. Jayabal
Department of Computer Science and Engineering, Sri Krishna College of Technology, Coimbatore, Tamil Nadu, India
e-mail: sandhya.g@skct.edu.in

M. S. Kaiser et al. (eds.), *ICT: Smart Systems and Technologies*, Lecture Notes in Networks and Systems 878, https://doi.org/10.1007/978-981-99-9489-2_2

Kong et al. done their research on detecting the fire on real time by using multifeatured fusion [2]. In this scenario, there are many threats regarding the lives of people and the property that gets destroyed by the fire. In order to address the problem, the alarm system has been proposed to detect the fire for serious cases. There are many features that are innovative for fire detection. This is done through the image detection. Therefore, this technique helps in time saving and decreases the computation time.

Yuan et al. did their research that was based on image detection where this technique uses the novel algorithm [3, 9]. This calculates the fire [8, 10] by moving the image as region by region to obtain the appropriate detection. This also segregates the features that are shape and area to detect the accuracy of fire region. Later this provided the best results by using the support vector machine [SVM] in order for training and moved to detect the fire images and received the accuracy of fire. This method is applied for the real-time monitoring [12] and also provides the alarm after detecting the fire.

Luo et al. done the research on sensing the fire area over few years. In case of alert the sensor provides the alarm [4]. They also used hardware and algorithm to detect the fire. This technique provides an alert to the fire department regarding the fire. This mainly detects fire, reduces the false positives, and provides the alert and control. This mainly uses convolutional neural network [CNN] network and detects the best accuracy with less time. These are done by using IoT to achieve the best results.

Jadon et al. had done the research on real-time applications such as by using IoT which helps to reduce the disasters and loss of lives and property of people [5]. They had done several attempts to detect the best accuracy for fire alert. Here they provided the best accuracy with less calculation time. They also found the difference between the performance and size of model that are installed on the devices. The detection of fire is done by using Raspberry Pi which conquers with the fire dataset and provides the best alert.

3 Existing Work

In the existing work, their research was based on detecting the fire by using deep learning. They observed them by merging both spatial and the temporal features in order to acquire the image most accurately in order to detect fire. In order to exploit the image properties one may use the lightweight CNN that is to get awareness about the fire detection system. The fire module is combined by two layers that is known to be squeeze and expand. The first and foremost layer is squeeze layer which is used by only one convolution filters whereas the second expand layer is done by one and three convolution filters. Therefore, the one-layer filter tends to play a vital role while comparing with the three-layer filters.

3.1 Drawbacks

- Different positions are used for classifying images.
- Examples are used by adversarial.
- Frames tend to be coordinate.
- Less performance.

3.1.1 Classification of Images with Different Positions

The major challenge is to know the computer vision to deal the variance that is present in the data. The data need to appear as per the real-world information.

Some of the images that human vision could have been:

- Difference in angles
- Difference in backgrounds
- Difference in lighting conditions.

3.1.2 Coordinate Frame

The network that is used is convolution network in order to recognize the images. These are also done by clustering the pixels that are available as patterns. They could not be able to understand the images that are present in the components. The images that are available in this could be visual only by CNN, and this could not have any internal representations. They tend to hold only the part of whole relationships. This type could not be able to form a basic component to human vision due to coordinate frames. This could also know to be as mental model because it keeps track only to the orientation and features of the object.

4 Proposed Work

Figure 1 shows the block diagram of the proposed system.

4.1 Haar Cascade Classifier

The machine learning object helps to detect the objects that are inside the image and video by using Haar cascade classifier. The feature used in this is from the trained set of positive and negative images. This could also detect other objects. All the features that appear in Haar feature tend to be irrelevant during the detection of object. The trained set helps to detect the image and video by using face recognition. This provides better fire detection than random prediction of images and video (Fig. 2).

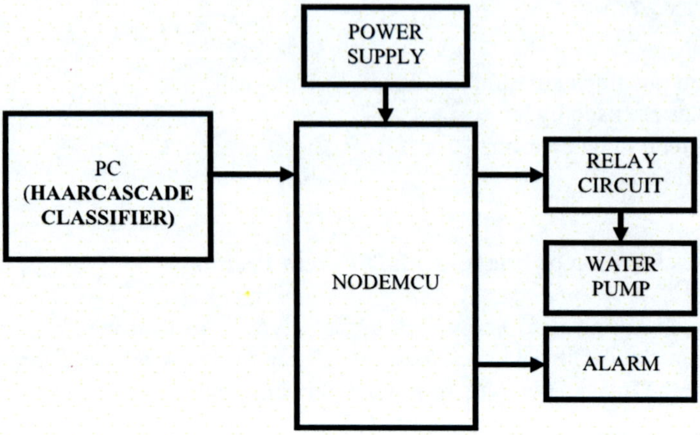

Fig. 1 Block diagram

Fig. 2 Haar classifier

4.2 Calculating Haar Features

The foremost step is to gather the features of Haar. In the detection window, the calculation of Haar feature is performed that appears on adjacent rectangular region. Here we need to calculate the sum of all the pixels that appear in each region and the difference between their sums.

4.3 Creating Integral Images

The calculation is need to be done fast; hence the integral images can be used. Here, we need to create some rectangles and also few sub-rectangles instead of computing each and every pixel. This helps to calculate the Haar feature faster. All the features that appear in Haar feature tend to be irrelevant during the detection of object. Hence many Haar features are to be used and known to be the AdaBoost.

5 Implementation Results

5.1 *Dataset Preparation and Training Set Creation*

In object recognition, the appropriate datasets are needed in every stage of calculation. This type of calculation begins from the initial phase of validation and provides the best result by using recognition algorithms. As the initial step, the images from the internet could be categorized according to their sets.

5.2 *Cascade Training*

The positive and negative images are used that are trained by using machine learning. They are extracted and combined using the detection object for fire. The pixel that can be used is in the scale of 232 × 232 pixel. Every block needs to be matched and combined with every feature. In the case of non-matching blocks, the process tends to be stopped. The rest of the process will be terminated when there is no fire in that block. Later, the next block will be tested and start the process once again. By using cascade classifier, the blocks are tested along with their pixels.

5.3 *Testing*

As, the final stage the fire is detected during the processing of input images by using trained classifier model. It shares the information through the NodeMCU to the water pump and alerts the system. This type of sensor needs to be ionized by smoke and fire. Moreover, the user receives the alert by using digital camera and video processing technique. In the case, large and open spaces are used which are based on video-based system. In the current scenario, the fire monitoring system is installed in many indoors and outdoors. In the current scenario breaking out fire is a common thing that occurs everywhere, and this may also cause many damages for both nature and humans. The major role of proposed work is to detect fire that provides more information which helps to compare using sensor. The detection process is done by using the method of image processing. The prediction using image detection provides more accuracy rather than any other detection process (Figs. 3 and 4).

Training data—Haar cascade needs less data to train than CNN, which requires thousands of images per class to achieve respectable accuracy. Inference periods—The CNN has an advantage in processing, training, and inference times. Cascades take longer to learn and provide an inference because they are larger models. SIFT—Scale-invariant feature transform method—is useful for cascade classifiers but not for CNN. It enables the classifier to perform well in a variety of situations where the object may be present. Accuracy—The CNN and Haar cascade algorithms both have

Fig. 3 Fire detection and automation

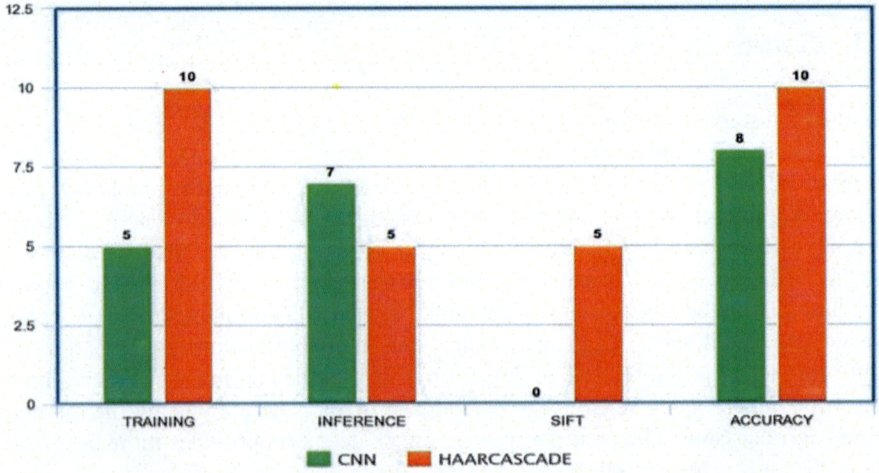

Fig. 4 Comparison of CNN and Haar cascade

reasonable accuracies. However, thanks to groundbreaking ongoing study in deep learning, the accuracy of Haar cascade models is now approaching 100%.

6 Conclusion and Future Work

The most dangerous event that occurs in our day-to-day scenario is fire, disaster. The fire needs to be controlled at the early stage before it results in the huge disasters. It may affect both the human and nature that provides huge losses. Hence, Haar cascade approach is used in this work. Each and every approach is under our camera surveillance [7] which provides more accuracy on fire detection almost around 90%. This type of sensor needs to be ionized by smoke and fire. Moreover, the user receives the alert by using digital camera and video processing technique. In this case, large and open spaces are used which are based on video-based system. In the current scenario, the fire monitoring system is installed in many indoors and outdoors. In case of fire, the system provides an alarm to avoid huge loses which helps in detecting the fire which is under observation.

References

1. Wang T, Bu L, Yang Z, Yuan P, Ouyang J (2019) A new fire detection method using a multi-expert system based on color dispersion, similarity and centroid motion in indoor environment. IEEE/CAA J Autom Sinica 7(1):263–275
2. Liu S, Kong L, Wang H (2018) Face detection and encryption for privacy preserving in surveillance video. In: Pattern recognition and computer vision: first Chinese conference, PRCV 2018, Guangzhou, China, proceedings, part III 1. Springer International Publishing, pp 162–172
3. Yuan C, Liu Z, Zhang Y (2017) Fire detection using infrared images For UAV-based forest fire surveillance. In: 2017 international conference on unmanned aircraft systems (ICUAS). IEEE, pp 567–572
4. Luo Z, Hsieh JT, Balachandar N, Yeung S, Pusiol G, Luxenberg J et al (2018) Computer vision-based descriptive analytics of seniors' daily activities for long-term health monitoring. Mach Learn Healthcare 2(1):1
5. Jadon A, Omama M, Varshney A, Ansari MS, Sharma R (2019) FireNet: a specialized lightweight fire & smoke detection model for realtime IoT applications. arXiv preprint arXiv: 1905.11922
6. Jain A, Srivastava A (2021) Privacy-preserving efficient fire detection system for indoor surveillance. IEEE Trans Indus Inf 18(5):3043–3054
7. Muhammad K, Khan S, Elhoseny M, Ahmed SH, Baik SW (2019) Efficient fire detection for uncertain surveillance environment. IEEE Trans Indus Inf 15(5):3113–3122
8. Kabra P, Singh BK (2019) Fire detection using infrared images for UAV-based forest fire surveillance
9. Valero MM, Rios O, Pastor E, Planas E (2018) Automated location of active fire perimeters in aerial infrared imaging using unsupervised edge detectors. Int J Wildl Fire 27(4):241–256
10. Ghassempour N, Zou JJ, He Y (2018) A SIFT-based forest fire detection framework using static images. In: 2018 12th international conference on signal processing and communication systems (ICSPCS). IEEE, pp 1–7
11. Kong SG, Jin D, Li S, Kim H (2016) Fast fire flame detection in surveillance video using logistic regression and temporal smoothing. Fire Saf J 79:37–43

12. Foggia P, Saggese A, Vento M (2015) Real-time fire detection for video-surveillance applications using a combination of experts based on color, shape, and motion. IEEE Trans Circ Syst Video Technol 25(9):1545–1556
13. Mueller M, Karasev P, Kolesov I, Tannenbaum A (2013) Optical flow estimation for flame detection in videos. IEEE Trans Image Process 22(7):2786–2797

Evolutionary Patterns in Modern-Era Cloud-Based Healthcare Technologies

Vishal Shrivastava, Vibhakar Pathak, Saumya Mishra, Ram Babu Buri, Sangeeta Sharma, and Chandrabhan Mishra

1 Introduction

Cloud computing stands as a significant milestone in the ongoing digital transformation of the healthcare industry, notwithstanding any apprehensions. Hospitals and various non-IT-centric businesses are increasingly capitalizing on this technology. In healthcare, cloud computing plays a vital role in managing digitized health-concern components which are responsible for accumulating data from myriad sources. Most promising developments in the health-concern point toward substantial growth in the global healthcare cloud computing market in the near future.

To cater to the diverse necessities and peculiarities in health-concern cloud-based service givers, provide enormous. This, in turn, facilitates the swift adaptation of healthier practices, ushering in new capabilities. Healthcare organizations' concerted efforts are poised to enhance clinical outcomes and individuals' quality of life. The overall well-being of the populace finds expression in the promotion of health, which also contributes to the financial stability of the healthcare system.

Many healthcare organizations continue to operate with outdated IT systems or non-interoperable technologies, impeding physicians' ability to access timely and accurate information, thereby hindering inferred deductions. Increasing adoption of cloud storage in healthcare data handling fostered improved physician-patient interactions and the seamless sharing of information among healthcare professionals.

Cloud computing, in essence, involves the storage, management, and processing of data over the Internet, diverging from traditional practices that entail setting up on-site data centers or housing data on local computers. Healthcare providers are required to make significant investments in infrastructure and maintenance to handle

V. Shrivastava (✉) · V. Pathak · S. Mishra · R. B. Buri · S. Sharma · C. Mishra
Computer Science and Engineering, Arya College of Engineering & I.T., Jaipur, India
e-mail: vishalshrivastava.cs@aryacollege.in

V. Pathak
e-mail: vibhakar@aryacollege.in

M. S. Kaiser et al. (eds.), *ICT: Smart Systems and Technologies*, Lecture Notes in Networks and Systems 878, https://doi.org/10.1007/978-981-99-9489-2_3

the various internal processes associated with storage, data processing, transmission, and collaboration. Cloud computing has reshaped the healthcare landscape, contributing to the expansion of infrastructure and facilities while paving the way for interdisciplinary medical research that benefits all.

Cloud computing's technological advancements have yielded numerous improvements in healthcare. Notable examples include the enhancement of patient care within hospitals through virtual consultations, even in the most remote locations.

Medical professionals and clinical researchers worldwide have harnessed cloud-based infrastructure to develop a range of vaccines and facilitate the creation of effective antigens. This technological advancement has not only enabled experts to acquire essential knowledge but has also streamlined information accessibility and exchange, thereby making significant contributions to the enhancement of healthcare systems. Cloud-based services offer a multitude of advantages for businesses, encompassing cost efficiency, simplified data retrieval, heightened productivity, and enhanced security measures. The positive outcomes achieved in terms of productivity and efficacy extends well beyond the realms of cloud computing and security measures. Such progress empowers healthcare institutions to elevate their productivity levels and enhance the quality of patient care, a pivotal aspect to consider for any organization. Delving into this comprehensive white paper sheds light on the vast potential that cloud computing holds within the healthcare sector.

1.1 Historical Context

The ancient context phase of the studies paper on "Evolutionary styles in current-generation Cloud-based Healthcare" gives background facts on the emergence and early ranges of cloud-primarily based healthcare structures. This phase plays a pivotal function in comprehending the origins and reasons behind the improvement of cloud-based healthcare, as well as how it fits into the broader historic panorama of healthcare generation.

Early Adoption of Electronic Health Facts (EHRs)

On this section, we delve into the preliminary steps taken to digitize healthcare facts. It elucidates the emergence of electronic health facts (EHRs) as a precursor to modern cloud-based healthcare systems. It's miles important to well known the demanding situations confronted at some stage in this era, extensively bearing on information security and interoperability. Those early challenges set the groundwork for the subsequent evolution of cloud-based solutions designed to cope with those issues.

Improvements in Connectivity

This subsection explores the technological development that laid the foundation for the integration of cloud generation into healthcare. It emphasizes the significance of fast advancements in excessive-velocity net connectivity and the great adoption of cell generation. These technological strides made it realistic to transition from conventional localized information storage to cloud-based answers, responding to the healthcare industry's growing need for greater efficient strategies of having access to and sharing affected person statistics.

1.2 Key Component of Cloud-Based Health Care System

Digital Health Records (EHRs)

EHRs constitute virtual renditions of sufferers' medical facts, encompassing comprehensive records regarding a affected person's scientific records. This encompasses data on diagnoses, prescribed medicinal drugs, treatment regimens, laboratory consequences, and extra. Inside the realm of cloud-based healthcare, EHRs discover secure website hosting and accessibility through the Internet. This allows healthcare companies' ability to access and amend affected person statistics from numerous locations, thereby improving affected person care and coordination.

Telemedicine and Faraway Patient Tracking

Telemedicine constitutes the utilization of videoconferencing, secure messaging, and analogous communiqué technology to supply remote healthcare services. Meanwhile, faraway patient tracking capitalizes on wearable devices and sensors for the purchase and transmission of actual-time health-associated data from sufferers to healthcare carriers via cloud-based totally channels. Those included components empower healthcare professionals to manage digital consultations and oversee patients' fitness statuses from afar. This augments healthcare accessibility and elevates affected person outcomes.

Fitness Facts Exchanges (HIEs)

HIEs are systems engineered to facilitate the comfortable exchange of affected person data throughout various healthcare corporations and structures. These systems play an instrumental function in ensuring interoperability by way of allowing healthcare carriers to proportion affected person statistics seamlessly throughout an array of healthcare settings. This not most effective ensures continuity of care however also mitigates the need for repetitive facts entry.

Records Analytics and Device Mastering Packages

Inside cloud-based totally healthcare systems, the arsenal includes records analytics and gadget getting to know tools designed for the processing and evaluation of huge volumes of affected person statistics. Those packages are able to detecting patterns, forecasting sickness outbreaks, helping in medical decision-making, and optimizing healthcare operations. Thru the strong capabilities offered with the aid of cloud computing, healthcare companies are enabled to conduct elaborate information evaluation, thereby extracting helpful insights pertinent to patient care and the management of populace fitness [1].

2 What Does Cloud Computing Mean?

Simplifying the concept, cloud computing employs the Internet for storing and accessing data and software, relieving reliance on local hard drives. In this model, users don't need to own the infrastructure; instead, they acquire it from external providers. Key features of cloud computing include self-service on-demand access, seamless network connectivity, resource pooling, and rapid adaptability. The ascent of cloud computing's popularity can be attributed to its advantageous attributes. Enterprises opt for cloud services to sidestep hefty software license fees and reduce dependence on local systems. The perpetual connectivity of cloud resources facilitates the use of diverse platforms.

Cloud computing can be categorized into three core types: infrastructure as a service (IaaS), platform as a service (PaaS), and software as a service (SaaS). IaaS forms the foundation, granting users access to computers, networks, and data storage. This affords comprehensive control over IT resources, akin to conventional setups in many IT departments and among developers. PaaS, the second type, empowers developers to focus on their applications without managing the underlying infrastructure intricacies. The absences of tasks like resource procurement, capacity planning, and software maintenance enhances productivity.

The third type, SaaS, delivers fully managed and maintained functional products by service providers. This often refers to end-user applications such as web-based email, where the provider handles the infrastructure, ensuring consistent service delivery.

3 Requirements of Cloud Computing for Healthcare

The healthcare sector generates a substantial volume of data on a daily basis. The challenge lies in efficiently and securely providing access to this data for both patients and healthcare providers. Healthcare cloud computing emerges as a solution, enabling businesses to transcend limitations and enhance patient outcomes. Cloud computing leads to personalized care for patients and cost reductions for healthcare providers. Furthermore, the incorporation of cloud technologies contributes to improved services and streamlined processes, ultimately leading to quicker responses from the medical community. Cloud technologies facilitate access to health information, empowering individuals to proactively manage their well-being.

Medical practitioners and healthcare professionals leverage cloud computing services to deliver high-quality solutions. This generates clear and readily accessible data that can be swiftly analyzed through user-friendly cloud technologies. This analysis aids experts in making informed treatment decisions. The push for expedited interactions between patients and medical experts drives online medical consultations and seamless remote medical support. The healthcare industry has long grappled with traditional systems, but recent years have witnessed a call for digital transformation. Wearable technology and virtual medicine hold the promise of tailored patient care. The demand for secure, fast, and cost-effective cloud-based solutions across various applications is on the rise, along with the need for seamless integration among digitized health-concern monitoring base [2].

4 Aim of Research Proposed

Distributed computing in the healthcare domain promotes efficient resource sharing and progression. This approach significantly reduces substantial labor expenses while establishing a highly efficient clinical monitoring and management system. The advent of cloud-based radio-frequency identification has revolutionized the secure, streamlined, and effective organization of clinical data. This advancement greatly contributes to superior outcomes in data transmission, intelligent health monitoring, and precise positioning. The latest innovation, referred to as IoT Cloud, amalgamates various Internet-connected technologies to offer continuous solutions across diverse contexts and environments. The burgeoning Internet of Things (IoT) technology finds applications ranging from constant health monitoring for CEOs to pre-empting various medical issues, delivering substantial benefits to healthcare services. Furthermore, the integration of cloud and web technologies into healthcare offers advantages like heightened reliability, efficiency, virtualization, and adaptability.

The paper delineates three primary objectives: (1) An exploration of cloud computing, its suitability for healthcare, and the supportive technologies. (2) Examination of the principal benefits of cloud computing within the healthcare sector. (3) Investigation into the challenges associated with implementing cloud computing in the healthcare industry.

4.1 Primary Key Advantages of Cloud-Based Solutions in the Medical Fields Green Digital Scientific File-Maintaining

The federal mandate for digital clinical information which took impact on January 1, 2014, turned into signed into regulation as a part of the American restoration and Reinvestment Act. The mandate calls for hospitals and healthcare centers to illustrate significant use of digital clinical information for storing statistics approximately affected person interactions.

The goals of cloud computing inside the clinical subject are to improve the fine, protection and efficiency of medical services, to better interact sufferers and family, enhance the coordination of care, and to keep affected person privacy and protection.

Nowadays, the majority of clinic and healthcare facilities have deserted the practice of paper document-preserving on the subject of health facts and are turning to cloud storage in healthcare. Electronic health data are saved inside the cloud and up to date electronically through physicians, nurses and different healthcare providers.

Streamlined Collaborative Affected Person Care

The implementation of cloud garage for electronic clinical information has streamlined the process of collaborative affected person care in the United States. Cloud storage makes it simpler for doctors to collaboratively view or percentage a patient's medical data.

Historically, a affected person likely had separate files of medical facts at every health practitioner, professional, or hospital they visited. This made it very hard for physicians to collaborate on the affected person's care.

The massive use of cloud garage in hospitals—mainly as it pertains to electronic health facts—makes it less difficult for medical doctors to share statistics with each other, see the outcomes of interactions among different physicians and the affected person and offer care that completely consider what the affected person has experienced with other medical doctors inside the past.

Reduced Information Garage Fees

Establishing on-site storage requires an up-front funding in hardware and calls for buying difficult drives to keep facts on, and further IT infrastructure to preserve that facts cozy and available always.

Vendors of cloud-based healthcare answers handle the administration, creation, and renovation of cloud statistics storage offerings, allowing healthcare companies to reduce their initial costs and attention efforts on the matters they do quality: being concerned for patients.

Superior Statistics Protection

In the beyond, physicians who used submitting shelves to save reams of patient statistics confronted good sized hazard of data theft or damage. Paper statistics are without difficulty misplaced or stolen, and will be absolutely destroyed with the aid of a flood, hearth, or some other herbal catastrophe. The lack of security surrounding those files become a full-size threat to patient protection.

As soon as the EMR mandate become established, healthcare providers ought to set up their personal on-website facts storage infrastructure, however once more, that could require the retention of IT personnel which are knowledgeable in information security to ensure that affected person statistics was blanketed.

As an alternative, healthcare companies had been capable of outsource records garage and safety to HIPAA-compliant cloud storage services. These offerings provide statistics garage of patient EMRs that complies with legally mandated necessities for information safety and privacy. "The cloud" has helped to make certain that every healthcare issuer can access a data storage solution with a view to correctly defend patient's sensitive facts.

Use of Cloud Computing in Healthcare Paves the Manner for Large Information Programs

The full-size adoption of cloud-based totally statistics storage solutions in healthcare has created new opportunities for "massive information" applications to improve patient effects.

In the past, medical doctors everywhere in the U.S.A. saved their patient information in paper documents. There was usually a huge quantity of potentially useful statistics in patient EMRs—statistics that might be used to are expecting when a deadly disease would possibly arise, to detect diffused correlations in affected person ailments that might monitor the causes of ailment or to clarify which treatment alternatives have been the simplest for a set of symptoms.

With the creation of cloud computing in hospitals and doctor practices, all the information that turned into formerly inaccessible in filing cabinets may be searched via and analyzed using the maximum complex computer algorithms available. This could allow healthcare carriers to locate and reply to public fitness threats that might formerly had been invisible till a lot later in their life cycle.

Cloud-Primarily Based Solutions Provide Flexibility and Scale Easily

Beyond the immediately monetary benefits associated with choosing cloud garage over an in-residence statistics garage solution, companies gain within the long-time period from simpler upgrades and reduced scaling costs. Companies of cloud garage solutions for healthcare use economies of scale to force down records control costs for his or her clients—hospitals and healthcare centers.

Cloud computing in healthcare additionally gives additional flexibility thru the standard pay-as-you-cross fee shape related to records storage. When healthcare centers build their personal facts garage answers, they must estimate how much capacity they need and invest their own money to growth that ability as they begin to run out of garage space. With cloud-primarily based solutions, a simple name to your service issuer is all that's needed to enlarge your records storage capacity to the ranges you want.

Cloud-based totally healthcare answers are fully-scalable and able to expanding in unison with your commercial enterprise [3].

More Desirable Affected Person Safety

Cloud-primarily based EMRs can play a full-size function in enhancing affected person safety.

For instance, a mentally sick patient in California visited numerous health center emergency rooms and stroll-in clinics hundreds of instances over the path of a year, on every occasion filing to exams and hoping to attain prescriptions for medicinal drug. Thanks to the adoption of cloud EMR solutions, healthcare vendors at every facility the affected person visited may want to glean direct perception into interactions between the affected person and physicians at other centers.

This included the affected person from being over-prescribed medicines that might have brought on her severe damage and stored the hospitals great money, as they had been able to avoid subjecting the affected person to trying out that she had these days performed at other facilities.

Drives Scientific Research

Inside the identical way that cloud computing will permit healthcare providers to leverage big facts and analytics in administration of their facilities, clinical researchers of the future will benefit appreciably from the digitization of healthcare records thru cloud-primarily based information garage.

Similarly, to the buildup of large records units, the cloud permits clinical researchers to leverage large quantities of computing strength that were previously inaccessible.

As these days as 2005, a clinical researcher who desired to research a large facts set would have to request time at one of the kingdom's supercomputing facilities, ready weeks on a waiting listing earlier than getting their hazard. With the cloud, researchers can leverage supercomputer-like analytical energy on their personal time, and at a fraction of the value.

Drives Records Interoperability

As we flow into the next decade of development in digital fitness, interoperability between linked medical gadgets, medical era, and the various systems and applications that save patient facts will become an more and more prominent difficulty.

A developing variety of product builders are constructing IoT-enabled gadgets for the healthcare enterprise, and without an regularly occurring well known for communiqué and facts transfer between devices, we're missing out on a few of the benefits of a related healthcare surroundings.

As developers work toward a global where wearable related devices, cell fitness packages and electronic health data can interface freely, facilitating rapid information transfer and analysis that drives affected person care results, cloud computing in healthcare will provide the primary platform for the garage and maintenance of all that information.

5 Pros of Cloud-Usage in Health-Concern Market

Cloud computing is revolutionizing the healthcare landscape in the modern age. By 2025, estimates from Global Markets Insights Inc. predict that the healthcare cloud computing sector could reach a valuation of $55 billion. Presently, healthcare institutions are swiftly embracing cloud technologies for their array of advantages, including enhanced collaboration, accessibility, efficiency, and security. Scalability

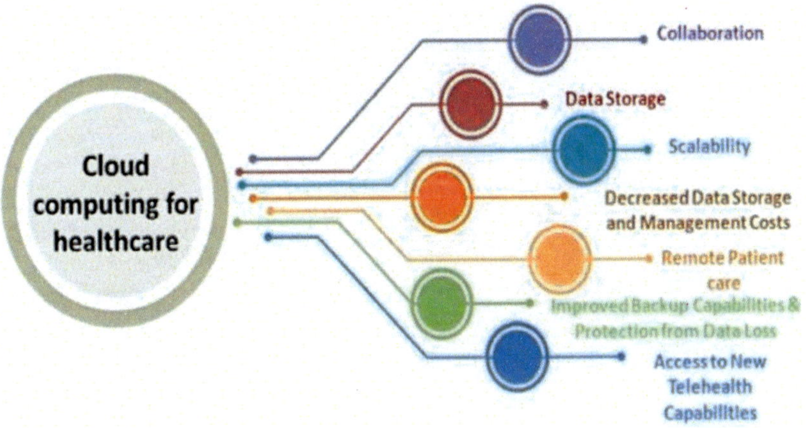

Fig. 1 Primary advantages of cloud-based solutions in the medical field

and storage capabilities further amplify its appeal. Infrastructure as a service (IaaS) offers healthcare establishments' substantial storage and on-demand computational power. Additionally, cloud-based platforms (PaaS) provide a secure environment for delivering web-based services and applications. Depicted in Fig. 1.

This shift toward cloud-based healthcare is more than just facilitating the exchange of medical information across various devices and locations; it forms the core of a transformative process. The advantages of seamlessly sharing patient health data between healthcare facilities and cloud users are also highlighted. Healthcare cloud computing not only proves cost-effective and easily implementable but also boasts numerous additional perks. Consequently, cloud-based services play a pivotal role in boosting interoperability and data accessibility within healthcare infrastructure.

This integration leads to cost reduction and facilitates the provision of high-quality personalized care. Healthcare Departments can now extend consistent levels of care to patients accustomed to such standards. The strengthening of patient concern along with streamlined medicinal help have contributed to improved patient outcomes through cloud computing. This technology presents a swifter, more efficient and cost-effective solution to healthcare challenges. It holds substantial potential in areas like clinical trials, disease prediction, epidemiological research, and medical analysis [4].

Cloud analytics, predictive analytics, and AI-driven tools assist hospitals in making informed decisions grounded in the latest evidence. Cloud analytics continuously monitor results by evaluating data from diverse interconnected medical devices. To fulfill their objectives in marketing, diagnostics, patient monitoring, prevention, and management, healthcare providers must strategically determine which cloud-based solutions to implement.

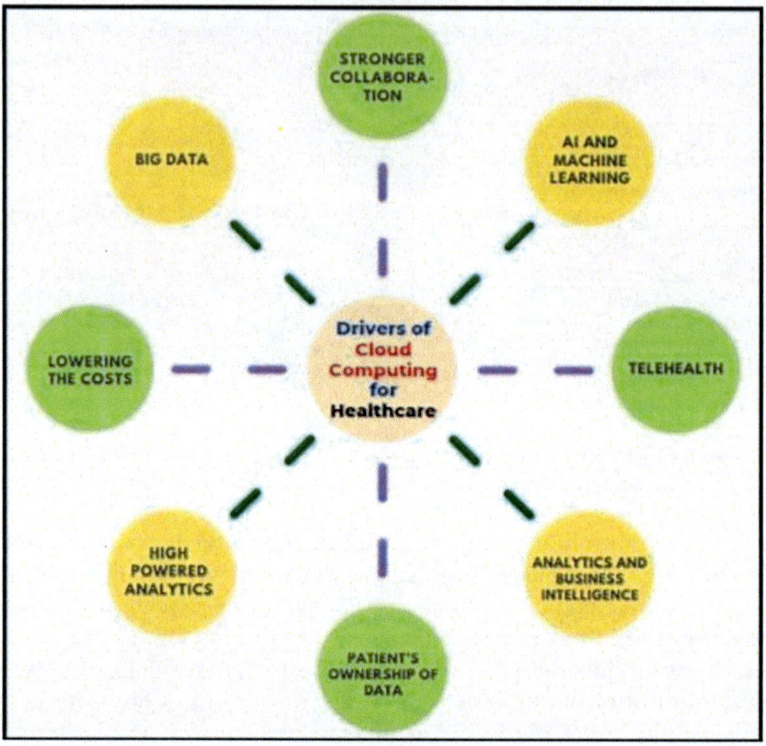

Fig. 2 Several key drivers that can profoundly enhance healthcare performance through cloud computing

6 Cloud's Utility in Health Concern

Figure 2 outlines several key drivers that can profoundly enhance healthcare performance through cloud computing. The integration of cloud technology in the healthcare sector brings about dual advantages, benefiting both patients and medical practitioners.

7 Constrains in Establishing Health-Concern Framework on Cloud

The concept of utilizing cloud computing within the healthcare sector has long been associated with challenges and limitations, particularly concerning security and potential system downtime. Figure 3 elucidates a range of concerns and complexities that require thorough consideration when contemplating the implementation of cloud computing strategies for healthcare. Further obstacles and procedural issues

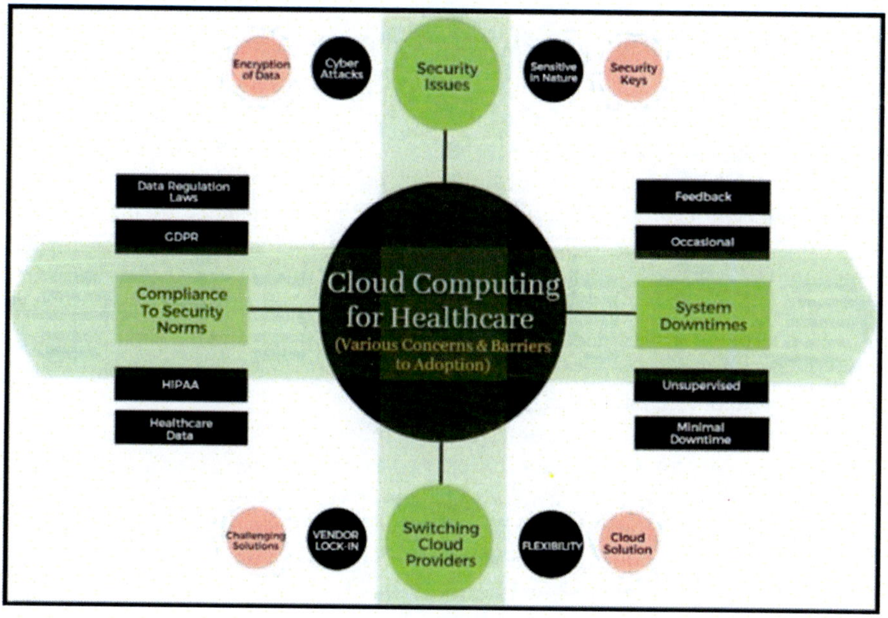

Fig. 3 Challenges and hurdles in adopting cloud computing for healthcare

in healthcare cloud operations encompass transitioning between cloud providers and ensuring compliance with uniform security protocols. Persistent concerns encompass the aftermath of cyberattacks, data privacy, effective use of encryption keys, overall system flexibility, and adherence to regulatory and legal standards. Despite these challenges, the adoption of cloud computing tools and technologies by healthcare organizations has led to notable and highly interactive outcomes.

A primary drawback of cloud computing lies in relinquishing control over the underlying infrastructure. While this poses a significant concern for businesses, service providers address this through assurances, warranties, and contractual agreements. Handling numerous clients concurrently can result in issues and support challenges for cloud providers. Although certain outages may impact the operational continuity and profitability of healthcare organizations, service providers are equipped with solutions to manage such situations. Notably, the healthcare industry, although initially hesitant, is rapidly embracing cloud computing as a prevailing trend across various sectors.

8 Cloud Applicability in Health Concern

Cloud computing serves as the backbone for various healthcare applications including electronic medical records, mobile apps, patient portals, Internet of Things medical research and predictive health benefit significantly from big data analytics in the cloud. Healthcare applications integrated with cloud technology function on a subscription-based model, offering a predictable cost structure that provides insights into monthly expenses. The concept of interoperability drives cloud platforms, enabling the aggregation, analysis, and sharing of data from various sources like IoT devices, pharmacies, insurance providers, hospitals, clinics, and the broader healthcare industry. Cloud solutions seamlessly connect central and distributed master data, enhancing collaborative medicine practices.

Since the inception of electronic medical records, collaborative medicine has evolved, reducing errors and delays in critical information exchange through tailored solutions. The integration of Internet of Things further streamlines information processes, augmenting healthcare efficiency. In today's landscape, technology plays a pivotal role across industries, healthcare being no exception. Cloud computing, with its on-demand computing capabilities, proves especially valuable for hospitals and healthcare entities, facilitating deployment, retrieval, and processing of network information. While embracing these technological advancements, adherence to healthcare standards is essential to prevent compromise and data loss [5].

9 Discussion

Medical cloud computing solutions offer remarkable dependability when accessing networks remotely. In contemporary times, hospitals universally employ this approach to ensure the security of patient data. This method enables medical personnel to electronically store patient records, substantially expanding the volume of securely stored information. Doctors today employ mobile, video, and application technologies to enhance patient interactions, consequently boosting communication and patient care through cloud computing.

Cloud-based Internet of Things (IoT) services have the potential to revolutionize entire healthcare systems. These devices, connected through various methods and locations, are linked to the hospital system.

Within hospitals, this technology allows for the electronic storage of patient records, encompassing images, documents, and videos. Authorized users can access critical data for research and service provision, fostering efficient real-time data exchange between hospitals and healthcare providers. Leveraging cloud-based IoT technology becomes increasingly pertinent for delivering effective healthcare services. By integrating various technologies and enabling data exchange, providers can amass substantial data pools, further enriched by more intricate systems.

Employing cloud computing technology elevates the efficiency and convenience of medical processes. Amidst heightened demands and expectations, healthcare systems grapple with novel challenges. Cloud computing empowers healthcare organizations to seamlessly scale to accommodate millions of electronic medical records while simplifying hospital visits, clinic appointments, and procedures through integrated medical data. Cloud-based healthcare solutions bridge geographical gaps among professionals, enabling case studies and perspectives to be shared effortlessly. Cloud-based healthcare computing, by democratizing data, empowers patients to manage their health proactively.

10 Future Reach

The future of healthcare applications with cloud computing appears promising. This technology is poised to empower doctors in crafting exceptional patient journeys, leveraging advanced areas like telemedicine and telemonitoring. As the global population ages, the evolution of cloud computing becomes pivotal for advancing healthcare in the years ahead. Recognizing the imperative to enhance healthcare through technological advancements for better treatment, administration, and efficiency is a consensus among clinicians and healthcare leaders.

Ongoing market developments are already paving the way for burgeoning healthcare opportunities. Clinical trial research strives to pinpoint optimal medications tailored to specific patient groups, while crucial details of procedures and therapies can be seamlessly preserved. This data not only fuels research but also serves as valuable case studies for aspiring medical professionals. The analysis of vast data sets further aids in precise data segmentation, aiding accurate, and insightful exploration.

11 Peroration

Cloud computing significantly enhances medical information technology by replacing paper work by digitization. A key advantage of cloud computing lies in its ability to provide on-demand access through computational resources such as data storage and processing power.

This increased capability in medical facilities has the potential to facilitate more affordable treatment for a greater number of patients, facilitate workforce expansion, and enable the addition of new hospital facilities. Cloud computing simplifies the delivery of services by outpatient caregivers within inpatient hospitals. The technique also leads to cost reduction by mitigating expenses associated with hardware and software procurement. These savings can accumulate over time, especially given the costs linked to replacing outdated equipment and acquiring new software and hardware.

Cloud interoperability ensures the easy accessibility of patient data, facilitating healthcare service planning, delivery, sharing, and insight development. Cloud-based technologies are fundamentally reshaping how healthcare enterprises function by streamlining data access for backup and recovery purposes.

References

1. Sultan N (2014) Making use of cloud computing for healthcare provision: Opportunities and challenges. Int J Inf Manage 34(2):177–184
2. Ahuja SP, Mani S, Zambrano J (2012) A survey of the state of cloud computing in healthcare. Netw Commun Technol 1(2):12
3. Griebel L, Prokosch HU, Köpcke F, Toddenroth D, Christoph J, Leb I, Sedlmayr M (2015) A scoping review of cloud computing in healthcare. BMC Med Inf Decis Mak
4. Doukas C, Maglogiannis I. (2012) Bringing IoT and cloud computing towards pervasive healthcare
5. In 2012 sixth international conference on innovative mobile and internet services in ubiquitous computing. IEEE, pp 922–926

The Development of the Semiconductor Supply Chain in India: Challenges and Opportunities

Manik Sadashiv Sonawane, Sanjay Shamrao Pawar, Jayamala Kumar Patil, and Vikas Dattatray Patil

1 Introduction

The semiconductor industry serves as the backbone of modern technology, enabling innovations that shape the way we communicate, compute, and interact with the world. India, with its burgeoning technology ecosystem, seeks to reduce its dependence on imported semiconductors and establish a self-reliant semiconductor supply chain. This paper aims to analyze the challenges and opportunities that India encounters in its quest for semiconductor supply chain development.

2 Historical Context and Current State

The semiconductor industry in India has witnessed gradual growth over the years, with a focus on design and software development rather than manufacturing. India's semiconductor market is largely dominated by imports, contributing to trade imbalances and vulnerabilities. However, recent initiatives like the National Electronics Policy and the Production Linked Incentive Scheme have laid the foundation for indigenous semiconductor manufacturing.

The semiconductor industry serves as a crucial driver of technological innovation, enabling advancements across diverse sectors. This literature review synthesizes

M. S. Sonawane (✉) · S. S. Pawar · J. K. Patil · V. D. Patil
Bharati Vidyapeeth's College of Engineering, Kolhapur, Kolhapur, India
e-mail: manik.sonawane@bharatividyapeeth.edu

S. S. Pawar
e-mail: sanjay.s.pawar@bharatividyapeeth.edu

J. K. Patil
e-mail: jayamala.k.patil@bharatividyapeeth.edu

M. S. Kaiser et al. (eds.), *ICT: Smart Systems and Technologies*, Lecture Notes in Networks and Systems 878, https://doi.org/10.1007/978-981-99-9489-2_4

insights from recent research papers, highlighting the challenges and opportunities associated with the development of the semiconductor supply chain in India.

Singh and Kumar [1] analyze the semiconductor supply chain in India, shedding light on its challenges and opportunities. The authors emphasize the need for a robust supply chain to reduce import dependency and establish a self-reliant semiconductor ecosystem. The paper identifies challenges related to infrastructure and policy framework, while also highlighting opportunities stemming from domestic demand and skill development [1].

Sinha and Mittal [2] present a roadmap for semiconductor manufacturing success in India. The paper delves into the key elements required for a thriving semiconductor manufacturing ecosystem. The authors outline strategies to address infrastructure challenges and regulatory hurdles, while emphasizing the importance of collaborations between industry and academia to drive innovation [2].

Mandal and Tripathi [3] investigate the challenges faced by the semiconductor industry in India. The paper comprehensively examines issues ranging from research and development constraints to policy impediments. The authors highlight the necessity of streamlined policies and increased R&D investments to overcome these challenges and foster industry growth [3].

Srivastava and Mishra [4] explore the opportunities within the Indian semiconductor sector. The paper underscores the immense potential for growth and innovation in the domestic market. The authors discuss avenues for collaboration between industry stakeholders, focusing on leveraging India's technological expertise and skilled workforce to position the country as a semiconductor hub [4].

Pandey and Singh [5] analyze semiconductor policy and regulations in India. The paper critically evaluates the existing regulatory framework, highlighting areas where reforms are required to attract investments and promote industry development. The authors stress the significance of a conducive policy environment for fostering a competitive semiconductor ecosystem [5].

In summary, these research papers collectively underscore the challenges and opportunities associated with the semiconductor supply chain development in India. They emphasize the need for focused efforts in infrastructure development, research collaboration, policy reforms, and skill enhancement to position India as a significant player in the global semiconductor industry.

3 Environmental Sustainability and the Challenges to Semiconductor Industry in India

The semiconductor industry is a major polluter, consuming large amounts of water and energy and generating significant amounts of waste.

One of the biggest challenges to environmental sustainability in the semiconductor industry is the high water usage. Water is used in various stages of the semiconductor manufacturing process, including cleaning, etching, and rinsing. The industry also consumes a lot of energy, both to power the manufacturing facilities and to cool the equipment.

Another challenge is the generation of waste. Semiconductor manufacturing produces a variety of hazardous wastes, including heavy metals, acids, and solvents. These wastes must be carefully disposed of to prevent environmental contamination.

Despite the challenges, there are also a number of opportunities for the semiconductor industry in India to become more sustainable. For example, the industry can adopt new water-saving technologies, such as recycling and rainwater harvesting. It can also use renewable energy sources, such as solar and wind power, to reduce its carbon footprint.

The industry can also work to reduce waste generation by using more efficient manufacturing processes and by recycling materials whenever possible. Additionally, the government can play a role in promoting environmental sustainability in the semiconductor industry by providing incentives for companies to adopt sustainable practices.

Here are some specific potential challenges and opportunities for India's semiconductor industry in the area of environmental sustainability:

Challenges:

- High water usage
- High energy consumption
- Hazardous waste generation
- Lack of awareness and expertise in sustainable manufacturing practices
- High upfront cost of implementing sustainable technologies.

By addressing these challenges and seizing these opportunities, India's semiconductor industry can become a global leader in sustainable manufacturing. This will not only benefit the environment, but it will also make the industry more competitive and attractive to investors.

Infrastructure: Semiconductor manufacturing is a highly complex and technologically intensive process that demands state-of-the-art infrastructure. However, India has a growing pool of technical talent and a strong manufacturing base, which can be leveraged to develop the semiconductor manufacturing infrastructure in the country.

Clean Rooms: India has several clean rooms that meet international standards. However, there is a need to increase the number of clean rooms and to upgrade the existing ones to meet the requirements of the semiconductor industry.

Advanced Fabrication Facilities: India does not have any advanced fabrication facilities (fabs) that can produce the latest generation of chips. However, there are several companies that are planning to set up fabs in India.

Reliable Power Supply: India's power infrastructure is improving, but there are still some challenges related to reliability and consistency. This can be a challenge for semiconductor manufacturing, which requires a consistent and high-quality power supply.

Way forward, to address the challenges in developing semiconductor manufacturing infrastructure in India, the government can take the following steps:

- Provide financial assistance to companies that are setting up fabs in India.
- Reduce import duties on semiconductor equipment and materials.
- Set up a national semiconductor institute to develop and transfer semiconductor technologies.
- Create a conducive environment for the growth of the semiconductor industry, such as by providing tax breaks and other incentives.

By taking these steps, India can develop a strong semiconductor manufacturing infrastructure that can meet the needs of the domestic industry and contribute to the global semiconductor supply chain.

In addition to the above, India can also focus on developing the following areas to strengthen its semiconductor manufacturing capabilities:

Semiconductor design: India has a growing number of semiconductor design companies. The government can support these companies by providing funding and other assistance.

Semiconductor packaging and testing: India has a strong manufacturing base for packaging and testing of semiconductor devices. The government can further strengthen this base by providing incentives to companies in this sector.

Semiconductor materials and equipment: India can also develop its own capabilities in the production of semiconductor materials and equipment. This will help to reduce the country's dependence on imports and make the semiconductor manufacturing industry more competitive.

By taking these steps, India can become a major player in the global semiconductor industry.

4 Research and Development

Semiconductor innovation requires substantial investment in research and development. India's R&D efforts need to be scaled up to foster breakthroughs in design, materials, and manufacturing processes, enabling the country to stay competitive in a rapidly evolving industry. India is developing capabilities for manufacturing mature nodes, such as 28 and 16 nm. These nodes are used in a wide range of applications, including smartphones, tablets, and laptops. Indian companies such as Tower Semiconductor and HCL Infosystems are involved in this work.

At the 40–22 nm level, India is developing capabilities for manufacturing leading-edge nodes. These nodes are used in high-performance applications, such as data centers and artificial intelligence. Indian companies such as the Indian Institute of Technology Bombay and the Centre for Development of Telematics are involved in this work.

At the 14–10 nm level, India is still in the early stages of development. However, there is significant interest in this area, as these nodes are used in the most advanced semiconductor chips. Indian companies such as the Indian Institute of Science and the Indian Institute of Technology Delhi are involved in this work.

At the 4–7 nm level, India is still in the research phase. However, there is a strong focus on this area, as these nodes are expected to be the basis for the next generation of semiconductor chips. Indian companies such as the Indian Institute of Technology Madras and the Indian Institute of Technology Kharagpur are involved in this work.

The Indian government is also investing heavily in semiconductor manufacturing, with the goal of making India a global hub for this industry. The government has set up several initiatives, such as the National Mission on Electronics (NME) and the Semiconductor Manufacturing Corporation of India (SMIC). These initiatives are aimed at developing the semiconductor ecosystem in India, from design and R&D to manufacturing and testing.

The progress made by India in semiconductor manufacturing is significant. However, there is still a long way to go before India becomes a global leader in this industry. The government and the private sector need to continue to invest in R&D and manufacturing capabilities. Only then will India be able to reap the full benefits of the semiconductor revolution.

Here are some of the challenges that India faces in semiconductor manufacturing:

Lack of skilled manpower: India lacks the skilled manpower required for semiconductor manufacturing. This is a major challenge that needs to be addressed.

High capital costs: Semiconductor manufacturing is a capital-intensive industry. This makes it difficult for small and medium-sized companies to enter this field.

Lack of government support: The Indian government needs to provide more support to the semiconductor industry. This could be in the form of tax breaks, subsidies, and other incentives.

Despite these challenges, India has the potential to become a major player in the global semiconductor industry. With the right investments and policies, India can become a global hub for semiconductor design, R&D, and manufacturing.

5 Detailed Resources Required

The type of equipment required for semiconductor manufacturing varies depending on the size of the features being manufactured. For micron-level manufacturing, the most common equipment includes:

Photolithography scanners: These machines are used to project patterns onto the semiconductor wafer.

Etch tools: These machines are used to remove material from the wafer to create the desired features.

Deposition tools: These machines are used to deposit material onto the wafer.

Deposition tools equipment for semiconductor manufacturing.

Cleaning tools: These machines are used to remove contaminants from the wafer.

Cleaning tools equipment for semiconductor manufacturing.

Metrology tools: These machines are used to measure the features on the wafer.

Metrology tools equipment for semiconductor manufacturing.

For nano-level manufacturing, the equipment becomes more complex and specialized. Some of the additional equipment that may be required includes:

Ion beam lithography: This technique uses a beam of ions to create patterns on the wafer. Ion beam lithography equipment for semiconductor manufacturing.

EUV lithography: This technique uses extreme ultraviolet light to create patterns on the wafer. EUV lithography equipment for semiconductor manufacturing.

Atomic layer deposition: This technique deposits material onto the wafer one atom at a time. Atomic layer deposition equipment for semiconductor manufacturing.

Spin coating: This technique applies a thin film of material to the wafer by spinning it at high speed. Spin coating equipment for semiconductor manufacturing.

India is still developing its semiconductor manufacturing capabilities, but it has made significant progress in recent years. In 2020, the Indian government announced a $10 billion plan to boost the semiconductor industry. This plan includes funding for the development of semiconductor manufacturing facilities, as well as for the training of engineers and technicians.

The Indian government has set up a National Semiconductor Policy, which aims to make India a global hub for semiconductor manufacturing. The government has also established the India Semiconductor Mission, which is a public–private partnership that will coordinate the development of the semiconductor industry in India.

The government is providing financial assistance to Indian companies to help them set up semiconductor manufacturing facilities.

The government is also providing tax breaks and other incentives to attract foreign investment in the semiconductor industry. The government is working to develop a skilled workforce for the semiconductor industry by setting up training programs and providing scholarships to students. These initiatives are helping India to develop its semiconductor manufacturing capabilities and position itself as a major player in the global semiconductor market.

6 India's Semiconductor Industry Compared to Other Countries

India's semiconductor industry is still in its early stages of development, compared to other countries such as Taiwan, South Korea, and China. These countries have been investing heavily in the semiconductor industry for many years and now have a significant competitive advantage. Taiwan is the world's leading semiconductor producer, accounting for over 60% of the global market share. Taiwan has a number of world-class semiconductor companies, such as TSMC and UMC, which have the capability to produce the most advanced chips.

South Korea is another major semiconductor producer, with a global market share of around 20%. South Korea is home to companies such as Samsung and SK Hynix, which are also leaders in the production of advanced chips. China is the world's largest consumer of semiconductors, but it is still developing its own semiconductor manufacturing capabilities.

China has a number of ambitious plans to boost its semiconductor industry, but it is still several years behind Taiwan and South Korea. Compared to these countries, India's semiconductor industry is relatively small and underdeveloped. India has a global market share of less than 1%. However, the Indian government is committed to developing the semiconductor industry and has launched a number of initiatives to attract investment and support the development of domestic chip design and manufacturing companies.

India's Competitiveness and Capabilities. Despite its relatively small size, India's semiconductor industry has a number of competitive advantages. India has a large pool of talented engineers and scientists, and the cost of doing business in India is relatively low. Additionally, the Indian government is providing a number of financial incentives to attract investment in the semiconductor industry.

India's semiconductor industry has also made significant progress in recent years. Several Indian companies have developed world-class semiconductor design capabilities. For example, MediaTek is a leading Indian fabless semiconductor company that designs chips for smartphones and other electronic devices.

However, India's semiconductor manufacturing capabilities are still in their early stages of development. India does not have any advanced chip fabrication plants, and it relies heavily on imports to meet its semiconductor needs.

Here are some specific areas where India's semiconductor industry lags behind other countries:

- *Chip manufacturing*: India does not have any advanced chip fabrication plants.
- *Research and development*: India invests less in semiconductor research and development than other countries.
- *Ecosystem*: India does not have a well-developed semiconductor ecosystem, including suppliers, customers, and research institutions.

7 Current Semiconductor Manufacturing Scenario in India

Here are some of the semiconductor devices that are currently manufactured in India:

- Power semiconductor devices
- Solar cells and modules
- LED displays
- Electronic and telecommunication equipment
- Medical devices
- Automotive electronics
- Industrial automation equipment.

As the semiconductor industry in India grows, it is expected that more and more types of semiconductor devices will be manufactured in the country. Currently, there are no full-fledged semiconductor fabrication plants (fabs) in India. However, there are several companies that manufacture semiconductor devices in India, including:

SPEL Semiconductor: This company is India's only semiconductor IC assembly and test (OSAT) facility. It provides a range of services, including wafer bumping, wire bonding, packaging, and testing.

MosChip Semiconductor Technologies: This is a fabless semiconductor company that designs and markets a range of analog and mixed-signal integrated circuits (ICs). Its products are used in a variety of applications, including power management, automotive electronics, and industrial automation.

Einfochips: This company provides semiconductor design services, including chip design, verification, and testing. It also offers a range of intellectual property (IP) blocks, which can be used to speed up the design process.

Tata Elxsi: This company provides a range of semiconductor design and engineering services, including system-on-chip (SoC) design, embedded software development, and test and validation.

HCL Technologies: This company provides a wide range of IT services, including semiconductor design and manufacturing. It has a strong track record in providing engineering services to the semiconductor industry.

In addition to these companies, there are a few other semiconductor-related companies operating in India, such as:

Vedanta: This company is planning to set up a semiconductor fab in Gujarat.

Foxconn: This company is partnering with Vedanta to set up the semiconductor fab in Gujarat.

Semiconductor Laboratory: This is a government-run research and development (R&D) institute that is involved in developing semiconductor technologies.

The Indian government is taking steps to promote the growth of the semiconductor industry in India. In 2020, the government announced the Production Linked Incentive (PLI) scheme for semiconductor manufacturing. This scheme provides financial incentives to companies that set up semiconductor fabs in India.

The government is also investing in semiconductor R&D. In 2022, the government announced the setting up of a National Centre for Electronics Design and Manufacturing (NCEDM) in Karnataka. The NCEDM will provide a platform for companies to collaborate on semiconductor design and manufacturing.

The Indian semiconductor industry is still in its early stages of development. However, the government's initiatives are expected to help the industry grow in the coming years.

8 Policy Framework

A conducive policy environment is crucial for semiconductor industry growth. Complex regulatory procedures, ambiguous intellectual property laws, and lack of incentives for research discourage investments in the sector. Some of the policy recommendations in more technical terms with company names and facilities:

- Invest in R&D
- Create a skilled workforce
- Reduce capital costs
- Attract foreign investment
- Encourage collaboration.

India can explore a number of international collaborations and partnerships to strengthen its semiconductor supply chain. Some potential partners include:

- The **USA**: The USA is the world's leading semiconductor market and has a number of world-class semiconductor companies. India can collaborate with US companies on chip design, manufacturing, and research and development.
- **Japan**: Japan is another leading semiconductor producer and has a number of strong relationships with Indian companies. India can collaborate with Japanese companies on chip design, manufacturing, and equipment supply.
- **South Korea**: South Korea is a major semiconductor producer and has a number of advanced chip fabrication plants. India can collaborate with South Korean companies on chip manufacturing and technology transfer.
- **Taiwan**: Taiwan is the world's leading semiconductor producer and has a number of world-class semiconductor companies. India can collaborate with Taiwanese companies on chip design, manufacturing, and research and development.
- **Germany**: Germany is a leader in semiconductor manufacturing equipment and materials. India can collaborate with German companies to develop its own semiconductor manufacturing capabilities.

In addition to these countries, India can also explore collaborations with other countries that are developing their own semiconductor industries, such as China, Brazil, and Israel.

9 Conclusion

India's journey toward semiconductor self-sufficiency is rife with challenges, but the potential rewards are substantial. By addressing infrastructure bottlenecks, boosting R&D, implementing favorable policies, and nurturing a skilled workforce, India can position itself as a prominent player in the global semiconductor supply chain. The success of these efforts will not only drive economic growth but also contribute to technological sovereignty and innovation.

References

1. Singh AK, Kumar S (2023) Semiconductor supply chain in India: challenges and opportunities. Microelectr Reliab 120:114–121
2. Sinha RK, Mittal AK (2023) Semiconductor manufacturing in India: a roadmap for success. Semicond Eng 2023(3):1–10
3. Mandal SK, Tripathi AK (2022) Challenges of the semiconductor industry in India. Int J Eng Res Technol 11(1):559–564
4. Srivastava AK, Mishra AK (2022) Opportunities in the Indian semiconductor sector. Microelectr J 79:105547
5. Pandey SK, Singh AK (2022) Semiconductor policy and regulations in India. J Indian Inst Sci 102(5):1159–1174

Water Quality Analysis of Major Rivers of India Using Machine Learning

Ashish Kumar Singh and Sanjay Patidar

1 Introduction

Water is considered to be one of the most important natural resource for all living organisms on earth. The existence of life without water is not possible. Water is obtained from various sources like rivers, lakes, etc. The rivers play a very important role in the lives of most of the Indian people living in various parts of the country as this river water is used for various kinds of purposes like irrigation, potable water, electricity, etc. that contribute to humans, creatures, and industries [1]. We need this water at a certain purity level because exceeding or crossing these limits can become life-threatening for living organisms. According to the study of various surveys on Indian rivers it has been evaluated that most of the rivers are very polluted and not fit for regular usage. The contamination of river can be because of man-made activity or because of some natural activity. The domestic sewage discharged into the water bodies cause various kinds of waterborne diseases. The animal dunghills also contaminate the water sources. Huge amount of heavy metals are added into the ground water source by various activities like mining and construction. The fertilizers and pesticides used by farmers in the fields contaminate the ground water. The harmful chemicals released from the industries also contaminate the water. Rapid industrial development is the major cause for degrading the water quality of all the rivers [2].

Some of the recent studies have shown that approximately 15–16 lakhs people die every year due to the consumption of such polluted water. World Health Organization (WHO) recently conducted a study and the results were alarming, it found that nearly 37% of people in the urban region and 64% people living in rural areas are living without access to pure water. About 80 percent of the illness and diseases are waterborne, which have resulted in death of about fifty lacs people and 250 crore

A. K. Singh (✉) · S. Patidar
Department of Software Engineering, Delhi Technological University, Delhi, India
e-mail: singhashish040897@gmail.com

M. S. Kaiser et al. (eds.), *ICT: Smart Systems and Technologies*, Lecture Notes in Networks and Systems 878, https://doi.org/10.1007/978-981-99-9489-2_5

infirmities [3]. The process of analyzing the quality of water is said to be one of determining the quality of the water by looking at several water quality criteria and seeing if they fall within the established limits. This complete process becomes very tedious and time consuming when it is done manually thus now machine learning is used to solve such classification problems by means of regression and forecasting. The water quality data containing several parameters of major rivers are collected in which the parameters that are investigated are BOD, pH, DO, temperature, and conductivity. Eventually this water quality data obtained is classified using different classification techniques as clean or not clean [1]. The required water quality of water is calculated using various links of different parameters to form a standardize index called Water Quality Index (WQI) [4]. Machine learning is one of the most efficient method in-order to evaluate huge dataset which is collected from a range of different sources in-order to find patterns in the data or to draw some conclusions about the data which was collected. Some suitable procedures must be chosen and the models need to be trained and validated in-order to apply various machine learning algorithms on the data collected. The choice of algorithms that needs to be applied on various data becomes important as one algorithm may give poor performance on some kind of data whereas the same algorithm might perform excellently on other data. The machine learning technology can be divided into two main supervised and unsupervised learning. The labeling of datasets is the main distinction between these two categories [5].

Some of the machine learning algorithms which are used are:

A. **Naïve Bayes**

One of the most well-known machine learning algorithms is naive Bayes. A type of probabilistic classifier is the naive Bayes. It is based on Bayes' theorem and assumes that the features used for classification purpose are not related or independent of each other such an assumption is referred to as the "naive" assumption. Although the algorithm is simple still it is very effective for various kinds of purposes and applications. For, e.g., spam filtering and text classification. The naïve Bayes algorithm calculates the probability of the data point belonging to each class. This is done by utilizing all the previous knowledge of the class distribution and the conditional probabilities of the features given each class. The training dataset is used to estimates the probabilities [6].

Naive Bayes learns the probabilities by counting the occurrences of each feature in each class and then computing the corresponding probabilities during the training phase, where the model is trained using the training data. The prior probability of each class is also estimated and predicted using the training data's frequency of occurrence. In the testing phase where the model is tested with the test data the algorithm uses the learned probabilities in the testing phase in-order to classify new, unseen data points. The Bayes' theorem is used by the naïve Bayes classifies to calculates the posterior probability of each class for a given data point and the class which have the highest posterior probability is assigned as the predicted class for the data point.

B. **Logistic Model Tree (LMT)**

Logistic Model Tree (LMT) algorithm is a machine learning algorithm which combines decision trees with logistic regression in-order to create a very powerful classification model. The LMT algorithm is basically known for its ability where it handle both numerical and categorical features very effectively. The creation of decision trees is the foundation of the LMT algorithm. By repeatedly splitting the data based on feature conditions that maximize the separation of the target classes, decision trees are created. This procedure keeps on until a stopping criterion is satisfied, such as when the depth is reached or a specific minimum number of samples are present in each leaf node. In traditional decision tree algorithms, the predicted class at each leaf node is determined by the majority class of training data samples in that node. However, in LMT, logistic regression is used to refine the predictions at the leaf nodes. Instead of relying solely on majority voting, logistic regression assigns probabilities to each class, giving a more nuanced and probabilistic prediction.

The decision tree and the logistic regression models are both constructed concurrently at the leaf nodes during the training phase of our LMT approach. The associations between the input features and the target variable within each leaf node are captured by the logistic regression models. In the testing phase, a new data point is passed through the decision tree to determine the appropriate leaf node. Then, the logistic regression model associated with that leaf node is used to calculate the class probabilities, which are further used to make the final class prediction. The LMT algorithm offers several benefits. First, it combines the strengths of decision trees and logistic regression, allowing for a more accurate and interpretable classification model. Additionally, LMT algorithm can manage numerical as well as the categorical features, making it versatile for a wide range of datasets. Its ability to capture nonlinear relationships and interactions between features enhances its predictive power.

C. **J48**

The J48 algorithm, also referred to as C4.5. It is a well-known decision tree-based machine learning technique that is primarily used for classification tasks. The J48 algorithm is created by further extending the ID3 algorithm. J48 creates a decision tree by iteratively dividing the data according to the characteristic that results in the greatest information gain or impurity reduction. Making a tree that maximizes the separation of the target classes is the goal here. Usual criteria for determining a node's impurity include entropy and the Gini index. The root node, which stands in for the entire dataset, is where the algorithm begins. Then, based on information gain or impurity reduction, it chooses the best feature to partition the data. Until a stopping requirement is satisfied, such as reaching a predetermined depth or having a minimum amount of samples in each leaf node, this process is repeated iteratively for each child node. The J48 algorithm's capacity to handle both categorical and numerical information is one of its advantages.

D. **Decision Tree**

A decision tree is one of the popular machine learning algorithm which can be used for both classification as well as the regression tasks. It is represented in some kind a tree structure in the form of flow chart in which each and every internal node in the tree represent a feature or attribute of the data, each branch of the tree structure represents a decision rule, and each leaf node represents the outcome or the prediction [2]. The process of constructing a decision tree involves recursively partitioning the data based on different feature values to create nodes and branches. The major aim is to find the best splits that maximize the separation of the target classes or we can say that minimize the prediction error. A feature is chosen at each node of the decision tree to divide the data according to a certain criterion, such as information gain, Gini index, or mean squared error. The chosen feature is the one that offers the greatest post-split increase in class purity or forecast precision. Until a stopping requirement is satisfied, such as reaching a maximum depth or having a minimum amount of samples in each leaf node, the process iteratively repeats itself. The decision tree model is very easy to interpret and understand because the resulting tree structure can be visualized very easily. Each decision rule along a path from the root to a leaf node represents a condition that guides the classification or regression process. The leaf nodes contain the final predictions or outcomes. The decision tree method learns the best splits by examining the training data and identifying the feature values that produce the best separation or prediction during the training phase, when the model is trained using the training data. In the classification situation, the majority class of the training samples that reach each leaf node is allocated. Regression assigns the mean or median value of the target variable to the leaf nodes. In the testing phase where the model is tested with the testing data, new data points are classified or predicted by traversing the decision tree based on their feature values. The data point follows the decision rules from the root to a specific leaf node, which provides the final prediction or outcome [2].

2 Literature Survey

Khullar et al. in 2023 [2], proposed a paper on the classification of river water quality by using some hybrid machine learning methodology. They proposed model for water quality classification of river Yamuna. The accuracy and the performance of the classification technique used in the study was compared and analyzed with respect to some of the other basic classification techniques that include the Support Vector Machine (SVM), Naïve Bayes classifier, and the bagged and boosted tree classifiers. The proposed approach in the paper by them obtains an overall accuracy of 99.65%. In their work, Bajpai et al. [7] offered a real time approach that will aid in classifying the river Ganga's water quality by looking at the information gathered from the mehndi ghat and the information from kannauj. They proposed a paper in which they wanted to identify and analyze the water quality of Ganga that whether it is pure and clean or not. They identified the water quality of river Ganges that whether the water is healthy and portable. Their major objective was to classify the water of river Ganga on some of the major water quality parameters using different classification algorithms of machine learning.

The accuracy of the proposed model is found to be around 99 percent, which is far ahead in terms of accuracy in compared to other approaches used for water quality prediction. In 2015 Chawla et al. [1] proposed a paper where they predicted the Pollution Potential of several Rivers in India with the help of empirical equation which consisted of water quality parameters. In 2014 Jaloree et al. [8] presented a paper where they studied the approach of decision tree to study the water quality. To analyze the water quality data of the Narmada River at a site in the Harda district, the research provided a classification model using a decision tree approach. They used the WEKA software to implement the data model. So as to classify the water quality data decision tree was applied which identify the class of water. The water quality parameters like Nitrogen, Temp, pH, COD, BOD plays a very important role in the assessment of the water quality. Gorde et al. presented a review study for the evaluation of water quality parameters in 2013 [9]. The parameters required to assess the quality of the water were the subject of a study. Salinity, phosphates, turbidity, temperature, pH, and nitrates are among the primary factors that determine the quality of water. In January 2019 Mahesh [3] proposed a review on various machine algorithms. As these machine leaning algorithms are used for various purposes which reduces the human effort and saves time. In 2022 Nanjundan et al. [10] presented a method in 2022 that will be utilized to compute and assess the water quality using a supervised machine learning methodology. The paper studied how to analyze and calculate the value of Water Quality Index (WQI) which can be further used to identify the quality of water. In 2019 Jalal et al. [11] proposed a paper where they analyzed the performance, i.e., the accuracy of various machine learning algorithms for a system which determines the water quality. Some of the machine learning algorithms include tree and SVM. Their performance was measured and evaluated further. In 2017 Shukla et al. [4] proposed a paper in-order to analyze the water quality of surface water of river Ganga using a methodology called as index mapping. To determine the water quality of Ganga river effectively they formed a method which can integrate Geographic Information System (GIS) with the Water Quality Index (WQI). They observed very poor water quality in the lower reaches of river whereas the Varanasi station had the worst water quality. In 2021 Nair et al [12] proposed an examination of predictive models for analyzing the water quality of rivers using different machine learning approaches. The performance of various water quality prediction models created utilizing machine learning and big data techniques is examined in this research. In 2023 Kavitha et al [13] proposed a survey on water quality prediction. The objective is to study various machine learning methodology and approaches so as to predict the waste water quality with most accuracy assess the water quality using a supervised machine learning methodology. The paper studied how to analyze and calculate the value of Water Quality Index (WQI) which can be further used to identify the quality of water. In 2015 Chawla et al. [14] presented a approach to develop an empirical equation to predict the pollution potential of river water in which they observer that the pollution potential predicted using empirical equation is congruence with the current pollution potential of the Indian rivers. In 2022 Abraham et al. [6] conducted an experimental study in which they used various machine learning algorithms to decide the river water quality of San Antonio river. In 2023 Brindha et al. [15] conducted a research whose main objective is to water quality using machine learning technique (Table 1).

A. Data Analysis Tools

In this survey WEKA tool is used to analyze the data collected. It is basically a open source software tool which is used for machine learning. It provides collection of various machine learning algorithms and other relevant tools which becomes handy for data classification, data preprocessing, data clustering, and data regression, etc. This software was developed in a university of New Zealand named Waikato.

Weka supports a diverse range of data formats and allows users to manipulate and preprocess their datasets efficiently. It offers a wide array of data preprocessing techniques, such as attribute selection, normalization, and missing value handling, which are essential for preparing data before applying machine learning algorithms.

One of the key strengths of Weka is its extensive library of machine learning algorithms. It includes several popular machine learning algorithms such as random forests, decision trees, k-nearest neighbors (k-NN), support vector machines (SVM), naive Bayes, and many other algorithms. This rich selection of algorithms caters to various types of machine learning tasks, enabling users to experiment and choose the most suitable algorithm for their specific needs.

B. Data Collection

In this study we have used data to classify and build the decision tree and the data was collected from Central Pollution Control Board which is sponsored by Ministry of Environment and Forests, Government of India.

C. Water Quality Parameters Used

Various water quality parameters can determine the level of water quality of any water source. In this study we have used five major water quality parameters to observe and analyze the quality of water of various rivers of India. The parameters which were used to determine whether the water is clean is pH level, Temperature, Conductivity, BOD, and DO.

D. Workflow (Fig. 1)

a. **Data Collection**

In this study we have used data to classify and build the decision tree and the data was collected from Central Pollution Control Board which is sponsored by Ministry of Environment and Forests, Government of India. Few data is also collected from Kaggle website.

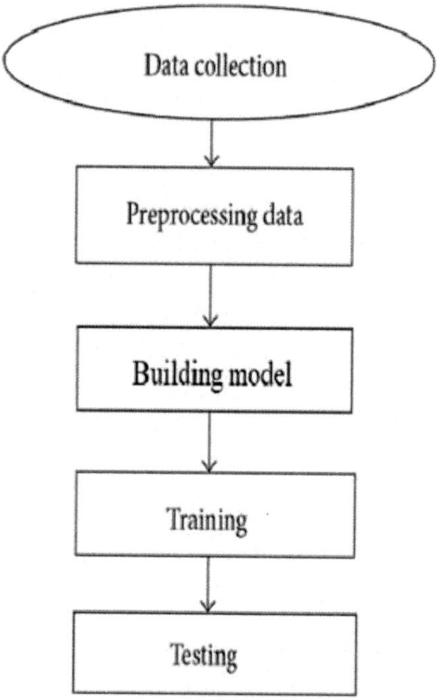

Fig. 1 Workflow

b. **Data Preprocessing**

The data raw which is collected needs to be further processed before it is fed to the algorithms for the analysis. This phase involves data cleaning, data integration, data transformation, and data reduction.

c. **Building Model**

To discover patterns in data, machine learning is utilized. We can perform supervised or unsupervised learning. Some of the machine learning include regression, classification, regression, and forecasting. In this phase a machine learning algorithm is chosen and trained which makes the prediction on the provided data.

d. **Training**

The model formulated is trained on the dataset. A subset of the complete dataset is used to train the model in this phase. We used a weak software for this purpose of training the model in which the data is uploaded first and then the model is trained.

Testing

The model trained on the training data is here tested with the test data. This phase cheeks the accuracy of the model.

3 Conclusion

An attempt to understand how to classify the river water depending upon its usability that whether it is clean or not. The data of various rivers containing several parameters like DO, pH, TC, BOD, and conductivity is collected and processed. An attempt to understand how to classify the river water depending upon its usability. This data is fed to several machine learning algorithms and we decide about the quality of the water by observing the result of the algorithm with the highest accuracy score. The Logistic Model Tree, J48, and Naïve Bayes are some of the algorithms that are used to build the model and the algorithm which shows highest accuracy among them is used to determine the water quality of the given dataset as clean or not clean. Some of the methodologies that are required for the above-mentioned procedure are also discussed in the paper including the workflow and the tool which is used for the analysis of data.

4 Result

The data of river Beas was tested with three classification algorithm—Naïve Bayes, J48, and LMT. Firstly 20% of data was tested with these algorithms but the accuracy was not on the higher side so the complete data was tested with the above-mentioned algorithm and results were on higher side. Naïve Bayes gave the highest accuracy of 90.47%. The data of river Godavari was tested with three classification algorithm—Naïve Bayes, J48, and LMT. Firstly 20% of data was tested with these algorithms but the accuracy was not on the higher side so the complete data was tested with the above-mentioned algorithm and results were on higher side. Highest accuracy was obtained by using J48 and LMT. Similarly LMT gave highest accuracy 94.64% with the data of river Ganga. The data of river Yamuna observed highest accuracy of 91.66 % using J48 algorithm. Sutlej river data on testing with the same three algorithms showed maximum accuracy of 93.75 with J48. Also the Brahmaputra river data shows maximum accuracy 83.33% using LMT algorithm.

Table 1 Summary table

Author	Year	Methodology	Dataset
Kullar et al. [2]	2022	Hybrid ML technique	Delhi based
			Central
			Pollution Control
			Board (CPCB)
Bajpai et al. [7]	2022	Random forest	CPCB
Chawla et al. [1]	2015	Empirical equation	Real time data
Jaloree et al. [8]	2014	Decision tree	Real time data
Gorde et al. [9]	2013	Analysis	Real time data
Mahesh [3]	2019	ML algo	Real time data
Nanjundan et al. [10]	2022	Unsupervised ML	Environment Information System
Jalal et al. [11]	2019	Classification algo	Water treatment station "Ghadir El Golla" of Tunisia
Shukla et al. [4]	2017	Index mapping	CPCB, UPPCB, NIH, CWC, ASTER, DEM, LPD, AAC
Nair et al. [12]	2021	ML and big data	Real time data
Kaviha et al. [13]	2023	Machine learning	Real time data
Chawla et al. [14]	2015	Machine learning	Real time data
Abraham et al. [6]	2022	Machine learning, data analysis	Kaggle
Dzeroski et al. [5]	1995	Analysis and classification	Real time data
Brindha et al. [15]	2023	Machine learning	Real time data

References

1. Chawla P, Hasteer N Prediction of pollution potential of Indian rivers using empirical equation consisting of water quality parameters. In: IEEE 2015 international conference on technological Innovation in ICT for agriculture and rural development
2. Khullar S, Singh N (2022) River water quality classification using a hybrid machine learning technique. In: 9th international conference on computing for sustainable global development. IEEE
3. Mahesh B (2019) Machine learning algorithms. Int J Sci Res 9 (a review)
4. Shukla AK, Ojha CSP, Garg RD (2017) surface water quality assessment of Ganga River basin. India using index mapping. IEEE
5. Dzeroski S, Grbovic J Knowledge discovery in a water quality database
6. Abraham A, Livingston D, Guerra I, Yang J (2022) Exploring the application of machine learning algorithms to water quality analysis. In: 7th international conference on big data, cloud computing and data science (BCD). IEEE
7. Bajpai A, Chaubey S, Patro B, Verma A (2022) A real time approach to classify the water quality of the river ganga at Mehandi Ghat, Kannuaj. In: International conference on artificial intelligence in engineering and technology. IEEE
8. Jaloree S, Rajput A, Gour S (2014) Decision tree approach to build a model for water quality. Binary J Data Mining Netw 4:25–28

9. Gorde SP, Jadhav MV (2013) Assessment of water quality. Int J Eng Res Appl 3(6):2029–2035. www.ijera.com. ISSN: 2248-9622
10. Nanjundan P, George JP, Vij A (2022) A reliable method of predicting water quality using supervised machine learning model. In: IEEE international conference on data science and information system (ICDSIS)
11. Jalal D, Ezzedine T (2019) Performance analysis of machine learning algorithms for water quality monitoring system. IEEE
12. Nair JP, Vijaya MS (2021) Predictive models for river water quality using machine learning and big data techniques—a survey. In: Proceedings of the international conference on artificial intelligence and smart systems (ICAIS-2021). IEEE Xplore Part Number: CFP21OAB-ART. IEEE. ISBN: 978-1-7281-9537-7
13. Kavitha D, Gayathri TR, Dhamini Devaraj Hasitha V (2003) Survey on water quality prediction. IEEE
14. Chawla P, Hasteer N (2015) Prediction of pollution potential of Indian rivers using empirical equation consisting of water quality parameters. In: International conference on technological Innovation in ICT for agriculture and rural development. IEEE
15. Brindha D, Puli V, Sobula BK, Mittakandala VS, Nanneboina GD (2023) Water quality analysis and prediction using machine learning. IEEE

Enhancing Trust in AI-Generated Medical Narratives: A Transparent Approach for Simplifying Radiology Reports

Vivek Kumar Verma and Bhavna Saini

1 Introduction

1.1 AI in Healthcare: Evolution, Benefits, and Transformative Impacts

Artificial intelligence (AI) has undergone remarkable evolution in healthcare over recent years, shifting from experimental applications to essential tools in various medical domains. Its integration into diagnostics, treatment planning, and patient management has revolutionized the way care is delivered, offering greater precision and personalized treatment modalities [1]. The benefits are manifold, with AI algorithms enhancing diagnostic accuracy, streamlining administrative tasks, and providing predictive analytics for patient outcomes. These transformative impacts not only improve clinical decision-making but also pave the way for more efficient, patient-centric care models, redefining the boundaries of what's achievable in modern medicine [2].

Tracing back to its nascent stages, AI's initial foray into healthcare was met with a mix of skepticism and curiosity. Over the decades, as computational capacities expanded and algorithms matured, AI transitioned from rudimentary applications to more complex roles in diagnostics, drug discovery, and patient care. Today, its prominence in the medical field is a testament to relentless advancements in machine learning and data analytics, catalyzing a paradigm shift in healthcare delivery [3].

V. K. Verma (✉)
Manipal University Jaipur, Jaipur, India
e-mail: vermavivek123@gmail.com

B. Saini
Central University of Rajasthan, Jaipur, India

M. S. Kaiser et al. (eds.), *ICT: Smart Systems and Technologies*, Lecture Notes in Networks and Systems 878, https://doi.org/10.1007/978-981-99-9489-2_6

One of the most profound impacts of AI in medicine is its capability to enhance diagnostic precision. Machine learning models, particularly deep learning, have showcased remarkable efficacy in imaging diagnostics, often matching, or even surpassing human expert performance. This precision, coupled with data-driven insights, has paved the way for personalized treatment plans, tailoring interventions to individual patient needs [4].

Beyond clinical applications, AI has been instrumental in addressing administrative burdens within healthcare systems. Automated algorithms assist in patient scheduling, billing, and claims processing, reducing errors and improving operational efficiency. Natural language processing (NLP) tools have revolutionized electronic health record (EHR) management, extracting meaningful insights from vast textual data, and aiding in clinical decision support [5].

Embracing the power of AI-driven predictive analytics, healthcare providers can now anticipate patient needs, potential complications, and likely outcomes. These models process vast datasets, from patient histories to genetic information, predicting disease risks and optimizing interventions. Such proactive care approaches, grounded in AI's predictive prowess, are redefining preventative medicine and chronic disease management [6].

1.2 Radiology Reports: Complexity

Radiology plays a pivotal role in modern medicine, offering critical insights into patient conditions through detailed imaging. However, the intricacies of radiology reports, laden with medical jargon and technical nuances, often pose significant comprehension challenges for patients. The dense terminology and the multifaceted nature of these reports can leave patients feeling overwhelmed and uncertain, hindering their ability to fully grasp their medical status and make informed decisions regarding their care [7–9].

This comparison Table 1 offers a theoretical highlight the core distinctions between traditional and patient-friendly radiology reports. While it provides a generalized view, it's important to note that the nuanced differences between the two types of reports can fluctuate depending on scenarios or settings. Table 1 designs to give an overarching idea, but specifics can differ in actual practice based on unique circumstances.

1.3 The Imperative for Transparent AI in Medical Narratives

As artificial intelligence steadily becomes a cornerstone in medical diagnostics and treatment planning, there's an ever-increasing emphasis on its transparency, especially when converting medical findings into narratives. Given the life-altering implications of medical decisions, there's little room for obscurity in AI outputs. According

Table 1 Traditional versus patient-friendly radiology reports

Criteria	Traditional radiology reports	Patient-friendly radiology reports
Language	Medical and technical jargon	Simplified
Details	Highly detailed with technical specifics	Summarized, highlighting main findings
Length	Typically, longer with full observations	Shorter with concise explanations
Target Audience	Physicians and medical professionals	Patients and public
Interpretability	Requires medical knowledge	Designed for easy comprehension
Accessibility	Often limited to medical portals	Potentially available in various formats

to authors [10], the interpretation of medical narratives necessitates both accuracy and clarity to ensure that healthcare professionals, patients, and other stakeholders can understand and trust the conclusions drawn by AI systems. The authors stress that without transparency, even the most advanced AI models might face resistance in clinical adoption due to potential mistrust or misinterpretations.

Some points underscore the significance of transparency in AI, especially when converting intricate medical information into narratives that guide decisions impacting human health and wellbeing as indicated in Table 2.

Table 2 Importance of transparency in AI when generating medical narratives

Aspect	Significance
Trust factor	Ensures users trust AI's results and recommendations in healthcare
Ethical considerations	Upholds standards, highlights biases, and ensures ethical integrity
Safety and accountability	Enables error tracing and understanding of decision-making processes
Facilitating communication	Enables error tracing and understanding of decision-making processes
Regulatory and compliance	Meets increasing regulatory standards and scrutiny
Stakeholder confidence	Boosts trust among insurance companies, regulators, investors
Interdisciplinary collaboration	Promotes cooperation between tech developers and healthcare professionals

2 AI-Generated Medical Narratives

2.1 Transparency Criteria for AI-Generated Medical Narratives

Interpretability: The AI model should provide outputs in a form that humans can understand. This is paramount in a clinical context where decision-making relies on comprehensible insights [11].

Traceability: It should be possible to trace back every decision made by the AI to a particular piece of data or a series of data processing steps [12].

Consistency: The AI model should behave predictably, producing consistent results when presented with similar data [13].

Fairness: Ensuring that AI-generated narratives don't reinforce existing biases, especially when interpreting medical data from diverse patient populations [14].

Robustness and Reliability: The AI system should be dependable, minimizing errors or inconsistencies, especially in varying operational conditions [15].

Transparency in Training Data: Stakeholders should have a clear understanding of the data that trained the AI model, especially considering the heterogeneity of patient data in healthcare [16].

Feedback Mechanism: There should be a system in place for users to provide feedback on the AI narratives, ensuring a loop for continuous improvement [17].

Documentation: Comprehensive documentation on the AI model's design, functioning, and validation is essential for its broader acceptance in the medical community [18].

2.2 Integration of Explainability Techniques

The seamless integration of explainability techniques into natural language processing (NLP) models is crucial, especially when these models generate medical narratives. Explainable AI (XAI) serves the dual purpose of enhancing trustworthiness and facilitating interpretability. According to authors [19], explainability in NLP is not just about understanding model predictions, but also about elucidating the underlying linguistic features and reasoning that contribute to the final narrative. By integrating these techniques, AI-generated medical narratives can offer insights into their decision-making processes, allowing both healthcare professionals and patients to understand and critically engage with the content. Such transparency not only enhances patient-provider communication but also ensures the ethical use of AI in clinical settings. Incorporating XAI into NLP is, therefore, both a technical and ethical imperative, paving the way for more reliable and patient-centric care.

When delving into the realm of artificial intelligence, one can liken the process to understanding a master chef's intricate culinary creations. LIME offers a taste-test, highlighting specific flavors but not encompassing the entire dish. SHAP, on the other hand, provides a comprehensive flavor profile, revealing how each ingredient contributes to the final taste, though it might take some time to dissect [20]. Attention mechanisms are akin to observing which ingredients the chef frequently glances at during cooking; it's indicative but not always definitive. Counterfactual explanations play out like a culinary experiment, where removing an ingredient like salt helps gauge its overall impact. Activation maximization is about discerning the chef's favorite tools, offering hints about their specialty dishes. Integrated gradients are like a behind-the-scenes tour, watching a dish evolve from a simple base to a gourmet masterpiece. Finally, rule extraction simplifies the chef's sophisticated techniques into basic cooking guidelines, but in doing so, some nuanced flavors might be overlooked. Together, these techniques try to demystify the AI "cooking process," helping us grasp how decisions are whipped up as shown in Table 3.

Table 3 Integration of explainability techniques with NLP models for medical narratives

Technique	Advantage	Limitation	Other considerations
LIME	Model-agnostic; highlights influential words	Local explanations might not generalize	Requires multiple perturbed samples
SHAP	Unified measure; consistent and accurate	Computationally intensive for deep models	Based on cooperative game theory
Attention mechanisms	Directly embedded in model; visually	Might not always align with true	Often used for model performance
Counterfactual explanations	Provides actionable insights by showing	Can be computationally expensive, might not cover all possible scenarios	Useful for "what if" scenarios
Activation maximization	Offers direct visualization of learned	Interpretations might not be semantically	More relevant for deep learning
Integrated gradients	Provides feature attributions; doesn't require multiple samples	Requires baseline input; might be computationally demanding for larger models	Suitable for deep models
Rule Extraction	Provides clear, rule-based explanations	Might oversimplify complex behaviors simpler, interpretable rules	Transforms complex models into

3 Methodologies for Simplifying Radiology Reports

3.1 Streamlining AI Narratives for Clarity

Distilling complex AI outputs into layman terms, especially in sensitive fields like medical narratives, is vital for comprehension by a broader audience. Here are some innovative techniques that could be employed:

Semantic Mapping: Utilize a dictionary or database of medical terms and their simpler counterparts. The AI model maps complex medical terms to more understandable synonyms when generating the narrative.

Interactive Visual Aids: Offer visual representations (like charts, graphs, or even infographics) that visually depict the AI's findings. For radiology, annotated image overlays can highlight and explain specific areas of interest.

Hierarchical Reporting: Present findings in layers. The top layer provides the simplest overview, while subsequent layers offer increasing detail, allowing readers to delve deeper as per their understanding and needs.

Natural Language Conversational Agents: Implement chatbots that allow users to ask questions about the report and get simpler explanations in real time.

Use Case-based Learning: Provide similar past case summaries alongside the AI output. This approach helps relate the current findings to a real-world context and offers simpler explanations derived from past cases.

Relevance Filtering: Remove redundant or less relevant information and present only the most critical findings in simpler terms. An advanced version of this would allow users to toggle the level of detail they want.

Analogical Reasoning: Use analogies common in everyday life to explain intricate medical concepts, making the narrative relatable and easier to understand.

Contextual Pop-ups: In digital versions of the report, incorporate pop-ups or hover-over tooltips that provide succinct explanations of technical terms.

Feedback Loop Integration: Allow users to provide feedback on confusing parts of the report. Use this feedback to continuously train the model to improve clarity in subsequent outputs.

Glossary Inclusion: Attach a glossary of complex terms at the end of the report, explaining each in layman's language.

Collaborative Filtering: Use algorithms that provide narrative simplifications based on what similar users found most comprehensible.

Sentiment Analysis: Check the generated narrative for potentially alarming phrases and reword them to convey the message without causing unnecessary distress, ensuring the meaning remains intact.

Adaptive Learning Systems: These systems remember users' preferences and questions from previous reports. Over time, they provide more tailored and understandable reports based on past interactions.

3.2 *Algorithmic Overview: Radiology Report Simplification Using BERT*

The process to simplify radiology reports using a pre-trained BERT model with attention mechanisms. Starting with an input raw radiology report, the code first preprocesses the report by tokenizing and cleaning the text. If the BERT model hasn't been fine-tuned for radiology data, the code will do so using available training data. BERT, with its inherent attention mechanism, then generates a simplified version of the report. Optional post-processing steps correct any grammar or language inconsistencies and ensure all crucial information remains. Finally, the simplified report is outputted for display or storage.

```
  // 1. Input Radiology Report
   INPUT: raw_report
   // 2. Pre-processing
   processed_report = Tokenize(raw_report)
   processed_report = TextCleaning(processed_report)
   // 3. Load Pre-trained BERT Model
   bert_model = LoadPretrainedBERT()
   // 4. Fine-tuning on Radiology Dataset (assuming training_data is available)
   IF NOT bert_model.is_finetuned_on_radiology:
      bert_model = FineTuneModel(bert_model, training_data)
   // 5. Attention Mechanism (inherent in BERT)
      // The attention mechanism will automatically be applied during the
simplification process
   // 6. Simplified Report Generation
   simplified_report = bert_model.GenerateSimplifiedReport(processed_report)
   // 7. Post-processing (Optional)
   simplified_report = GrammarCorrection(simplified_report)
   simplified_report = EnsureInformationCompleteness(simplified_report)
   // 8. Output Simplified Radiology Report
   OUTPUT: simplified_report
   RETURN simplified_report
END FUNCTION
// Main Execution
raw_report = GET_REPORT_FROM_DATABASE_OR_INPUT()
simplified_report = SimplifyReport(raw_report)
DISPLAY_OR_STORE(simplified_report)
END
```

The pseudocode provides a structured algorithm for transforming intricate radiology reports into simpler versions using the BERT model, a state-of-the-art tool in natural language processing. The process commences with the receipt of a raw radiology report. Before employing the model, the report undergoes preprocessing—this involves converting the report into individual tokens (like words or phrases) and removing any extraneous or irrelevant details. This ensures that the data fed into the BERT model is clean and organized. Next, a pre-trained BERT model, which already knows a lot about language from its previous training, is loaded. However, to make it adept at handling radiology-specific terms and nuances, the model is fine-tuned on a dataset specific to radiology reports. This means it is further trained on radiology data to adapt its knowledge.

The heart of the process lies in the utilization of BERT's inherent attention mechanisms. Attention mechanisms allow the model to weigh the importance of different parts of the report when generating a simplified version. This ensures that crucial information remains intact even in the simplified output. After the model produces a simpler version of the report, there's an optional post-processing stage. Here, any minor grammatical errors made by the model are rectified, and a secondary check ensures that no essential information has been unintentionally left out.

3.3 Case Studies of Successful Simplification Tools in the Medical Domain

IBM Watson Health: IBM's Watson has been employed in various healthcare scenarios, including assisting oncologists in treatment planning. While not directly related to radiology reports, its natural language processing (NLP) capabilities have been utilized to distill vast medical literature into actionable insights [21].

Outcome: Clinicians were provided with summarized, relevant information to assist in decision-making.

Zebra Medical Vision: Zebra's imaging analytics platform uses AI to read medical imaging and generate findings. While initially focused on the detection aspect, there's potential in integrating simplified report generation based on its analyses [22].

Outcome: Increased diagnostic speed and potential for clearer reporting.

PathAI: PathAI uses advanced machine learning techniques to assist pathologists in diagnosing diseases from medical images. Their platform could potentially be extended to create patient-friendly summaries of findings [23].

Outcome: While primarily aiding in diagnosis accuracy, there's room to leverage its capabilities to provide simplified patient reports.

SigTuple: SigTuple, a Bengaluru-based startup, focuses on AI-driven solutions for pathology. Their platform, Manthana, processes medical data (like blood samples) and converts them into machine-readable formats. These are then analyzed to provide insights and results [24].

Outcome: Quicker, standardized pathology reports that could potentially be simplified for patient understanding with additional tools.

3.4 Evaluation Metrics and Validation Strategies for Simplified Narratives

Evaluating and validating simplified narratives, especially those generated by AI, requires a combination of quantitative and qualitative methods. Evaluating AI-simplified medical narratives requires a blend of quantitative and qualitative measures. Readability metrics, such as the Flesch Reading Ease and Gunning Fog Index, assess textual complexity, while comprehension can be gauged through question-answering or paraphrasing. Engagement with the narrative can be measured through attention time and drop-off rates, and feedback metrics like Likert scales can capture user sentiment. To validate the effectiveness of these narratives, strategies like user testing, A/B testing, and expert reviews are essential. Moreover, aligning the content with established guidelines, such as plain language standards, and monitoring real-world outcomes like patient compliance can further attest to the narrative's clarity and impact.

4 Ethical and Regulatory Considerations

When deploying AI-generated medical narratives in radiology, it's imperative to uphold patient privacy and data security, ensuring compliance with standards. Patients should be informed about the AI's role and provide consent. The AI's methods must be transparent and explainable, with a focus on maintaining accuracy and minimizing biases. Liability lines should be clear in case of discrepancies, with continuous monitoring of the AI's performance and feedback mechanisms in place. Healthcare professionals should receive training to navigate these narratives, and the AI system should adhere to local and global regulatory standards [25]. Engaging with all stakeholders, including healthcare providers, patients, and regulators, will ensure an ethically sound and holistic approach.

5 Discussion

Artificial intelligence (AI) is making a big impact in medicine, especially in turning complex radiology reports into simple explanations that patients can understand. However, our study shows that for AI to be truly useful in healthcare, it's not enough for it to be just technically good. It must be transparent. This means that patients, doctors, and others need to trust it, and it should be clear about how it makes decisions. Our AI model doesn't just simplify reports; it also explains how it decides to simplify them.

One might argue about the necessity of such a granular level of transparency, especially when outcomes (simplified reports) meet the desired criteria. The rationale is twofold: First, healthcare, in its essence, is profoundly personal. A patient's comprehension of a medical report isn't merely academic; it can influence decisions, perceptions, and emotional states. Ensuring that an AI's simplification can be trusted becomes paramount. Second, for medical practitioners, understanding an AI's decision-making can be integral to validating its outputs, ensuring that nuances aren't lost in translation. Our study, in its scope and findings, makes a compelling case for integrating transparency into AI tools in healthcare. Simplification, while pivotal, is just one dimension. An AI tool's effectiveness in the medical domain isn't just measured by its operational efficiency but significantly by its ability to engender trust. And trust, as our research highlights, is as much about clarity and justification as it is about outcomes. In conclusion, the future of AI in healthcare, as envisioned through our research, is not of isolated, inscrutable algorithms churning out results but of transparent models working in tandem with humans, fostering trust, ensuring understanding, and jointly navigating the intricate labyrinth of healthcare narratives. The path ahead is clear: AI tools in healthcare should not only be about intelligence but crucially about "intelligibility."

References

1. Topol EJ (2019) High-performance medicine: the convergence of human and artificial intelligence. Nat Med 25(1):44–56
2. Rajkomar A, Dean J, Kohane I (2019) Machine learning in medicine. N Engl J Med 380(14):1347–1358
3. Miotto R, Wang F, Wang S, Jiang X, Dudley JT (2020) Deep learning for healthcare: review, challenges and the future. Brief Bioinform 21(1):123–136
4. Rajpurkar P, Irvin J, Ball RL, Zhu K, Yang B, Mehta, H, Ng AY (2018) Deep learning for chest radiograph diagnosis: a retrospective comparison of the CheXNeXt algorithm to practicing radiologists. PLoS Med 15(11):e1002686
5. Jiang F, Jiang Y, Zhi H, Dong Y, Li H, Ma S, Wang Y, Dong Q, Shen H, Wang Y (2019) Artificial intelligence in healthcare: past, present and future. Stroke Vasc Neurol 4(4):230–243
6. Obermeyer Z, Emanuel EJ (2016) Predicting the future: big data, machine learning, and clinical medicine. J Am Med Assoc 316(10):1061–1062
7. Johnson AJ, Chen MY, Swan JS, Applegate KE (2011) Patient understanding of radiation risk from diagnostic imaging: a multi-institutional survey. J Am Coll Radiol 8(12):865–872

8. Gunn AJ, Gilcrease-Garcia B, Mangano MD, Sahani DV, Boland GW, Pandharipande PV (2018) Structured reporting in radiology. Acad Radiol 25(1):66–73
9. Bossuyt PM, Reitsma JB, Linnet K, Moons KG (2012) Beyond diagnostic accuracy: the clinical utility of diagnostic tests. Clin Chem 58(12):1636–1643
10. Shortliffe EH, Sepulveda MJ (2018) Clinical decision support in the era of artificial intelligence. J Am Med Assoc 320(21):2199–2200
11. Carvalho DV, Pereira EM, Cardoso JS (2019) Machine learning interpretability: a survey on methods and metrics. Electronics 8(8):832
12. Bello F, Bhatia S, Ienca M (2022) Artificial intelligence in healthcare: ethics, transparency, and accountability. Springer, Switzerland
13. Ding D, Wang Z, Li X (2022) Towards consistent and interpretable artificial intelligence for healthcare. Springer, Switzerland
14. Buolamwini J, Gebru T (2020) Gender shades: intersectional accuracy disparities in commercial gender classification. Proc Natl Acad Sci 117(15):7464–7473
15. Rajkomar A, Oren E, Chen K, Dai AM, Hajaj N, Hardt M, Liu PJ, Liu X, Marcus J, Sun M, Sundberg P (2018) Scalable and robust deep learning with implicit regularization. Springer Briefs Comput Sci 105–114
16. Kearns M, Roth A (2019) The ethical challenges of machine learning. Nature 575(7781):354–358
17. Lipton J, Berkowitz J, Sutskever I (2016) Learning to diagnose from medical images with deep neural networks. Adv Neural Inf Process Syst 256–265
18. Lipton J, Steinhardt J (2018) Does my model understand what it's doing? A critical assessment of interpretability methods for machine learning. Mach Learn Knowl Discov Databases 56–71
19. Alonso M, Llinares J (2022) Explainable artificial intelligence: concepts, taxonomies, opportunities, and challenges. Springer, Switzerland
20. Patel V, Wachter R (2022) The ethics of artificial intelligence in healthcare. Springer, Switzerland
21. Kumar A, Joshi S (2022) Applications of AI in healthcare sector for enhancement of medical decision making and quality of service. In: International conference on decision aid sciences and applications (DASA). IEEE
22. Harry A (2023) Transforming patient care: the role of artificial intelligence in healthcare; a mini-review. Bull Jurnal Multidisiplin Ilmu 2(3):530–533
23. Khanna NN et al (2022) Cardiovascular/stroke risk stratification in diabetic foot infection patients using deep learning-based artificial intelligence: an investigative study. J Clin Med 11(22):6844
24. Thakur A (2022) Market determinants impacting distributed ledger technology, and AI-based architectures in the healthcare industry. Int J Bus Anal Intell 10(10)
25. Jeyaraman M et al (2023) Unraveling the ethical enigma: artificial intelligence in healthcare. Cureus 15(8)

Green Construction Project Management: A Bibliometric Analysis

T. Gunanandhini, S. Sivakumar, Aswathy Sreenivasan, and M. Suresh

1 Introduction

The term "green construction" refers to the methods and ideologies that aim to create and use buildings as environmentally friendly as possible. The green building focuses on minimizing harmful effects on the environment and even introducing some beneficial effects, from the design phase to assembly to the functionality of the project after completion. To effectively handle green construction projects, project managers need to grasp the critical qualities they need to have as the green building phenomena continue to spread and become more popular [1, 2].

Given the increased focus on sustainability on a global scale, it is the obligation of the construction sector to act decisively to reduce their environmental impact when erecting a new structure and to design that structure to run sustainably for many generations to come. The construction industry needs to be a pioneer in implementing sustainable practices and reducing the impact on the environment because of its significant social impact. Green buildings will keep expanding and establishing themselves as the new norm as the long-term implications of our activities come into clearer light. One of the examples that can be provided for green construction project is Masdar City in Abu Dhabi. Masdar City is a leader in sustainability and a center for research and development, driving the advancements toward more environmentally friendly, sustainably oriented urban living. With attributes including a pedestrian-friendly design, the utilization of renewable energy, and garbage recycling, Masdar City aspires to be one of the most sustainable urban developments in the world. It illustrates how sustainable communities can be built using a comprehensive approach to green construction. This paper refers to the various research conducted on green project management, especially in construction. We seek to identify the major themes and subjects within the green construction project management field by examining the current literature. This study aims to show how the sector has changed over

T. Gunanandhini (✉) · S. Sivakumar · A. Sreenivasan · M. Suresh
Amrita School of Business, Amrita Vishwa Vidyapeetham, Coimbatore 641112, India

M. S. Kaiser et al. (eds.), *ICT: Smart Systems and Technologies*, Lecture Notes in Networks and Systems 878, https://doi.org/10.1007/978-981-99-9489-2_7

time by identifying emerging trends and patterns in research on green construction projects. Examining studies' geographic and chronological distribution can help us understand how green building project management has changed across time and space. Finally, by identifying potential research gaps, this analysis will aid current and future researchers and practitioners in addressing topics that need more investigation. To achieve the aim of this paper, the following questions are developed and answered: "RQ1: How have the concept of green implementation in construction been investigated in terms of a year, region, research technique, context, and outcomes?" "RQ2: What are the limitations of the existing literature?" "RQ3: What are the possibilities for future research for lean implementation in the construction industry?."

2 Literature Review

Hwang and Ng [1] state that the top 10 difficulties that project managers face when handling green construction were (1) "the pre-construction process takes longer than it should;" (2) "there are difficulties in choosing subcontractors who offer green construction services;" (3) "there is uncertainty about the use of green materials and equipment;" (4) "there is a high cost associated with using green materials and equipment;" (5) "more discussions and cooperation with green consultants and engineers required;" (6) "more frequent design changes and variations during construction;" (7) "difficulty in understanding green requirements in contract details;" (8) "unplanned events that arise when carrying out green projects;" (9) "planning for unconventional construction processes;" and (10) "preparing for various methods of constructing." Hwang et al. [3] reveal that the major components of job satisfaction as "job content," "resource adequacy," "work context," and "personality and competence of stakeholders." Also, eight measures were devised in this study to improve the job satisfaction of project managers, including the increasing possibilities for the project managers recognition, government financial aids and introduction and training courses on green construction laws and procedures.

Rawai et al. [4] explore the most beneficial cloud computing technologies for construction collaboration and the key ideas of sustainable project management. The research shows that using cloud computing in the construction sector will streamline project collaborations, communications, relationships, and networks. This study will improve collaboration within the construction sector, which could ultimately strengthen and expand the sector's competitive advantages. It also provides innovative ideas for collaboration and integration within the industry, increasing productivity and boosting efficiency and effectiveness in the transition to a greener construction sector. Robichaud and Anantatmula [5] researched on "Green Project Management Practices for Sustainable Construction" to make specific recommendations for modifying traditional construction techniques to maximize the execution of green building projects that are both affordable and environmentally friendly. It summarizes research on the expenses and prospects of green building and then uses the findings to suggest procedures for greening project management in construction. The findings of this study indicate that using green project management techniques can significantly improve a sustainable building project while completing it within reasonable financial limitations. With the premise that a green project has a better chance of being financially successful if an inter-discipline team is involved from the beginning to the end of the project, this study presents certain improvements to conventional project management techniques. Hwang et al. [2] state that green construction projects took 8% longer on average compared to a similar-sized conventional building project. Also, these projects had a 4.8% delay relative to their intended schedules. Finally, a list of suggestions to improve the schedule's performance was also presented. It will provide an industry benchmark for measuring the total length and effectiveness of green construction projects.

According to Wu et al. [6], the most effective green performance building plans will consider various risks, standards, design norms, and financial constraints. It is discovered that eight variables account for 77.5% of the variation in the standards and quality of highway construction. The top five difficulties that project managers have while managing green construction projects including (1) "the lengthier time needed during the planning phase before construction;" (2) "trouble in the selection of partners who offer green services;" (3) "ambiguity concerning sustainable materials and equipment; (4) the high cost of required materials;" (5) "regular consulting with the green experts." The vital domains were human resource management, governance structure, communications and cost management and scheduling and planning. According to Arif et al. [7], buyers in India are ready to pay more to go green because of the nation's current energy problem. However, the cost of going green is unclear because there are no trustworthy lifecycle cost assessment methodologies. Luo et al. [8] identify the most recent developments in the field of green construction as well as the knowledge gaps in the earlier studies. A list of significant recommendations for potential future study areas was also put out to address the biggest knowledge gaps. Therefore, the information obtained from this study will aid those working on construction strategies and carrying out green construction projects. Table 1 synthesizes the past literature.

3 Research Methodology

We chose a total of 1597 papers pertinent to green implementation in construction project management for our study's thorough literature evaluation. We used precise search phrases, such as "Green AND Construction," "Green AND Project Management," and "Green AND Construction Project Management" in the TITLE-ABSTRACT-KEYWORDS field, to find these articles. We used the Scopus and Biblioshiny analysis tools to make our analysis easier. Our study period covers over 40 years and a wide spectrum of pertinent literature from 1983 to 2022. We gathered data from 758 different sources during the data-gathering procedure, enabling us to assemble a sizable dataset for analysis. We found and examined a total of 1597

Table 1 Synthesis of past literatures

Paper type	References	Aim
Review	[5, 8–12]	The objective is to offer thorough guidance on how to modify "traditional building methods" to facilitate the implementation of green building projects that are both economically advantageous and environmentally responsible. Before making suggestions for greening project management in the construction industry, it evaluates studies on the costs and possible advantages of green construction. To comprehend the significance of project management in reaching sustainable goals, it also includes a list of the elements influencing green building construction efficiency and sustainability
Case study	[13–15]	It attempts to endorse a technique for building known as "Green Construction Assessment" (GCA). "Management performance indicators (MPIs)" and "operational performance indicators (OPIs)" are the two recognized categories of environmental indicators. It also creates a framework for assessing a construction project's environmental performance. The framework offers a dynamic and engaging procedure that supports ongoing site operation monitoring and enhancements
Modeling	[1–4, 7, 9, 16–30]	These studies seek to pinpoint the challenges faced by project managers when working on green building projects as well as the crucial disciplines and skills needed to overcome these challenges. Through a review of the literature, surveys, and interviews with project managers, this study will provide a knowledge base for project managers to be competitive and successful and successfully accomplish sustainable project work

distinct documents in this collection. These documents were generally published around 7.91 years ago, demonstrating a varied and current array of sources.

4 Results and Discussion

We used the *R* programming language's Biblioshiny tool to extensively study prior literature. The major phrase used frequently in building and green literature is represented visually in Fig. 1. Notably, the phrases "green building," "sustainability," and "sustainable development" were most frequently used in this context. The word cloud in Fig. 1 displays these keywords in varied sizes to show their frequency, with the most widely used ones put prominently in the center for emphasis. To further examine the connections between these phrases and concepts, we also created a thematic map, shown in Fig. 2, divided into four geographic zones based on "centrality and density." The high relevance with high density is "construction, risk management, green," and "green building, and sustainable construction." Figure 3 shows the thematic

Fig. 1 Word cloud

evolution of terms. For example, prior to 2017, the most discussed concept was sustainable building. Post 2017, the importance has shifted to focus more on green building. Figure 4 shows the country collaboration map. The most collaborative country is China, with Australia, Hong Kong, the USA, and Singapore having the highest frequency. Hwang, B.G. is the most influential author with 14 documents (Fig. 5). Hwang, B.G. most cited article was "Green building project management: Obstacles and Solutions for Sustainable Development," with a citation of 283. The research attempts to identify typical challenges faced in managing green construction projects and, at the end, suggests some methods to address the challenges. The study's results indicate that even though project costs are the biggest obstacle to managing the construction of green buildings, there is still a wealth of sustainable expertise in the construction sector. The scope of government subsidies should be expanded to cover the use of green technologies and products in order to address the cost-related issue. To overcome these obstacles and perhaps encourage the use of sustainable construction in upcoming projects, a project management framework for green building construction should also be established (Fig. 5).

5 Findings

According to Hwang and Ng [1], the top 10 difficulties that construction project managers identify when handling green construction were identified as (1) "the pre-construction process takes longer than it should;" (2) "there are difficulties in choosing subcontractors who offer green construction services;" (3) "there is uncertainty about the use of green materials and equipment;" (4) "there is a high cost associated with using green materials and equipment;" (5) "more discussions and cooperation with green consultants and engineers required;" (6) "more frequent design changes and variations during construction;" (7) "difficulty in understanding green requirements in contract details;" (8) "unplanned events that arise when carrying out green projects;" (9) "planning for unconventional construction sequences;" and (10) "preparing for various construction methods." Hwang and Tan [16] found that

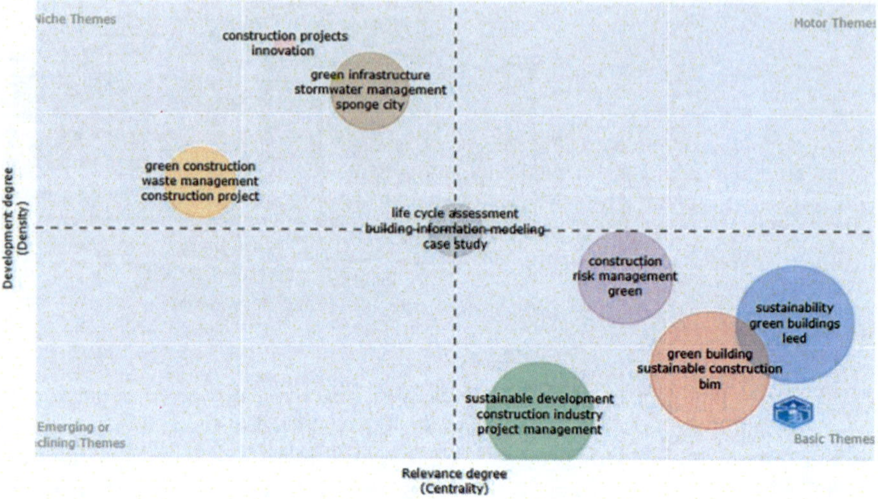

Fig. 2 Thematic map

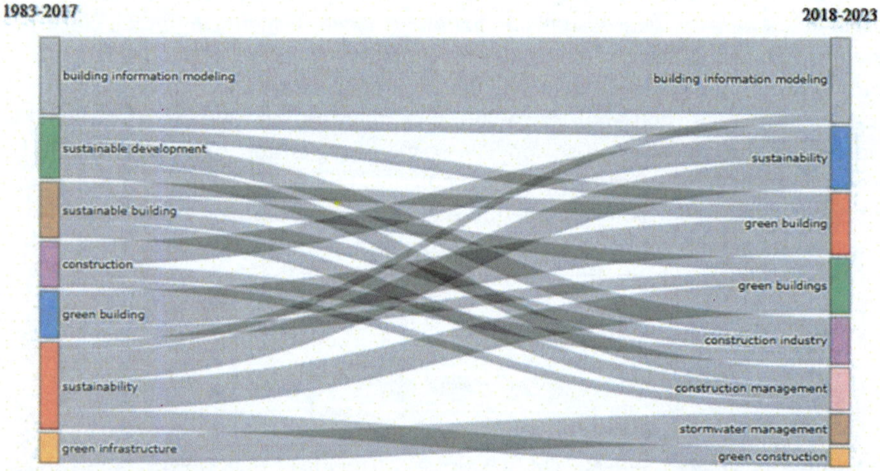

Fig. 3 Thematic evolution

the results of this study show that: (1) "there are significant differences between conventional and green construction projects, particularly in the amount of detail and communication needed;" (2) "there is no shortage of sustainable knowledge in the Singapore construction industry, but difficulties against executing the idea exist, and the absence of investment in green construction;" and (3) "a project control framework has to be created, perhaps encouraging the use of sustainable management techniques for future projects." According to Zhao et al. [17], the perceptions of "job content" and "personality and ability of employees" varied between

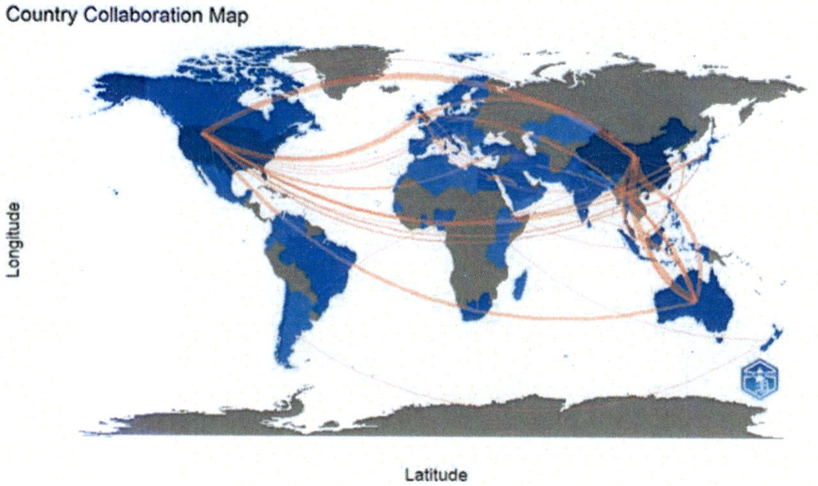

Fig. 4 Country collaboration map

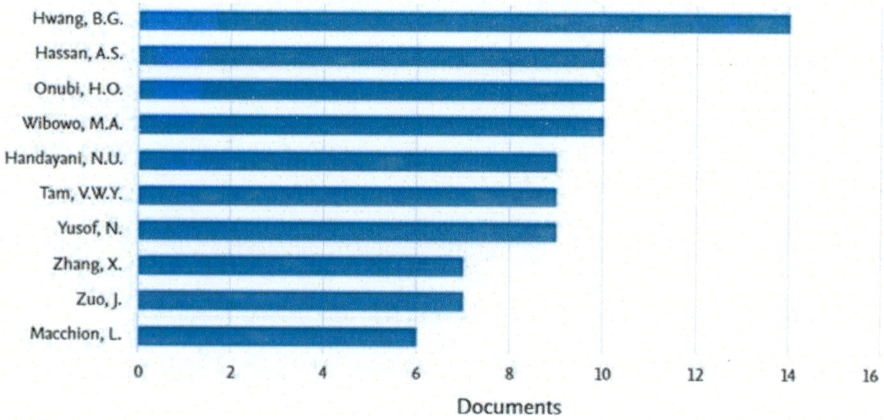

Fig. 5 Document by authors

green and conventional construction projects. The main factors that prevent project managers from being satisfied with their jobs in green construction projects are "technical demand," "split incentive situation," "project managers' competence and understanding," and "requirement and assistance from stakeholders." Mohd-Rahim et al. [12] highlight how healthy buildings that are constructed with consideration for the environment, energy emission, and emission reduction are referred to as "green buildings." The structural empirical review determines four fundamental aspects of retrofitting: (1) Improvements in energy efficiency, (2) technological change, (3) organizational difficulties, and (4) behavioral shifts.

According to Hwang and Tan [18], Singapore's construction industry is well-versed in sustainable practices, even if project costs pose the biggest barrier to controlling the development of green buildings. To solve the issue of costs, the range of incentive schemes has to be broadened to involve the application of green technologies and products. A project management framework for the construction of green buildings has to be designed to address these challenges and promote the usage of sustainable construction in upcoming projects. According to Hwang et al. [3], the significant elements of job satisfaction are "job substance," "resource sufficiency," "Work setting," and "stakeholders' personalities and skills." Rawai et al. [4] show that using cloud computing in construction will streamline project collaborations, communications, relationships, and networks. It is intended that this study will help to improve collaboration within the construction sector, which could ultimately strengthen and expand the sector's competitive advantages. It also provides innovative ideas for collaboration and integration within the industry, increasing productivity and boosting efficiency and effectiveness in the transition to a greener construction sector. According to Robichaud and Anantatmula [5], using green project management techniques can significantly improve a sustainable building project while completing it within reasonable financial limitations. With the premise that a green project has a better chance of being financially successful if an inter-discipline team is involved from the beginning to the end of the project, this study presents certain improvements to conventional project management techniques.

Luo et al. [8] conducted a scientometric research on "A Systematic Review of Green Construction Research using Scientometric Methods." The scientometrics analysis revealed the most recent developments in the field of green construction as well as the knowledge gaps in the earlier studies. A list of significant recommendations for potential future study areas was also put out to address the biggest knowledge gaps. Therefore, the information obtained from this study will aid those working in the construction strategies and carrying out green construction projects. Zhao et al. [21] revealed 13 leadership traits, with "strive for job performance and productivity" coming first. These traits were divided into two categories: relationship-oriented leadership and directive and task-oriented leadership. The findings showed that the project managers' leadership styles tended to be more directive and task-oriented but did not ignore their relationships with subordinates. Hasan and Zhang [22], in an attempt to learn about the concerns with adopting green construction and how these activities influence the environment and community significantly, conducted study on "Critical Barriers and Challenges in Implementation of Green Construction in China." The results of this study show that most stakeholders believe there are obstacles that the relevant authorities should consider. Higher costs and lack of knowledge of the technologies provide the biggest threat to possible challenges. Another significant conclusion is that supervisors differ in their opinions of top management's support for green construction. Senior management must address this issue and satisfy supervisors' concerns regarding green building.

Shi et al. [23] demonstrated that significant obstacles include increased expense, more effort, and a shortage of green providers and information. To remove these constraints and promote green construction practices, discussions were held. This study is a helpful resource for industry professionals and government decision-makers. Onubi et al. [30] intended to evaluate the effects of implementing particular green construction site practices on the performance of projects regarding health and safety. The results show that while storm-water management has little impact, energy and waste management methods considerably impact the health and safety performance of the construction projects. The research on "Green Building Project Management: Obstacles and Solutions for Sustainable Development" by Khodadadzadeh [25] has analyzed recent green building design and implementation developments in this survey. It has described how buildings contributed to global warming by using around 32% of the total final energy produced worldwide and emitting 19% of the greenhouse gases. There is a need for the globe to take the appropriate steps since global warming will soon become the biggest issue on the planet. The findings of this survey will assist academics, governmental organizations, and policymakers in developing new guidelines for promoting green building design. According to Shooshtarian et al. [14], future study is strongly encouraged to demonstrate GC programs; capability in the efficient management of construction; and demolition waste in various construction projects. The findings might help design policies and motivate the construction and waste recovery sectors to implement the management practices established in these programs.

6 Conclusion

Green construction project management is the process of overseeing building projects in a way that is sustainable and environmentally friendly [31–33]. This strategy aims to reduce the damaging effects of construction operations on the environment and encourage the use of green building products, processes, and technology. The literature review was used to determine the various impacts of green construction project management and how it affects the job satisfaction of the project managers. Through bibliometric analysis of the chosen journals on green construction, this paper provides the required suggestions and recommendations for construction managers to improve efficiency, including cost management and environmental performance. This study also enlists the skill set and knowledge areas demanded from the project managers in construction project management.

This study also found how knowing client satisfaction helps construction project managers to compete in the construction sector. Green construction is not merely the usage of renewable energy resources and recyclable materials, it also involves energy efficiency using sustainable practices and techniques that do not cause any damage to the environment. This could enhance not only the environmental performance but also the economic performance of the project. It was also revealed that green construction projects demand an integrated strategy for communication. Though it might increase construction project managers' challenges, green construction projects need competent and skilled managers. Building information modeling (BIM) has emerged as a crucial technology for managing green building projects. As a vital tool for the future of sustainable building, it may support sustainability objectives, improve collaboration, guide decision-making, and optimize resource management. BIM will become more and more important in determining the effectiveness of green building programs as the construction industry prioritizes environmental responsibility.

In conclusion, as people become more conscious of the effects of construction activities on the environment, green construction project management is becoming increasingly crucial in today's society. Deep knowledge of sustainable building techniques and the capacity to effectively convey and apply these techniques throughout the construction project are prerequisites for effective green construction project management. This study's limitation was the paucity of limitations and future research suggestions in previous studies. This study analyzed the available literature due to several restrictions, including time and restricted outreach. Researchers can look at case studies to see how various nations, states, or businesses have successfully used green construction techniques because this industry is big and is expected to impact India's GDP significantly. Additionally, a study may be required to identify the problems and offer remedies.

References

1. Hwang BG, Ng WJ (2013) Are project managers ready for green construction? Challenges, knowledge areas, and skills. Smart Sustain Built Environ 25
2. Hwang BG, Ng WJ (2013) Project management knowledge and skills for green construction: overcoming challenges. Int J Project Manage 31(2):272–284
3. Hwang BG, Zhao X, Lim J (2019) Job satisfaction of project managers in green construction projects: influencing factors and improvement strategies. Eng Constr Archit Manag 27(1):205–226
4. Rawai NM, Fathi MS, Abedi M, Rambat S (2013) Cloud computing for green construction management. In: 2013 Third international conference on intelligent system design and engineering applications. IEEE, pp 432–435
5. Robichaud LB, Anantatmula VS (2011) Greening project management practices for sustainable construction. J Manag Eng 27(1):48–57
6. Wu X, Zhao W, Ma T (2019) Improving the impact of green construction management on the quality of highway engineering projects. Sustainability 11(7):1895
7. Arif M, Egbu C, Haleem A, Kulonda D, Khalfan M (2009) State of green construction in India: drivers and challenges. J Eng Des Technol 7(2):223–234
8. Luo W, Sandanayake M, Hou L, Tan Y, Zhang G (2022) A systematic review of green construction research using scientometrics methods. J Clean Prod 132710
9. Wu P, Low SP (2010) Project management and green buildings: lessons from the rating systems. J Prof Issues Eng Edu Pract 136(2):64–70
10. Onubi HO, Yusof NA, Hassan AS (2020) Adopting green construction practices: health and safety implications. J Eng Des Technol 18(3):635–652
11. Dong Y (2020) Project management affecting the productivity and sustainability of a green building: a literature review. In: Sustainable development of water and environment: proceedings of the ICSDWE2020, pp 251–257
12. Mohd-Rahim F, Pirotti A, Keshavarzsaleh A, Zainon NI, Zakaria N (2017) Green construction project: a critical review of retrofitting awarded green buildings in Malaysia. J Des Built Environ 11–26

13. Tam CM, Tam VW, Tsui WS (2004) Green construction assessment for environmental management in the construction industry of Hong Kong. Int J Project Manage 22(7):563–571
14. Shooshtarian S, Maqsood T, Wong PS, Khalfan MJ, Yang R (2019) Green construction and construction and demolition waste management in Australia. In: Proceedings of the 43rd annual Australasian university building educators association conference (AUBEA 2019), Central Queensland University, pp 18–25
15. Zou X, Moon S (2014) Hierarchical evaluation of on-site environmental performance to enhance a green construction operation. Civ Eng Environ Syst 31(1):5–23; Eason G, Noble B, Sneddon IN (1955) On certain integrals of Lipschitz-Hankel type involving products of Bessel functions. Phil Trans Roy Soc Lond A247:529–551
16. Hwang BG, Tan JS (2012) Green building project management: obstacles and solutions for sustainable development. Sustain Dev 20(5):335–349
17. Zhao X, Hwang BG, Lim J (2020) Job satisfaction of project managers in green construction projects: constituents, barriers, and improvement strategies. J Clean Prod 246:118968
18. Hwang B, Tan JS (2012) Sustainable project management for green construction: challenges, impact and solutions. In: Global challenges in construction industry, pp 171–179
19. Hwang BG, Leong LP, Huh YK (2013) Sustainable green construction management: schedule performance and improvement. Technol Econ Dev Econ 19(sup1):S43–S57
20. Sang P, Liu J, Zhang L, Zheng L, Yao H, Wang Y (2018) Effects of project manager competency on green construction performance: the Chinese context. Sustainability 10(10):3406
21. Zhao X, Hwang BG, Lee HN (2016) Identifying critical leadership styles of project managers for green building projects. Int J Constr Manag 16(2):150–160
22. Hasan MS, Zhang RJ (2016) Critical barriers and challenges in the implementation of green construction in China. Int J Curr Eng Technol 6(2):435–445
23. Shi Q, Zuo J, Huang R, Huang J, Pullen S (2013) Identifying the critical factors for green construction–an empirical study in China. Habitat Int 40:1–8
24. Hwang BG, Leong LP (2013) Comparison of schedule delay and causal factors between traditional and green construction projects. Technol Econ Dev Econ 19(2):310–330
25. Khodadadzadeh T (2016) Green building project management: obstacles and solutions for sustainable development. J Proj Manag 1(1):21–26
26. Onubi HO, Hassan AS (2020) How environmental performance influence client satisfaction on projects that adopt green construction practices: the role of economic performance and client types. J Clean Prod 272:122763
27. Onubi HO, Yusof NA, Hassan AS, Bahdad AAS (2021) Analyzing the mediating effect of economic performance on the relationship between green construction practices and health and safety performance in Nigeria. Environ Sci Pollut Res 28(27):36598–36610
28. Wang Y, Chong D, Liu X (2021) Evaluating the critical barriers to green construction technologies adoption in China. Sustainability 13(12):6510
29. RezaHoseini A, Noori S, Ghannadpour SF (2021) Integrated scheduling of suppliers and multi-project activities for green construction supply chains under uncertainty. Autom Constr 122:103485
30. Onubi HO, Hassan AS (2020) Understanding the mechanism through which adoption of green construction site practices impacts economic performance. J Clean Prod 254:120170
31. Suresh M, Antony J, Nair G, Garza-Reyes JA (2023) Lean-sustainability assessment framework development: evidence from the construction industry. Total Qual Manag Bus Excellence 1–36. https://doi.org/10.1080/14783363.2023.2222088
32. Nair G, Suresh M (2021) Challenges faced by construction organizations during covid-19 era. In: IOP conference series: earth and environmental science, vol 796, no 1. IOP Publishing, p 012004
33. James PM, Ramesh S, Harish MT, Bhavani RR (2020) Semi-automation system for construction industry with emphasis on humanitarian technology. In: ICDSMLA 2019: proceedings of the 1st international conference on data science, machine learning and applications. Springer, Singapore, pp 1450–1462

Illuminating Agriculture: Crafting a Strategy IoT-Based Architectural Design for Future Growth

M. Pavithra, S. Duraisamy, and R. Shankar

Abbreviations

IoT	Internet of Things
ICTs	Information and Communication Technologies
No. of	Number of
MRWC	Medium Range Wireless Communication
SRWC	Short-Range Wireless Communication
LRWC	Long-Range Wireless Communication
LPWA	Low Power Wide Area
PA	Precision Agriculture
LTE	Long Term Evolution
Wi-Fi	Wireless Fidelity
NB-IoT	Narrow Band IoT's
SM	Soil_Moisture
TH	Threshold
TH_min	Minimum Threshold
TH_max	Maximum Threshold
SMC	Soil Moisture Content Percentage

M. Pavithra (✉) · S. Duraisamy · R. Shankar
Chikkanna Government Arts College, Tirupur, India
e-mail: pavithra02.09.98@gmail.com

S. Duraisamy
e-mail: sdsamy.s@cgac.in

R. Shankar
e-mail: shankarcgac2020@cgac.in

M. S. Kaiser et al. (eds.), *ICT: Smart Systems and Technologies*, Lecture Notes in Networks and Systems 878, https://doi.org/10.1007/978-981-99-9489-2_8

1 Introduction

India is a country of farmers, and it emphasizes farming more so since it is vital to the country's economic growth. We must increase agricultural productivity through effective water and energy usage because agriculture's contribution to the national economy is now declining. Agriculture will become more conscious of water conservation, automated, and the agricultural sector in the twenty-first century must be highly intelligent, and it must also be top-notch high-yielding, and pollution-free. To achieve this, IoT is an emerging technology and its instrumentation that is best suited to fields of study like agriculture.

The IoT is now a widely researched topic among academics, industry professionals, and specialists. The IoT is already affecting people all over the world, and from the perspective of the average user, it is setting the groundwork for the establishing a wide range of goods, such as smart medical facilities, smart decor, smart academic institutions lecturing, and computerization. Also, it is employed commercially in a variety of industries, including manufacturing, transportation, agriculture, and corporate management [1].

Agricultural IoT's "human–machine-things" linkage enables more sophisticated and dynamic human recognition, management, and control of numerous agricultural aspects, processes, and systems. It may also considerably increase human understanding of important parts relating to farming species' and crops' life, aid in the capacity to manage intricate ways of farming, and help with the management of agricultural emergencies. Agricultural IoT technology is now the subject of considerable and active global research, albeit the majority of the software continues to be in the experimental proof-of-concept stage [2].

As society and technology continue to advance, the a proliferation of the IoT offers strong technological backing for green farming advancements, which encourages the growth of the latter as a significant component of contemporary agriculture [3, 4] The IoT architecture is not subject to a single globally accepted agreement. Many scholars have presented several architectural designs. Architectures of three and five layers, respectively. Three layers constitute the most fundamental architecture [5]. The sensing layer comes first. It is made up mostly of different sensors that may utilize data to communicate with sensing devices. This allows for the collection of the necessary data, intelligent analysis, and eventually the establishment of a link between both the sensor and the network. The fundamental objective of the network layer, which comprises mostly the administration of the network system, the remote computing system, and the Internet system, is to observe the handling of information and transmission. The application layer is capable of fully completing the identification and detection between persons and commodities, commodities and things, is the last layer, and it allows for the completion of data research and analysis as well as the IoT's intelligent function [6, 7].

The sole objective of this research is to explore the importance of IoT and propose its architecture. Section 2 describes market momentum, value, and adoption of Internet of Things (IoT) around the globe. Agricultural IoT's current trends and applications are briefly discussed in Sect. 3. (Irrigation management system, Water quality monitoring system, Crop Monitoring and disease and pest management system) Sect. 4: We investigate and integrate Internet of Things (IoT) infrastructures designed specifically for smart farms, which include sensor, communication, and data analytics technologies that assist in operating farming equipment. For a smart agricultural system, we create our general IoT system architecture. Section 5 displays the first working of our research prototype of self-contain irrigation technology. Section 6 concludes with findings and suggests more study.

2 Current Trend of IoT

Until recently, an Internet connection was primarily used for interacting with devices at any point in time and from any point in the world, however, this required human interaction and control. The IoT is an innovative model that builds on existing ICTs. The Internet of Things allows any-time accessibility to objects that exist in the real world by linking them to the digital realm, everywhere connections for anything. In the Internet of Entities universe, real and virtual things coexist in the same location and time [8]. By 2025, there may be 27 billion linked IoT devices, an 18% increase from the current 14.4 billion. By 2030, the number of hooked up IoT gadgets is estimated to reach 25.4 billion. There are presently around 400 active IoT platforms. [9] Every minute, 152,200 IoT devices are expected to link to the World Wide Web by 2025. IoT services have the potential to generate $4–11 trillion in financial benefits by 2025 [10]. Figures 1 and 2 show the IoT active connections statistics [11]. A short outline of the available IoT applications is given below [12, 13].

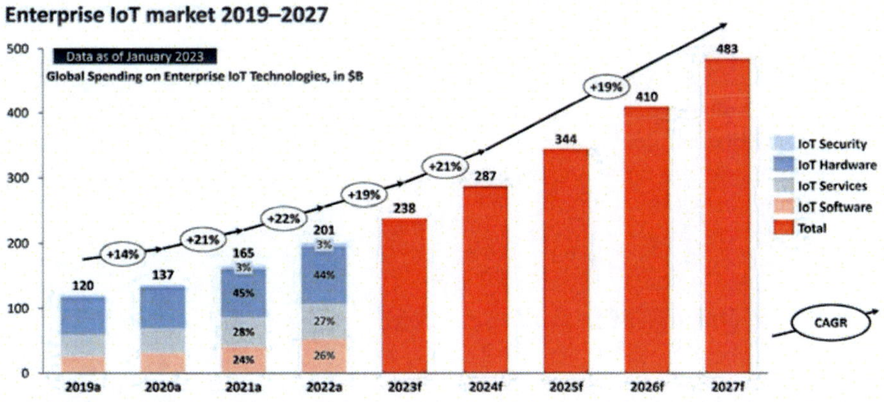

Fig. 1 Analytical view of IoT market [12]

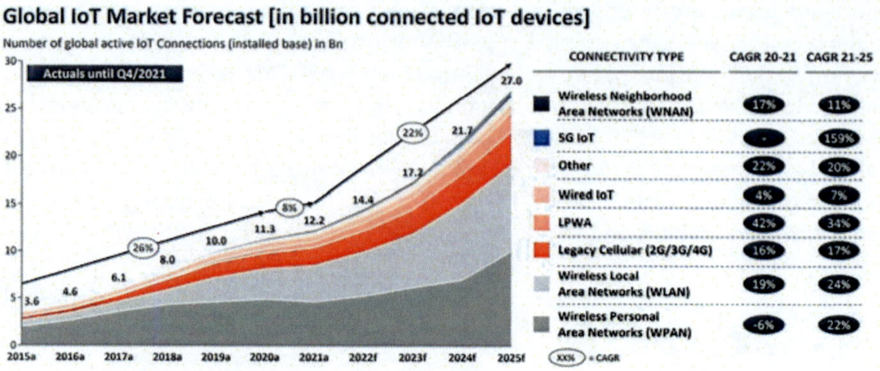

Fig. 2 IoT connected IoT devices [13]

3 Current Trend of Agricultural IoT

Farmers that use smart farming have a better awareness of important factors including water, topography, aspect, vegetation, and soil kinds. Farmers are then able to determine how to manage scarce resources in their producing region in a way that is both environmentally and economically viable. The use of additional technology in conjunction with agricultural production methods in order to decrease waste and boost output is known as "smart farming" [14].

3.1 Irrigation Management System

By improving the efficiency, productivity, and sustainability of irrigated agriculture and irrigation infrastructure, irrigation management aims to boost food production while simultaneously promoting economic growth. Everyone is focusing on smart irrigation technology and irrigation schedulers [15] since managing irrigation is becoming more and more important owing to water scarcity [16] and water waste from outdated irrigation systems [17].

3.2 Water Quality Monitoring System

Data acquired from smart sensors may be used to create a smart water distribution system. Real-time control of the water supply is provided through distribution lines equipped by pressure sensors, IoT flow of water meters, and electrically controlled valves. [18] Utilizing smart water distribution system, feasible can obtained following business benefits by applying smart water distribution service offering achieve:

- Cut down on water waste with features [19], like remote control, real-time monitoring, and predictive maintenance.
- To effectively distribute water across the distribution lines, balance water supply and demand [20].
- Monitor and manage water quality using smart water sensors that have been deployed at various places along distribution lines [19].

Smart water distribution by monitoring water flow and direction with water sensors built into the pipes [20, 21]. IoT solutions for the water sector may also alert water distributors to any pipeline leaks [22]. The main advantage of deploying an IoT water solution is that it can alert water distributors about any pipeline leaks. Intelligent water usage is made possible by IoT smart water solutions in homes, farms, and cities [23].

3.3 Smart Farming

Smart farming is controlled by technology and kept track of by instruments [24]. The escalating need for greater production for potential [25], the necessity to make better use of the environment's assets, to increasing sophistication and utilization of technological advancements in communication and information, and [26] the growing need for environmentally conscious technological developments. The agricultural sector contributes to the growth in the use of innovative technology [27].

3.4 Crop Monitoring and Disease and Pest Management System

Sensors, drones, and satellites are used in crop monitoring to track crop health and locate areas that need care. Sensors can identify issues early and assist farmers in choosing [28] the optimal times to plant and harvest crops [29]. Crops are regularly examined carefully during the growing season as part of monitoring [30]. Pesticides are released as needed to protect crops when sensors identify the presence of pests [31]. Data derived based on the environment's temperature, humidity, and pigment parameters are used to determine the presence in a plant pathology [32]. The method may [33] be improved to be more precise and efficient in determining values and defining whether or not leaves are unhealthy or sick by using image processing techniques in conjunction with it [34].

4 System Architecture

An IoT framework is an amalgamation of technologies spanning from detectors, protocols, and controllers toward cloud accommodations and layers. An IoT device's design includes sensors to gather data from the [35] surroundings, actuators with either wired or wireless connections, as well as an embedded machine incorporating a central processing unit, storage, connectivity modules, data entry and output interfaces, and rechargeable batteries a lot, in that sequence [36]. There is no tremendously approved Internet of Thing design. Diverse configurations have been proposed by various researchers, [37] Architectures of three and five layers [38]. A three-layered design is the one with the most rudimentary framework [39], as seen in Fig. 3. It was originally employed during the beginning stages of research in this sector. It is divided into three stages: perception, network, and app.

- Perception layer—The perceptual tier is a core structural aspect of IoT, which includes devices such as sensors and controllers, as well as linked gadgets that enable the creation of an IoT environment. It detects physical characteristics or recognizes other smart things in the surroundings [40].
- Network Layer—The network tier is accountable for linking devices with intelligence, network sections, and servers. The transport layer uses networks [41] like Wi-Fi, 3G, LAN, Bluetooth, RFID, NFC, and others to the data gathered by the sensor travels from the detection layer to the data processing tier and vice versa.
- Application layer—The IoT architecture's topmost level, the advantages and uses of IoT are most readily apparent at the software layer. The application layer is in charge of providing application-specific amenities to clients [42]. The three-tiered design illustrated the essential notion of the Internet of Things, however, IoT research typically focuses on its more sophisticated elements, making it insufficient.

Fig. 3 Basic architecture of IoT

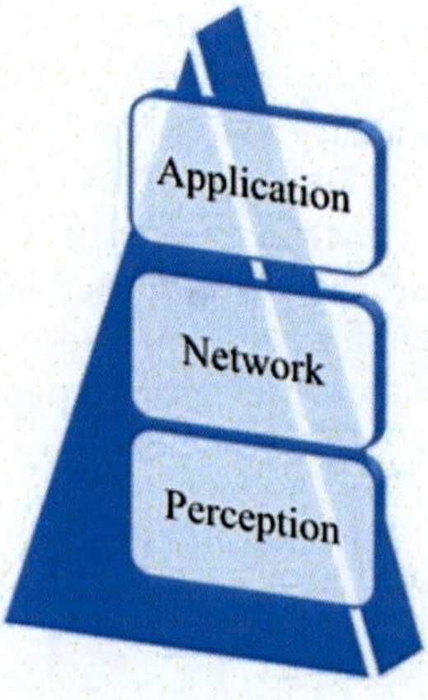

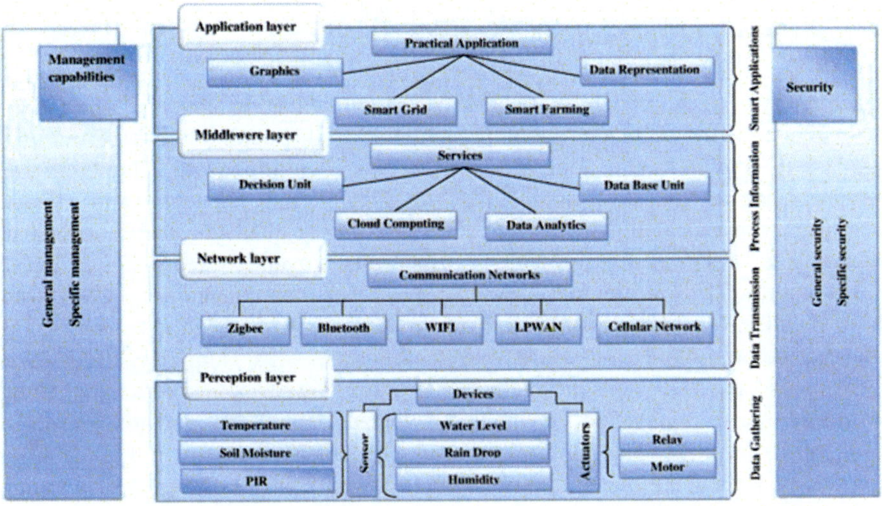

Fig. 4 Proposed IoT architecture model

Figure 4 illustrates how each tier of the IoT stack is examined in our work. We have included a fresh contribution to our poll by talking about various IoT designs. In earlier Internet of Things polls, this was not covered.

4.1 Perception Layer

The IoT is unable to perform without contextual consciousness, which has reliant on sensor technology [40]. The perceiving layer, the IoT architecture's lowest layer, is in charge of gathering data from various sources [43]. Most Internet of Things (IoT) sensors are small, reasonably priced, and energy-efficient. In our proposed model, this layer consists of sensors as well as actuators that are positioned in surroundings to capture data on humidity, ambient temperature, water level, illumination, wind speed, vital nutrients level, illnesses of plants, insect pests, and other physical characteristics [44]. The data is processed using embedded devices, and after further processing and analysis it is routed toward a upper tier via the network layer. WSNs, for instance, are often used to monitor storage and logistics facilities and manage the climate [41]. The network layer is connected to these devices through wired or wireless communication methods.

4.2 Network Layer

Based on universally accepted and functional communication standards, the IoT is a dynamic network architecture with the potential to self-configure [45]. Providing connectivity and communication amongst IoT system components falls within the purview of the network layer of an IoT design. Devices may connect to and interact with one another as well as with the rest of the Internet thanks to its protocols and technologies [41]. Depending on the distance they traverse, three different types of IoT communication methods exist. They are IoT communication systems for short, medium, and long distances. IoT uses a variety of network technologies, such as Bluetooth, Wi-Fi, ZigBee, and CN-cellular networks. Numerous issues have been addressed in the extensive scientific literature on wireless networks, including lowering energy consumption, enhancing networking capabilities, and boosting scalability and robustness [41]. Figure 5 provides a graphical representation of the division into communication technologies. The following categories are used to group these standards according to the range of transmission:

- Technologies such as Bluetooth, RFID, and UWB are examples of wireless communications with SRWC ranges (Højst 10 m) [40]. IrDa,
- This category includes wireless communication technologies with a MRWC (10–100 m) such as ZigBee [46], Wi-Fi [47], and WiMAX.
- LRWC systems (and over 100 m): Cellular networks and LPWA technologies classified as Non-3GPP (Sigfox [48, 49], LoRa [50], and Weightless) and

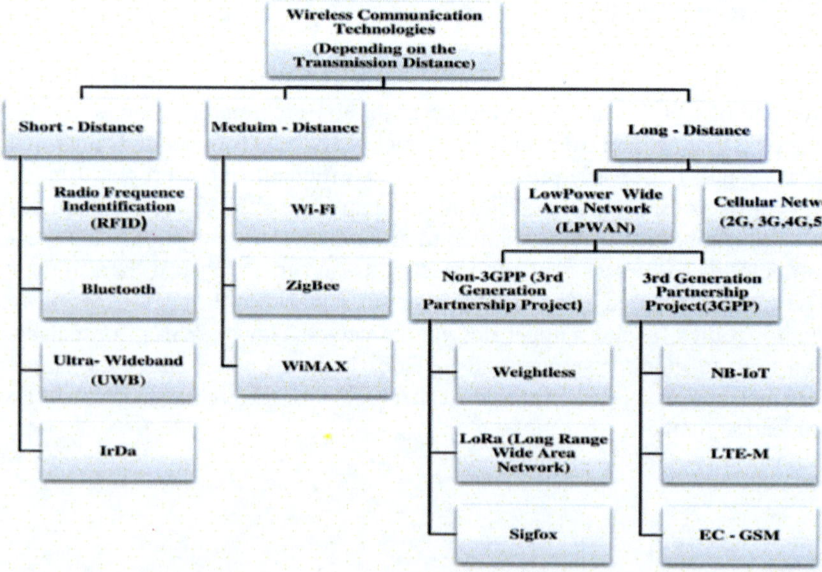

Fig. 5 Graphical representation on communication technologies

3GPP (NB-IoT [51] EC-GSM, LTE-M) technologies grouped as long-distance communication range.

Table 1 compares the many wireless communication technologies that are currently in use [45]. Technologies for tying agricultural systems together are also looked at. Cellular communication technologies [48], namely [52] 3G, 4G, LTE, and 5G, have proven to be the most appropriate and dependable modes of communication for precision agriculture, where a substantial [53] volume of real-time data necessitates transmission and analysis. NB-IoT, Bluetooth, ZigBee LoRa, RFM69, SigFox, Wireless Fidelity, and WiMAX are some samples of wireless technologies. Their primary uses include smart irrigation, plant protection, soil sensing, insect detection, and environmental management. Shi et al. [54] proposed a wireless sensor network and controller for a self-contained irrigation system that collects information on moisture of soil and temperature, and other factors and transmit that intel to the controller to control watering or not. Which has action in our model by using ZigBee, Wi-Fi, and cellular communication technology (like 3G, 4G, LTE, and 5G).

Table 1 Comparison of the wireless communication technologies

IoT technology	Standard	Power consumption	Network type	Speed	Range	Frequency spectrum	Mesh
Bluetooth (BLE)	IEEE 802.15.1	10 mW	WPAN	1 Mbps	50 m	2.4 GHz	No
ZigBee	IEEE 802.15.4	Very low	WPAN	20–250 Kbps	100 m	2.4 GHz	Yes
Z-Wave	Z-Wave Alliance	Very low	WPAN	100 Kbps	30 m	908.42 MHz	Yes
6LoWPAN	IEEE 802.15.4	Very low	WPAN	250 Kbps	10–100 m	2.4	Yes
Wi-Fi	IEEE 802.11a, 11b, 11 g, 11n, 11ac, 11ad	High	WLAN	100–250 Mbps	100 m +	2.4 GHz/5 GHz	No
LoRa/ LoRaWAN	IEEE 802.15 g	High	LPWAN	27 Kbps	10 km +	470–510 MHz (China), 865–925 MHz	No
WiMAX	IEEE 802.16	N/A	WMAN	70 Mbps	50 km	2–11 GHz	No
GSM/ GPRS	ETSI	Very high	WAN	Moderate	35 km +	850 MHz/ 1.9 GHz	No
LTE	3GPP	Very high	WAN	0.1–1 Gbps	28/10 km	700–2600 MHz	No
LTE-M	3GPP	Moderate	LPWAN	1 Mbps	Long	Various	No
NB-IoT	3GPP	Moderate	LPWAN	250 Kbps	20 km +	Various	No

4.3 Data Processing Layer

This layer is likewise referred to as a layer of middleware. It accumulates, trials, and analyses enormous amounts of information via the transport layer [55]. To the bottom tiers, it could supervise and offer a multitude of services [56]. The collecting, analyzing, and interpreting Intel obtained from IoT gadgets is liable to the software and hardware components of IoT architecture known as middleware. This facilitates the provision of numerous services and enhances the interoperability of smart devices. There are a variety of open source and commercial middleware delivery methods available for IoT devices. Examples include Oracle Fusion Middleware [57], OpenIoT [58], MiddleWhere [59], Hydra [60, 61], and FiWare [62]. In our proposed system, this layer is accountable for gathering unorganized information from the gadgets, [56] processing it, and making it accessible for further inquiry or action. Algorithms, analytics platforms, data management systems, and other tools are all included in the data processing layer. These technologies are used to glean insightful information from the data and base judgments on it.

4.4 Application Layer

The uppermost layer in an IoT architecture that communicates directly with end users is the application layer. This layer has a composition of a variety of applications and software, such as mobile apps and web portals [63], and other user interfaces that are intended to communicate with the underlying IoT infrastructure. Additionally, it has middleware services that facilitate easy data sharing and communication across various IoT systems and devices. In our suggested paradigm, processing, and analytics tools are also present on the application layer [42], allowing data to be reviewed and transformed into useful information. The software for data visualization, machine learning algorithms, and complex analytics tools are all included in this. The benefits and uses of IoT are best evident in this layer.

5 Basic Prototype for Self-Contain Irrigation System

Figure 6 shows our initial working model for intelligent farming, for that we used hardware components in this base model are Arduino Uno microcontroller, humidity, and temperature sensor—DHT11/DHT22, soil moisture sensor, ESP8266 Wi-Fi Module, battery (3.5 and 5 V), LCD display (20 × 4), pump relay. Only based on soil moisture and temperature which is operating based on moisture percentage that is the condition is if moisture percentage is low or high the watering is on or off respectively [64]. First and foremost, our prototype is accountable for gathering

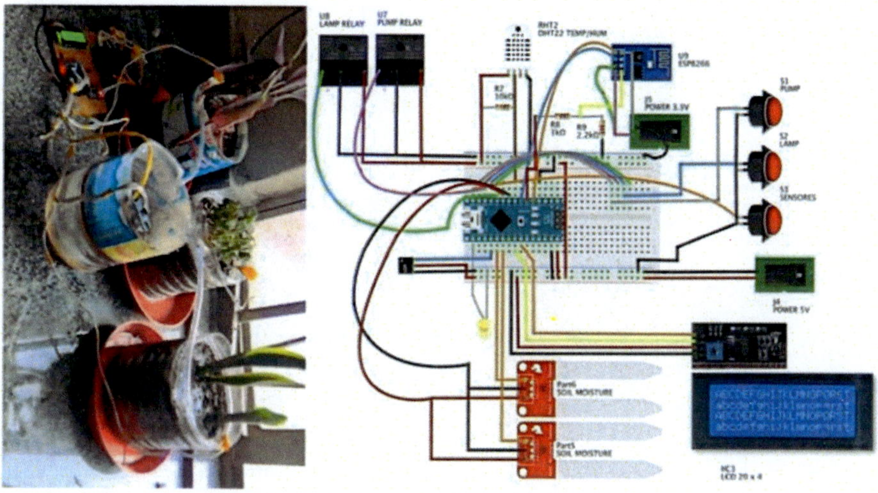

Fig. 6 Basic prototype for self-contain irrigation system

information on soil moisture and automating irrigation system operation in accordance with the Soil's Moisture Content ($SMC\ \% = \left(1 - \frac{D}{W}\right) \times 100$) at necessary intervals throughout the day [65]. A moisture sensor will be required to detect the state of the soil [64]. Both dry and wet soil types are possible. This sensor is near a body of water. The pump will automatically turn on when the soil's dry level reaches a certain level in such a situation, and this information is captured and kept in the cloud.

An algorithm for self-contain irrigation based on soil moisture:

Step 1. Set TH_min of SM to Dry Conditions to start the irrigation operation and TH_max of SM to Wet Conditions to halt it
Step2. Select the watering mode (auto)
Step 3. If (dial = auto)
{
Determine the soil moisture percentage (SMC)
If (SMC < = TH_min) // parameter assessment of current SM from user-specified threshold value
{
Send 1 via relay to begin irrigate// Signal to Switch on the Hydraulic Motor
}
Else
{
Send 0 via the relay to halt watering //Signal to shut the Hydraulic Motor
}
}

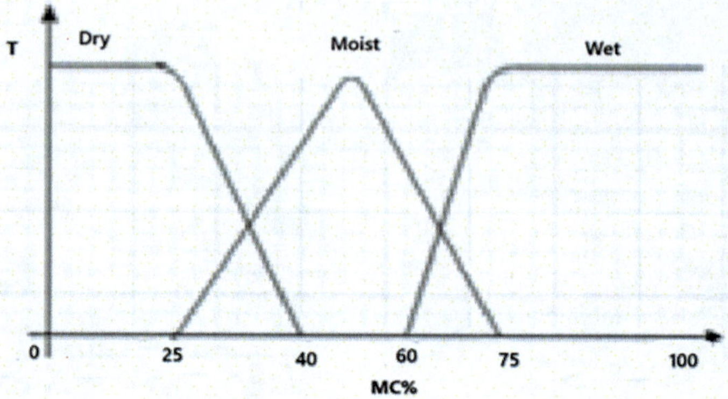

Fig. 7 Soil Moisture Content Percentage (SMC%) versus Temperature (T)

In the interim, the user will get information about the pump's status as well as information about the soil's dry level. The process of prototype shows in the graph Fig. 7 [Soil moisture content percentage (SMC%) versus Temperature (T)] [15]. It is our duty to improve agricultural output via efficient use of water and energy [66]. The architectural model's focus will be on how to effectively utilize resources on agricultural land without requiring human contact. Table 2 shows many IoT platforms currently available in the world. This table helps to many researcher to choice their IoT platform based on their work.

Table 2 List of IoT platform [67]

IoT platform	Real-time data capture	Data visualization	Cloud service type	Data analytics	Developer cost
Ubodots [68]	YES	YES	Public	YES	Free of cost
Thing Speak [69]	YES	YES (Matlab)	Public	YES	Free of cost
ThingWork	YES	YES	Public (IaaS)	YES	Pay per use
Xively	YES	YES	Public (IoTIaaS)	YES	Free of cost
Piotly	YES	YES (Matlab)	Public	YES	Free of cost
Nimbits	YES	YES (Matlab)	Hybrid	YES	Free of cost
Connecterra	YES	YES	Private (IaaS)	YES	Pay per use
Axeda	YES	YES	Private (IaaS)	YES	Pay per use
Phytech	YES	YES	Private (IaaS)	YES	Pay per use
Arkessa	YES	YES	Private	YES	Pay per use
Yaler	YES	YES	Private	YES	Pay per use
AWS [70]	YES	YES	Private	YES	Pay per use
AZURE [71]	YES	YES	Private	YES	Pay per use
SmartThings	YES	YES	Private	YES	Pay per use
Kura	YES	YES	Private	YES	Pay per use

6 Conclusion

India is one of the world's most imaginative and precise countries, and it has achieved significant advancements in every industry, including agriculture. This is fantastic news for our country and was made possible by improvements in science and technology like the IoT. The use of agriculture IoT is discussed in a compilation of earlier works. The identification and description of recommendations pertinent to IoT applications in intelligent farming are made possible by an in-depth examination of these papers. Already we seen architecture of our model in this article which consists four layer, among these first layer gives Intel of physical parameters of field like humidity, ambient temperature, water level, illumination, etc. the second layer offers. IoT communication protocols and connection between sensor networks and gateways or servers to address the expanding literature and errant information about them. Additionally, heterogeneous systems and other popular communication technologies like ZigBee, Wi-Fi, Bluetooth, and many others are investigated.

The findings show that every technology has unique benefits and restrictions and that various methods of wireless communication are appropriate in various contexts. We will leverage cellular, Wi-Fi, and ZigBee networks in our suggested paradigm to send data to higher layers. Algorithms, analytics platforms, data management systems, and other tools are all a part of the data processing layer, which is where the Intel received from the 2nd layer is kept and processed. The IoT infrastructure is meant to be communicated with by a variety of software and programs found in the higher layer of our architecture, such as mobile applications, Internet portals, and other user interfaces. This includes tools for complicated analytics, machine learning techniques, and data visualization applications.

In this study, we also explain our initial working prototype for intelligent farming, which is just dependent on soil moisture and temperature and operates according to moisture percentage, with irrigation being on or off depending on whether moisture % is low or high. In such a case, when the dry level of the soil reaches a specified level, the water supply will turn on immediately, and facts will be captured and preserved in the cloud. In the interim, the user will receive information on the condition of the pump and the amount of soil dryness on an LCD display. Also this article discusses the cost of development for a list of IoT platforms used in real time. Finally, we provided a number of pertinent suggestions that might help IoT applications in intelligent farming overcome their problems and create new research possibilities. A IoT of room exists for developing innovative, effective new systems in IoT-based agriculture with the arrival of contemporary technology. Inexpensive-cost systems with characteristics such as independent functioning, simple to care for, energy efficiency, ease of use, and durable design are in high demand. Future works will be released with all of our architectural design's characteristics.

References

1. Nayyar A, Puri V (2016) Smart farming: IoT based smart sensors agriculture stick for live temperature and moisture monitoring using Arduino, cloud computing & solar technology. In: The international conference on communication and computing systems (ICCCS-2016). https://doi.org/10.1201/9781315364094-121
2. Xu J, Gu B, Tian G (2022) Review of agricultural IoT technology. Elsevier B.V. on behalf of KeAi Communications Co., Ltd
3. Xie T, Zhang C, Zhang Z, Yang K (2019) Utilizing active sensor nodes in smart environments for optimal communication coverage. IEEE Access 7:11338–11348
4. Zhang J, Liu P, Xue W, Rui Z (2018) Farmland intelligent information collection system based on NB-IoT. In: The international conference on cloud computing and security, Springer, Haikou, China, June 2018. pp 331–343 (978-3-030-00017-2)
5. Sethi P, Sarangi SR (2017) Internet of Things: architectures, protocols, and applications. https://doi.org/10.1155/2017/9324035
6. Yan-e D (2011) Design of intelligent agriculture management information system based on IoT. In: Proceedings of the 2011 fourth international conference on intelligent computation technology and automation, vol 01, pp 1045–1049
7. Zhang Z, Zhang C, Li M, Xie T (2020) Target positioning based on particle centroid drift in large-scale WSNs. IEEE Access 8:127709–127719
8. Köksal Ö, Tekinerdogan B (2018) Architecture design approach for IoT-based farm management information systems. Springer. https://doi.org/10.1007/s11119-018-09624-8
9. https://techjury.net/blog/how-many-iot-devices-are-there/#gref (Online)
10. https://dataprot.net/statistics/iot-statistics (Online)
11. https://iotbusinessnews.com/2023 (Online)
12. https://iotbusinessnews.com/2023/02/09/06465-global-iot-market-size-to-grow-19-in-2023/ (Online)

13. https://iotbusinessnews.com/2022/05/19/70343-state-of-iot-2022-number-of-connected-iot-devices-growing-18-to-14-4-billion-globally/ (Online)
14. O'Grady MJ, O'Hare GM (2017) Modelling the smart farm. Inf Process Agricult 4:179–187. https://doi.org/10.1016/j.inpa.2017.05.001
15. Kamaruddin F, Abd Malik NN, Murad NA, Latiff NM, Yusof SK, Hamzah SA (2019) IoT-based intelligent irrigation management and monitoring system using arduino. TELKOMNIKA (Telecommun Comput Electron Control) 17:2378–2388. https://doi.org/10.12928/telkomnika.v17i5.12818
16. Rajkumar MN, Abinaya S, Kumar VV (217) Intelligent irrigation system—an IOT based approach. In: 2017 International conference on innovations in green energy and healthcare technologies (IGEHT). IEEE, pp 1–5
17. Vinoth M, Vithiya G (2018) Farm field monitoring and irrigation automation using IoT. Int J Eng Technol 53–56. https://doi.org/10.14419/ijet.v7i2.26.12534
18. Kumar G (2021) Precision water irrigation system for agriculture using IoT framework. Int J Mod Agric 10:718–724
19. Obeid AM, Karray F, Jmal MW, Abid M, Manzoor Qasim S, BenSaleh MS (2016) Towards realisation of wireless sensor network-based water pipeline monitoring systems: a comprehensive review of techniques and platforms. IET Sci Measur Technol 10:420–426. https://doi.org/10.1049/iet-smt.2015.0255
20. Sadeghioon AM, Metje N, Chapman DN, Anthony CJ (2014) SmartPipes: smart wireless sensor networks for leak detection in water pipelines. J Sens Actuator Netw 3:64–78. https://doi.org/10.3390/jsan3010064
21. Singh P, Kumari SK (2021) Smart irrigation system using IoT. CRC Press, pp 137–141
22. Khan FQ, Ahmad I, Khairullah EF, Bajaber O, Ullah A, Jan S (2022) Real-time consumer aware and alert system based on smart water tank. Mater Today Proc 63:790–797. https://doi.org/10.1016/j.matpr.2022.05.506
23. Manikandan D, Prabhu S, Chakraborty P, Manthra T, Kumaravel M (2022) IoT-based smart irrigation and monitoring system in smart agriculture. Springer, Singapore, pp. 45–56.
24. Patle KS, Saini R, Kumar A, Palaparthy VS (2022) Field evaluation of smart sensor system for plant disease prediction using LSTM network. IEEE Sens J 22:3715–3725
25. Vangala A, Das AK, Kumar N, Alazab M (2021) Smart secure sensing for IoT-based agriculture: blockchain perspective. IEEE Sens J 21
26. Citoni B, Fioranelli F, Imran MA, Abbasi QH (2019) Internet of Things and LoRaWAN-enabled future smart farming. IEEE IoT Mag 2:14–19. https://doi.org/10.1109/IOTM.0001.1900043
27. Kumar R, Mishra R, Gupta HP, Dutta T (2021) Smart sensing for agriculture: applications, advancements, and challenges. IEEE Consum Electron Mag 10:51–56. https://doi.org/10.1109/MCE.2021.3049623
28. Balaji GN, Nandhini V, Mithra S, Priya N, Naveena R (2008) IoT based smart crop monitoring in farm land2018. Imperial J Interdisc Res (IJIR) 4:88–92. https://www.researchgate.net/publication/322508245_IOT_Based_Smart_Crop_Monitoring_in_Farm_Land
29. Anil Kumar DB et al (2023) IoT-based water harvesting, moisture monitoring, and crop monitoring system for precision agriculture. In: 2023 International conference on distributed computing and electrical circuits and electronics (ICDCECE). IEEE, pp 1–6
30. Pandey RK, Kumar S, Jha RK (2017) Crop monitoring using IoT: a neural network approach. In: Soft computing: theories and applications: proceedings of SoCTA 2017. Springer, pp 123–132
31. Nayagam MG, Vijayalakshmi B, Somasundaram K, Mukunthan MA, Yogaraja CA, Partheeban P (2023) Control of pests and diseases in plants using IOT Technology. Measur Sens 26:100713. https://doi.org/10.1016/j.measen.2023.100713
32. Ntihemuka M, Inoue M (2018) IoT monitoring system for early detection of agricultural pests and diseases. In: 12th South East Asian technical university consortium (SEATUC). vol 1, pp 1–5
33. Salehin I, Noman SM, Baki-Ul-Islam IJ, Bishnu P, Habiba U, Nessa N (2021) IFSG: intelligence agriculture crop-pest detection system using IoT automation system. Indonesian J Electr Eng Comput Sci 1091–1099. https://doi.org/10.11591/ijeecs.v24.i2.pp1091-1099

34. Yashodha G, Shalini D (2021) An integrated approach for predicting and broadcasting tea leaf disease at early stage using IoT with machine learning--a review. Mater Today Proc 37:484–488. https://doi.org/10.1016/j.matpr.2020.05.458
35. Mishra D, Zema NR, Natalizio E (2021) A high-end IoT devices framework to foster beyond-connectivity capabilities in 5G/B5G architecture. IEEE Commun Mag 59:55–61. https://doi.org/10.1109/MCOM.001.2000504
36. Javed F, Afzal MK, Sharif M, Kim BS (2018) Internet of Things (IoT) operating systems support, networking technologies, applications, and challenges: a comparative review. IEEE Commun Surv Tutor 20:2062–2100. https://doi.org/10.1109/COMST.2018.2817685
37. Mashal I, Alsaryrah O, Chung TY, Yang CZ, Kuo WH, Agrawal DP (2015) Choices for interaction with things on Internet and underlying issues. Ad Hoc Netw 28:68–90. https://doi.org/10.1016/j.adhoc.2014.12.006
38. Said O, Masud M (2013) Towards internet of things: survey and future vision. Int J Comput Netw 5:1--17. https://cs.brown.edu/courses/csci2270/archives/2017/papers/Towards_Internet_of_Things_Survey_and_Fu.pdf
39. Wu M, Lu TJ, Ling FY, Sun J, Du HY (2019) Research on the architecture of Internet of Things. In: 2010 3rd international conference on advanced computer theory and engineering (ICACTE). IEEE, pp V5–484
40. Feng X, Yan F, Liu X (2019) Study of wireless communication technologies on Internet of Things for precision agriculture. Wirel Pers Commun 1785–1802. https://doi.org/10.1007/s11277-019-06496-7
41. Tzounis A, Katsoulas N, Bartzanas T, Kittas C (2017) Internet of Things in agriculture, recent advances and future challenges. Biosyst Eng 164:31–48. https://doi.org/10.1016/j.biosystemseng.2017.09.007
42. Phasinam K, Kassanuk T, Shabaz M (2022) Applicability of internet of things in smart farming. J Food Qual 1–7. https://doi.org/10.1155/2022/7692922
43. Shi X, An X, Zhao Q, Liu H, Xia L, Sun X, Guo Y (2019) State-of-the-art internet of things in protected agriculture. Sensors 1833. https://doi.org/10.3390/s19081833
44. Thorat A, Kumari S, Valakunde ND (2017) An IoT based smart solution for leaf disease detection. In: 2017 International conference on big data, IoT and data science (BID). IEEE, pp 193–198
45. Kumar A, Hancke GP (2014) Energy efficient environment monitoring system based on the IEEE 802.15. 4 standard for low cost requirements. IEEE Sens J 14:2557–2566. https://doi.org/10.1109/JSEN.2014.2313348
46. Gislason D (2008) ZigBee wireless networking. Newnes
47. Liu Y, Hassan KA, Karlsson M, Pang Z, Gong S (2019) A data-centric internet of things framework based on azure cloud. IEEE Access 53839–53858. https://doi.org/10.1109/ACCESS.2019.2913224
48. Trilles S, González-Pérez A, Huerta J (2018) A comprehensive IoT node proposal using open hardware. A smart farming use case to monitor vineyards. Electronics 7:419. https://doi.org/10.3390/electronics7120419
49. Centenaro M, Vangelista L, Zanella A, Zorzi M (2016) Long-range communications in unlicensed bands: the rising stars in the IoT and smart city scenarios. IEEE Wirel Commun 60–67
50. Yazar D, Taskin S (2020) A long-range context-aware platform design for rural monitoring with IoT in precision agriculture. Int J Comput Commun Control 15. https://doi.org/10.15837/IJCCC.2020.2.3821
51. IoT Enabling New Business Opportunities (2015) HUAWEI Technologies Co. Huawei, NB
52. Idbella M, Iadaresta M, Gagliarde G, Mennella A, Mazzoleni S, Bonanomi G (2020) Agrilogger: a new wireless sensor for monitoring agrometeorological data in areas lacking communication networks. Sensors 20:1589. https://doi.org/10.3390/s20061589
53. Shi X, An X, Zhao Q, Liu H, Xia L, Sun X, Guo Y (2019) State-of-the-art internet of things in protected agriculture. Sensors 19:1833. https://doi.org/10.3390/s19081833

54. Xiang X (2010) Design of fuzzy drip irrigation control system based on ZigBee wireless sensor network. In: Computer and computing technologies in agriculture IV: 4th IFIP TC 12 conference, CCTA 2010, Nanchang, China, October 22–25, selected papers, part I 4, pp 495–501. https://doi.org/10.1007/978-3-642-18333-1_58
55. Boonchieng E, Chieochan O, Saokaew A (2018) Smart farm: applying the use of NodeMCU, IOT, NETPIE and LINE API for a Lingzhi mushroom farm in Thailand. IEICE Trans Commun 101:16–23. https://doi.org/10.7546/ijba.2019.23.1.71-86
56. Adenugba F, Misra S, Maskeliūnas R, Damaševičius R, Kazanavičius E (2019) Smart irrigation system for environmental sustainability in Africa: an Internet of Everything (IoE) approach. Math Biosci Eng 16:5490–5503. https://doi.org/10.3934/mbe.201927
57. Oracle Fusion Middleware (2011) Cache, oracle web, pp 67–68
58. Soldatos J, Kefalakis N, Hauswirth M, Serrano M, Calbimonte JP, Riahi M, Aberer K, Jayaraman PP, Zaslavsky A (2015) Openiot: open source Internet-of-Things in the cloud. In: Interoperability and open-source solutions for the Internet of Things: international workshop, FP7 OpenIoT project, held in conjunction with SoftCOM 2014, split, Croatia, September 18, 2014, Invited Papers. Springer, pp 13–25
59. Ranganathan A, Al-Muhtadi J, Chetan S, Campbell R, Mickunas MD (2004) Middlewhere: a middleware for location awareness in ubiquitous computing applications. In: Middleware 2004: ACM/IFIP/USENIX USENIX international middleware conference, Toronto, Canada, October 18-22, 2004. Proceedings 5. Springer, pp 397–416
60. Reddy PC, Gungi A, Unni M (2019) Cellular and molecular mechanisms of Hydra regeneration. Evo-Devo Non-Model Species Cell Dev Biol 259–290 (2019)
61. FIWARE: The open source platform for our smart digital future. Accessed: Sep 2018
62. Cirillo F, Solmaz G, Berz EL, Bauer M, Cheng B, Kovacs E (2019) A standard-based open source IoT platform: FIWARE. IEEE Internet of Things Magazine, pp 12–18. https://doi.org/10.1109/IOTM.0001.1800022
63. Ray PP (2018) A survey on Internet of Things architectures. J King Saud Univ Comput Inf Sci 30:291–319. https://doi.org/10.1016/j.jksuci.2016.10.003
64. Goap A, Sharma D, Shukla AK, Krishna CR (2018) An IoT based smart irrigation management system using machine learning and open source technologies. Comput Electron Agric 155:41–49. https://doi.org/10.1016/j.compag.2018.09.040
65. Dahane A, Benameur R, Kechar B (2022) An IoT low-cost smart farming for enhancing irrigation efficiency of smallholders farmers. Wirel Pers Commun 127:3173–3210. https://doi.org/10.1007/s11277-022-09915-4
66. El-Shirbeny MA, Ali AM, Savin I, Poddubskiy A, Dokukin P (2021) Agricultural water monitoring for water management under pivot irrigation system using spatial techniques. Earth Syst Environ 5:341–351. https://doi.org/10.1007/s41748-020-00164-8
67. Tara N, Uddin MM, Hyder KS, Sharmin S (2022) A Survey on the inclusion of IoT in agriculture for soil monitoring. https://www.researchgate.net/publication/361115512
68. Foughali K, Fathallah K, Frihida A (2018) Using Cloud IOT for disease prevention in precision agriculture. Procedia Comput Sci 130:575–582. https://doi.org/10.1016/j.procs.2018.04.106
69. Aliev K, Jawaid MM, Narejo S, Pasero E, Pulatov A (2018) Internet of plants application for smart agriculture. Int J Adv Comput Sci Appl 9(4). https://doi.org/10.14569/IJACSA.2018.090458
70. Vincent DR, Deepa N, Elavarasan D, Srinivasan K, Chauhdary SH, Iwendi C (2019) Sensors driven AI-based agriculture recommendation model for assessing land suitability. Sensors 19(17):3667. https://doi.org/10.3390/s19173667
71. Gupta B, Mittal P, Mufti T (2021) A review on amazon web service (aws), microsoft azure and google cloud platform (gcp) services. In: Proceedings of the 2nd international conference on ICT for digital, smart, and sustainable development, ICIDSSD 2020, 27–28 February 2020, Jamia Hamdard, New Delhi, India

Video-Based COVID-19 Monitoring System

Devesh Shetty, Fayeq Zaidi, Asfaq Parkhetiya, Abhishekh Gangwar, and Deepali Patil

1 Introduction

The global spread of COVID-19 has prompted various countries to implement measures like mask mandates and social distancing to control the pandemic. Recent research by the World Health Organization highlights the effectiveness of masks in reducing virus transmission. However, not everyone complies with mask mandates, especially in densely populated areas.

To address this, we aim to establish a video surveillance system that automatically detects individuals not wearing masks and alerts authorities. Our approach involves training a deep learning model using transfer learning and Mask R-CNN with CCTV cameras for better accuracy in identifying mask adherence.

Our goal aligns with recent advancements in object detection, particularly in instance segmentation, which is more complex than traditional object detection. Our approach with Mask R-CNN simplifies implementation while offering flexibility in architectural design.

D. Shetty (✉) · F. Zaidi · A. Parkhetiya
Shree L. R. Tiwari College of Engineering, Mumbai, India
e-mail: deveshetty.2201@gmail.com

A. Gangwar
Department of Biometrics, Centre for Development of Advanced Computing, Pune, India
e-mail: abhishekg@cdac.in

D. Patil
Dwarkadas J. Sanghvi College of Engineering, Mumbai, India
e-mail: deepali.patil@djsce.ac.in

M. S. Kaiser et al. (eds.), *ICT: Smart Systems and Technologies*, Lecture Notes in Networks and Systems 878, https://doi.org/10.1007/978-981-99-9489-2_9

2 Literature Review

2.1 *Face Detection*

One innovative approach, the Selective Refinement Network (SRN) [1], combines two-step classification and regression operations to improve face detection accuracy and reduce false positives. It includes the Selective Two-step Classification (STC) module and the Selective Two-step Regression (STR) module, which have been validated as effective through extensive experiments.

Additionally, a study [2] focuses on enhancing glaucoma detection using deep learning. It uses pretrained CNN models like ResNet50V2, VGG16, InceptionV3, and Xception, along with data augmentation techniques to improve accuracy. VGG16 stands out as the most accurate model for clinical glaucoma detection.

2.2 *Masked Face Detection*

In [3], the paper explores using transfer learning with deep convolutional neural networks and image processing to identify herbal plants. The study focuses on five herbal species, creating an image dataset and retraining established models like Inception-v3, ResNet, MobileNet, and Inception ResNet V2 through transfer learning.

On the other hand, [4] introduces a Masked Face Detection Approach in Video Analytics. The authors employ a multi-step technique, including estimating distance from the camera, eye line detection, facial part detection, and eye detection. They use tools like Analog Devices Inc.'s Cross Core Embedded Studio (CCES) and HOGSVM for person detection and distance estimation, along with the Viola–Jones algorithm in MATLAB for detecting facial features. This study explores diverse techniques for masked face detection.

2.3 *Mask Segmentation*

In [5], researchers focus on enhancing instance segmentation performance in remote sensing images. They achieve this by combining Mask R-CNN, a well-known instance segmentation model, with Softer-NMS, a target detection algorithm. This novel approach aims to surpass traditional Mask R-CNN in terms of accuracy and efficiency. The authors tackle the challenges of precisely segmenting objects in remote sensing images, including variations in object size, environmental conditions, and instances overlapping. To overcome these hurdles, they enhance Mask R-CNN by incorporating Softer-NMS, which improves candidate frame generation and selection.

On the other hand, [6] explains the concept of mask segmentation, involving the detection and masking of regions of interest to facilitate improved object identification. A notable technique in this field is Mask R-CNN [7], a versatile framework for object instance segmentation developed by Facebook AI researchers. It efficiently detects objects in images while simultaneously generating high-quality segmentation masks for each instance. As an extension of Faster R-CNN, Mask R-CNN adds an additional branch to the existing structure, enhancing accuracy and enabling object masking for refined instance segmentation. This research sheds light on the synergy between object detection and image segmentation, providing valuable insights into this domain.

3 Proposed System

The proposed system primarily operates within the realm of computer vision, a burgeoning domain. Within this field, object detection stands as a pivotal area of advancement. Moreover, the detection of humans or faces, critical applications, garners widespread acceptance globally. Within this context, our focus lies on the application of face detection. The system not only identifies individuals but also discerns whether they are donning masks. This facet gains immense significance given the prevailing circumstances. Furthermore, the system goes beyond binary detection; it classifies the type of mask being worn and delineates the exact mask-adorned regions. A pivotal ingredient for the system's success is an extensive dataset, pivotal for achieving accurate and effective detection. Our exposition of the proposed system follows a structured approach, detailing each component comprehensively.

3.1 Part I

Dataset Preparation Dataset preparation became apparent that no readily available dataset existed for masked faces, which prompted us to embark on the task of creating our dataset from the ground up. This endeavor required a multifaceted approach, combining various techniques to ensure dataset completeness and quality. In the early stages, we harnessed the power of Google Chrome's "Download All Images" extension, enabling us to collect images from an array of websites efficiently. However, we did not stop there. Recognizing the need for a diverse and comprehensive dataset, we turned to web scraping techniques. The utilization of Beautiful Soup, a Python library, allowed us to traverse websites, extract images, and integrate them into our growing dataset. Acquiring a comprehensive dataset for training and testing a machine learning system to detect and classify different types of masks, such as N95 and surgical masks, the steps involved are:

1. Defining the Objective: In this case, the goal is to develop a system that can accurately detect and classify various mask categories. This step helps in determining the types of data needed and the criteria for dataset selection.
2. Collecting Specific Images: Collect images specific to different mask categories from diverse sources.
3. Meticulous Curation: Review and filter collected images to ensure they meet quality and relevance criteria. Exclude low-quality or irrelevant images.
4. Acquisition of 13,022 Masked Face Images: After curation, the dataset comprises 13,022 high-quality masked face images from various sources.
5. Incorporating External Data: Add 5000 images from the LFW Face Dataset on Kaggle to augment the dataset.
6. Dataset Preparation: Prepare the dataset for machine learning by resizing images, normalizing pixel values, and ensuring consistency.
7. Annotation: Label each image with the corresponding mask category (e.g., N95, surgical) for supervised learning.
8. Data Splitting: Divide the dataset into training, validation, and testing sets for model development and evaluation.
9. Training Machine Learning Models: With the curated and prepared dataset in place, machine learning models are trained using algorithms and techniques suitable for image classification tasks.
10. Model Evaluation and Fine-Tuning: Evaluate model performance using the testing dataset. Fine-tune models based on evaluation results to improve accuracy.
11. Deployment: Once a satisfactory model is developed, it can be deployed in real-world applications for mask detection and classification.

Dataset Preprocessing and annotation

1. Data collection:

While making the dataset, we did not have many images available of people wearing mask, so we had to create our own dataset by scrapping the webpages of different websites using Python beautiful soup library; apart from that, we did use some Chrome extensions to download data in bulk from the image section of the google. Apart from scrapping and downloading data, we used dlib face detector and leveraged the 68 facial landmarks to place mask artificially over the faces by developing a Python script that could place mask over the face.

2. Data Cleaning:

An important step in ensuring the dataset's quality and dependability was to eliminate any unneeded, grainy, or low-quality images. This was accomplished by carefully examining each image one at a time. For cleaning purposes, both automated and manual techniques were used. Automatic filtering removed unwanted images, and manual inspection assisted in locating and removing any residual unwanted elements. The thorough cleaning procedure was designed to improve the dataset's overall quality and suitability for further analysis.

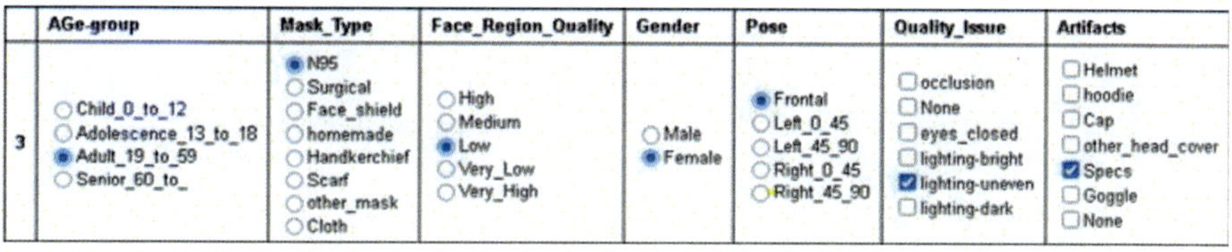

Fig. 1 Via Image annotation tool

3. Annotation:

Each image had many faces, so we had to annotate each region according to the attributes assigned for, e.g., if he or she is wearing a hat or specs etc. By collecting these attributes, it can be leveraged for future purposes. Using bounding boxes to mark precise boundaries of faces and recording annotations in structured JSON files as shown in Fig. 1

4. Categorization by Mask Type:

Categorize images based on the type of mask being worn to distinguish between various mask categories, such as surgical masks, N95 masks, cloth masks and assign labels to images according to the type of mask, ensuring the dataset includes different mask variations.

5. Region Extraction:

Extract bounded regions from annotated images, representing different mask types, and save them separately for each mask category. It can be done by developing a Python script to automate this process, creating separate files for different mask types in a uniform resolution.

6. Dataset Summary:

In this phase of our project, we summarize the dataset, which comprises 25,599 images categorized into various mask types, including cloth masks, face shields, N95 masks, other masks, scarfs, surgical masks, and without masks. We prepare the dataset for machine learning by normalizing pixel values, resizing images, and splitting it into training and testing sets. Subsequently, we train machine learning models to detect and classify masked faces using annotated images. We evaluate model performance with a testing dataset and fine-tune the models to enhance accuracy. To address class imbalance, we incorporate additional images for refinement. Finally, the trained models are deployed for real-world masked face detection and classification applications, integrated into practical systems or applications for immediate use.

Mask segmentation Mask segmentation surpasses basic detection by offering precise delineation of mask boundaries, essential in light of the diverse mask types during the COVID-19 pandemic. We adopted "edge-to-edge marking," a meticulous technique for contouring masks, far more precise than conventional bounding boxes.

We departed from conventional bounding box annotations, opting for "edge-to-edge marking" to capture the intricate mask shapes accurately. Utilizing the VIA

Fig. 2 Segmentation images Part II

Image annotation tool, we marked masks in detail, encompassing every curve, fold, and edge. Our commitment to thoroughness led us to categorize masks into distinct types, including N95 masks, cloth masks, surgical masks, scarves, and others. However, we primarily focused on four main categories: N95 masks, cloth masks, surgical masks, and scarves, given their prevalence during the pandemic. Our extensive annotation efforts resulted in a substantial dataset confirming mask presence and categorizing them by type. Importantly, this dataset included pixel-level segmentation data, allowing our system to precisely identify mask locations within images. To seamlessly integrate and analyze this data, we systematically organized the annotation results into structured JSON files. Each JSON file served as a comprehensive record, connecting image names with corresponding mask type annotations and pixel-level segmentation data. To enhance annotation consistency and clarity, we devised a Python script, replacing specific "Mask Type" annotations with a more streamlined "mask" annotation within the JSON files.

This standardization ensured data alignment with our system's overarching objectives. In total, our mask segmentation efforts culminated in a dataset comprising approximately 3169 meticulously annotated images, each accurately categorized by mask type and intricately marked to pinpoint mask positions on individuals' faces. Figure 2 shows mask segmented images.

3.2 Part II

3.2.1 Detection Using Transfer Learning

1. Leveraging Transfer Learning

Transfer learning is employed, which is a deep learning technique that utilizes knowledge learned from similar problems. In this case, the pretrained model has learned features from data related to mask detection tasks.

2. Defining Mask Categories

Recognize and classify seven distinct mask categories: cloth, face shield, N95, surgical, scarf, other mask, without mask.

3. MobileNet Architecture Selection

MobileNet is selected due to its lightweight architecture, which makes it well-suited for this task. It incorporates depth-wise separable convolutions to optimize the convolution process.

4. Depth-Wise Separable Convolutions

MobileNet uses depth-wise separable convolutions, a technique that improves efficiency by processing each input channel individually, reducing computational complexity. This is followed by an 11 convolution to further refine features, creating an effective combination for feature extraction.

5. Training and Performance Monitoring

The model is trained using a curated and annotated dataset. Throughout the training process, the performance is monitored using the loss metric.

6. Understanding Loss as a Key Metric

Loss is used as a critical metric during training and validation. Unlike accuracy, which measures the percentage of correct predictions, loss quantifies how far the models predictions are from the actual values. A lower loss indicates that the model predictions are closer to the ground truth. It helps ensure that the model is learning and generalizing effectively.

7. Loss Curve and Model Tuning

The loss curve, as shown in Fig. 3, provides a graphical representation of how the model performance evolves during training and validation. The goal is to minimize the loss over time, which signifies that the model is improving and converging toward accurate predictions.

3.2.2 Performance Evaluation Using Confusion Matrix

We visualize the confusion matrix to acquire more insights into the model performance. The confusion matrix indicates the model ability to categorize instances correctly into their appropriate classes. The confusion matrix provides a detailed analysis of the model predictions, allowing us to find areas where the model excels and regions where it may be improved further. The performance of our classification model is positive, with competitive precision, recall, and $F1$-score values for each class. The categorization outcomes of the model are represented visually in the confusion matrix, assisting in the evaluation of the model overall efficacy as shown in Fig. 4.

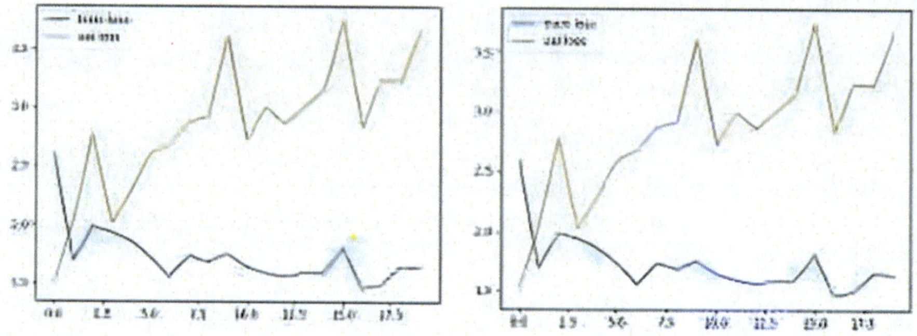

Fig. 3 Loss curve and model tuning

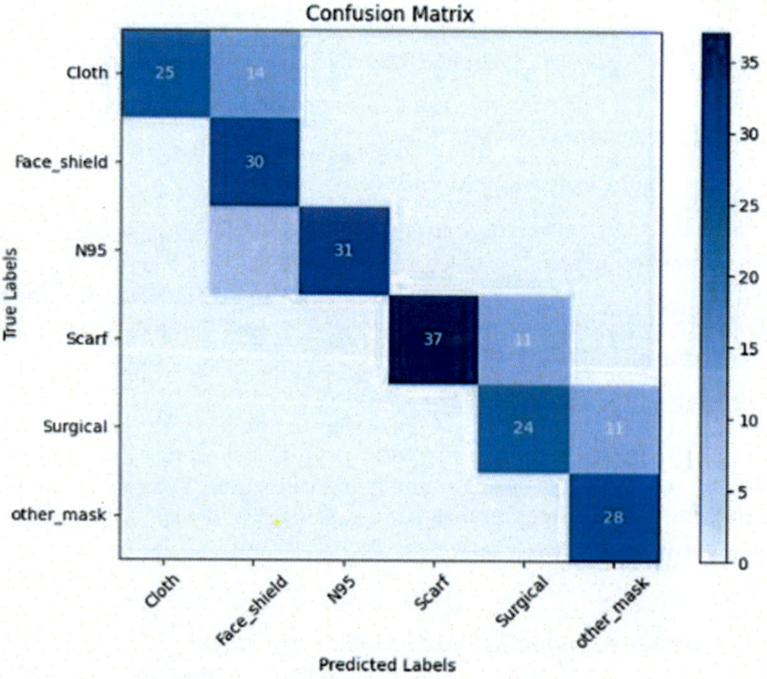

Fig. 4 Confusion matrix

3.3 Part III

Mask R-CNN As seen from related work [7, 8] an objective of Mask R-CNN is to simultaneously perform both object detection and pixel-level instance segmentation within images. In our scenario, we have acquired a dataset consisting of around 3169 segmented images. These images are unique because they contain masks that delineate different types of objects or regions within each image. To create a labeled dataset for training Mask R-CNN, we have employed a meticulous annotation process. This process involves generating JSON files that meticulously document the coordinates and characteristics of the segments within each image.

To ensure effective model training and evaluation, we have judiciously split the dataset into two subsets: a training set comprising 70% of the data and a validation set containing the remaining 30%. This data splitting is a standard practice in machine learning to assess how well the model generalizes to unseen data and to prevent overfitting. To further enhance the model's performance and speed up training, we have leveraged a pretrained model. This pretrained model is initially trained on a vast dataset like Common Objects in Context (COCO). By using this pretrained model as a starting point, we capitalize on the knowledge it has already acquired about objects and their features, which can significantly expedite and improve the training process for your specific segmentation task.

During this phase, the model learns to predict various aspects of the objects within the images, including their bounding boxes (to locate them within the image), object classes (to identify what the objects are), and object masks (to precisely segment each object's pixels). This training process is iterative and aims to minimize the difference between the model's predictions and the ground truth annotations in your dataset.

The outcome of this extensive training effort is a valuable asset: the model.h5 file. This file encapsulates both the architecture and the learned weights of the trained Mask R-CNN model. With this file, we can perform object detection and instance segmentation tasks on new, unseen images, providing you with accurate and detailed information about the objects present in the images. The result is a model capable of delivering granular insights into image content, making it an invaluable tool for a wide range of computer vision applications, from medical imaging to object recognition in various industries. Figure 5 shows result of mask R-CNN.

Fig. 5 Mask R-CNN output

4 Conclusion

This paper presents a system designed to address COVID-19 challenges by promoting preventive measures like mask usage and social distancing. Using advanced machine learning and computer vision techniques, our system detects mask adherence with precision through CCTV cameras and convolutional neural networks (CNNs). Our goal is to create an accurate and efficient system that integrates seamlessly with existing infrastructure, benefiting public health authorities. In these challenging times, our system serves as an innovative tool to safeguard public health and safety. Beyond the pandemic, it offers versatility and adaptability for applications in computer vision, public health, and safety, making it valuable for monitoring safety protocols, ensuring workplace safety, and contributing to broader public health initiatives. It demonstrates the power of technology in enhancing our collective well-being and resilience.

5 Future Work

One avenue is integrating temperature detection, a crucial step post-COVID-9, which could be achieved through thermal imaging or similar technologies. Additionally, a dedicated module for real-time monitoring and enforcement of social distancing is a promising addition, ensuring adherence to safety guidelines. Expanding our system to include contact tracing capabilities, leveraging anonymized data for outbreak identification, and integrating with healthcare systems could revolutionize pandemic response. Furthermore, scaling the system globally through collaboration with international organizations and governments is within reach.

References

1. Chi C, Zhang S, Xing J, Lei Z, Li SZ, Zou X (2019) Selective refinement network for high performance face detection
2. Phankokkruad M (2021) Evaluation of deep transfer learning models in glaucoma detection for clinical application. In: 2021 the 4th international conference on information and communications technology
3. Azeez YR, Rajapakse C (2019) An application of transfer learning techniques in identifying herbal plants in Sri Lanka. In: 2019 international research conference on smart computing and systems engineering (SCSE)
4. Deore G, Bodhula R, Udpikar V, More V (2016) Study of masked face detection approach in video analytics. In: 2016 conference on advances in signal processing (CASP)
5. Wang Y, Rao Y, Huang C, Yang Y, Huang Y, He Q (2021) Using the improved mask R-CNN and softer-NMS for target segmentation of remote sensing image. In: 2021 the 4th international conference on pattern recognition and artificial intelligence

6. Bamne B, Shrivastava N, Parashar L, Singh U (2020).Transfer learning-based object detection by using convolutional neural networks. In: 2020 International conference on electronics and sustainable communication systems (ICESC)
7. CHe K, Gkioxari G, Dollar P, Girshick R (2017) Mask R-CNN. 2017 IEEE international conference on computer vision (ICCV)
8. Ribani R, Marengoni M (2019) A survey of transfer learning for convolutional neural networks. In: 2019 32nd SIBGRAPI conference on graphics, patterns and images tutorials (SIBGRAPT)

Analyzing User Profiles for Bot Account Detection on Twitter via Machine Learning Approach

Deepti Nikumbh, Anuradha Thakare, and Deep Nandu

1 Introduction

People from around the world use online social networking services (also known as social media) to build relationships with like-minded people who share similar personal and professional interests and hobbies. Online social media platforms are favored by the masses all over the world. The extensive usage of social media has advantages as well as disadvantages on the society. It allows users to share their experiences, interests, and connect with friends and strangers. It is a medium through which businesses engage with their customers. However, on social media, anybody can share anything including materials that are unverified. Bot or fake accounts participate in the spread of such unverified information.

Online fraud and misinformation circulated by fake accounts on social media are a major concern and are increasing at a rapid pace. Popularities indicators such as likes, shares, and followings gained by business organizations and social media influencers must verified since bots and fake accounts play a major role in increasing the count. Such usage of important networking platforms can have a negative impact on society. Enemy countries maintain a network of bots that they use to spread spam messages, junk news, and even sway public opinion related to political figures. In the 2016 US presidential election, bots produced around 3.8 million tweets [15]. A Russian Internet Research Agency shared more spam links and junk news tweets compared to any other Twitter account and pretended to be political right and left wing to garner more likes and retweets [16].

The proposed system will analyze the user profile features of Twitter accounts and automatically detect fake or bot accounts. This will help the user and the platform

D. Nikumbh (✉) · A. Thakare
Pimpri Chinchwad College of Engineering, Pune 411044, India
e-mail: deepti.nikumbh@gmail.com

D. Nandu
Shah and Anchor Kutchhi Engineering College, Mumbai, India

M. S. Kaiser et al. (eds.), *ICT: Smart Systems and Technologies*, Lecture Notes in Networks and Systems 878, https://doi.org/10.1007/978-981-99-9489-2_10

creators to get rid of such accounts thereby reducing spams and hatred-filled fake content on Twitter.

The remainder of this paper is organized as follows. Section 2 presents prior and related work on bot detection Sect. 3 discusses the proposed methodology in detail mainly focusing on the derived features, EDA on the given feature set, feature engineering approaches used in the work, machine learning techniques, and detailed performance evaluation. Section 4 provides the conclusion of the work along with future research directions.

2 Related Work

In recent years, Twitter has experienced a proliferation of bot/Fake accounts. These accounts are mainly responsible for spamming, offering clickbait, compromising security via malware, and skewing public opinion [1, 2]. Around 9–17% of Twitter accounts are bots which contribute 16–56% of tweets [3, 4]. Reference [5] presented a detailed survey of social bots and their behavior. They also discussed various bot detection mechanisms and bot coordination strategies.

In literature, much research is carried out in detecting bot accounts by analyzing various social media features. Reference [6] developed a real-time Twitter credibility analysis model considering text, user, and social credibility features. Spam, misspelling, and bad word filters were used for text credibility analysis, verified account and account creation features for user credibility and followers impact, and follower-following ratio for social credibility. The authors further want to extend their model for bot detection and include text semantic analysis and multimedia data analysis in the framework.

The bots are also involved in circulating misinformation on social media. Reference [7] worked on the set of implicit and explicit features of user profiles and demonstrated that a correlation exists between user profiles and news spread (fake or real) on social media. Users who tend to believe in fake news have different characteristics as compared to the one who believes in real news. Comparative feature analysis in terms of explicit and implicit user profile features was presented. Authors further want to integrate their work into fake news detection models and incorporate bot detection techniques to discriminate bots from normal users.

Bot are also spammers. Reference [8] identified spam profiles by considering user-based, content-based, and graph-based features. A combination of the above features was learned by using J48 and Naïve Bayes classifiers, and a comparative analysis is presented. The authors further intend to evaluate the effectiveness of their hybrid features on various social media platforms such as Facebook and Myspace. Reference [9] along with content-based features also considered account metadata and interaction-based features for detecting spammers. Reference [10] addressed the spam drift problem on Twitter by studying 12 statistical features of user profiles. They developed a novel model that first discovers changed spam tweets from unlabeled tweets and includes it in a random forest classifier trained on labeled tweets. The authors further highlighted that techniques to eliminate "too old" or unuseful data from the training process should be explored.

From the literature, it is evident that lot of features are available for study hence careful feature engineering is needed. Machine learning (ML) and deep learning (DL) techniques are the best way to process the features. Reference [11] performed feature extraction using correlation selection and univariate analysis. Reference [12] performed dimension reduction of features using PCA, rank order correlation, and wrapper methods. The best features were trained using SVM, NN, and hybrid SVM-NN classifiers. Reference [13] surveyed various ML and DL techniques for tweet-based bot detection. The authors highlighted open issues which need attention which include feature selection, dataset labeling, dataset balancing, reducing false positive, and language diversity support.

Reference [14] focuses on detecting human-created fake accounts. Their study concluded that engineered features used to identify fake bot accounts cannot be used to identify human-created fake accounts. When used, the machine learning algorithm gave $F1$ score of only 49.75%. The authors highlighted that feature sets used in their work can be enriched with features from social science domains like psychology can be a way ahead.

The literature review carried out related to bot detection indicates that supervised-based machine learning techniques are best. The most widely and commonly used techniques to train a machine learning model remain support vector machines (SVM), artificial neural networks, the decision tree, and the random forest algorithm. Features engineering is essential to get optimal results.

Based on the research, we found out that the traditional data attributes were not enough to get better results, so we tried to derive a few more attributes/features which were based on the existing attributes from the dataset. Based on this derived attribute, we tried to train the model to achieve better accuracy. Table 1 gives a summary of all the features used for bot detection in the literature.

3 Proposed Methodology

The flowchart of the proposed work is shown in Fig. 1. First data is collected in the JSON format, and data preprocessing, cleaning, and exploratory data analyses are performed over it. Further new features are derived from traditional features mentioned in the dataset. Since the number of features in the dataset is high, feature selection and dimension reduction techniques are applied to get the best set of features.

Supervised machine learning techniques are used for final bot classification. These supervised classification algorithms learn from the labeled data. Using previous knowledge, the algorithm labels the new data by correlating patterns to the unlabeled data. The trained model is finally evaluated on various performance metrics like precision, accuracy, recall, and false positive rate.

Table 1 Summary of various features and machine learning model used in literature

Work	Features considered					ML algorithm
	Text	User/profile	Content-based	Graph/network based	Metadata	
[6]	• Spam words • Bad words • Misspelled words	• Verified account • Account creation date	• Followers impact • Following-follower proportion	–	–	–
[7]	–	• Verified user • RegisterTime • Gender • Age • Personality	• StatusCount • FavorCount	• FollowerCount • FollowingCount	–	–
[8]	• Spam words	• Followers count • Reputation • Account age • FF ratio • Following count	• Total number of tweets • Tweet frequency • Mentions ratio • URL's ratio • Hashtag ratio	• In/out degree • Betweenness	–	• J48 • Decorate • Naïve-Bayes
[9]	–	–	• Unique URL ratio • Unique mention ratio • Content and hashtag similarity • URL, mention, hashtag ratio • Automated tweet URL ratio	• Follower ratio • Mean FF to follower ratio • Reputation • Follower reputation • Clustering coefficient • Community reputation • Community clustering coefficient	• Retweet proportion (RP) • Automated RP • Tweet time standard deviation • Tweet time Interval standard deviation	• Random forest • Decision tree • Bayesian network
[10]	Account age, followers count, following, favorites count, no_list, no. of tweets tweets, retweets count, hashtag count, mentions counts, URL count, characters count, digits counts					• Random forest • C4.5 • BayesNet • SVM

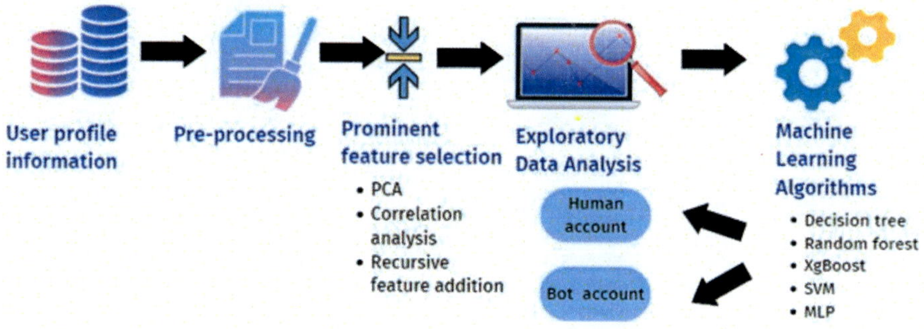

Fig. 1 Flowchart of proposed work

3.1 Data Collection

The dataset considered for the problem has a total of 41,560 entries of data, out of which around 22,970 entries are of bot accounts and 18,590 entries are of human accounts. A total of 40 features are provided in the dataset for bot detection. From the existing features of the dataset, few new features are derived. These newly derived features can help in increasing the accuracy of the model. The derived parameters are:

age_of_account—It is calculated using the Twitter account created date and current real-time date.

tweet_frequency—It is the number of tweets that occurred since account creation. It is calculated by the number of tweets divided by the age of the account

$$\text{tweet\_frequency} = \frac{\text{Total number of tweets}}{\text{age\_of\_account}} \tag{1}$$

follower_growth—It measures the new followers on the account since account creation. It is calculated as the ratio of the number of followers to the age of the account.

$$\text{follower\_growth} = \frac{\text{Number of followers}}{\text{age\_of\_account}} \tag{2}$$

friend_growth—It measures the new friends on the account since account creation. It is calculated as the ratio of number of friends to the age of the account.

$$\text{friend\_growth} = \frac{\text{Number of friends}}{\text{age\_of\_account}} \tag{3}$$

ratio_of_followers_to_friends—It is calculated by dividing followers_count with friends_count.

$$\text{ratio\_of\_followers\_to\_friends} = \frac{\text{Total number of followers}}{\text{Total number of friends}} \tag{4}$$

3.2 Exploratory Data Analysis

Valuable insights can be made by performing initial investigations on the dataset. Figure 5 shows cat plot analysis for followers_count, friends_count, status_count, and favourites_count. It can be seen that for accounts created by humans, the plots are scattered indicating that human have different followers and friends count, and they often update their status and have varying favorite action on Twitter, whereas for bot accounts, the cat plots do not show much variation.

Figure 2 shows box plot analysis of account age for human and bot accounts. From the plot, it can be seen that median, max values for accounts created by human are higher than those of bot accounts. Thus, the age of human accounts is much older than that of bot accounts. This is because most of the time bot accounts are intentionally created to spread spam, and fake news, and once their work is done, they are discontinued.

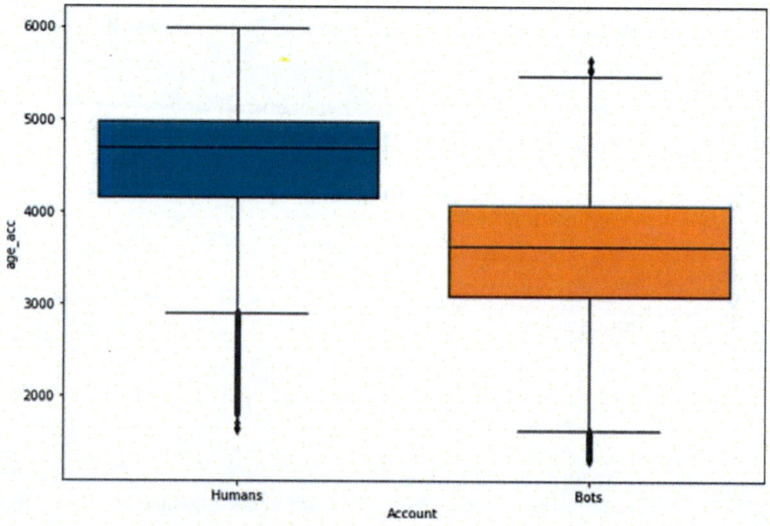

Fig. 2 Analysis of account age with respect to account type (bot, human)

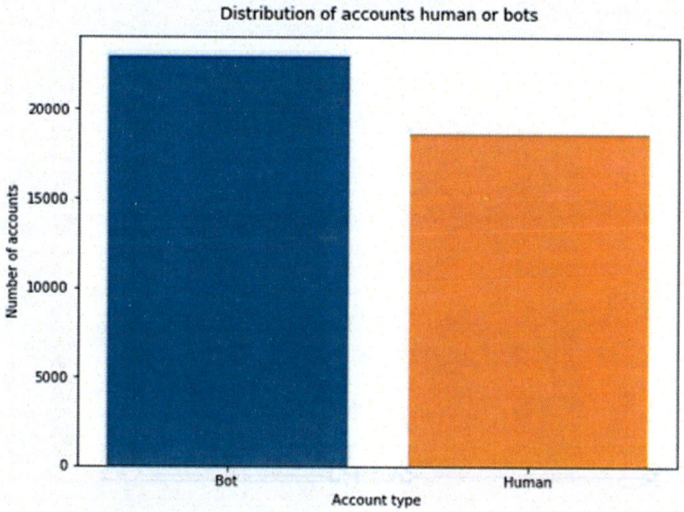

Fig. 3 Before applying SMOTE

3.3 Oversampling via SMOTE

Imbalanced datasets can affect the performance of the model, as the results might be biased toward a single majority class. The dataset used for the problem is imbalanced, i.e., it has 41,561 records out of which 22,970 are of bot accounts and 18,591 are human accounts. Thus, the minority class data is balanced using SMOTE method. SMOTE uses KNN algorithm to generate new synthetic data tuples for the minority class. Figures 3 and 4 show dataset distribution before and after SMOTE analysis.

3.4 Feature Engineering Techniques

Higher dimensional data requires high processing power, and more execution time, resulting in higher computational cost. Feature engineering techniques like feature selection and dimension reduction will remove noisy and redundant features resulting in an improved classification model. Three feature reduction techniques are applied on the dataset to obtain the most promising feature set.

PCA (Principle Component Analysis)

PCA is used in cases where the dataset is huge and has many features. It is used to reduce the dimension of such datasets, wherein a new smaller set of features is derived from the original feature set. PCA captures variance in the data; the higher the variance, more information the feature contains. The first principal component

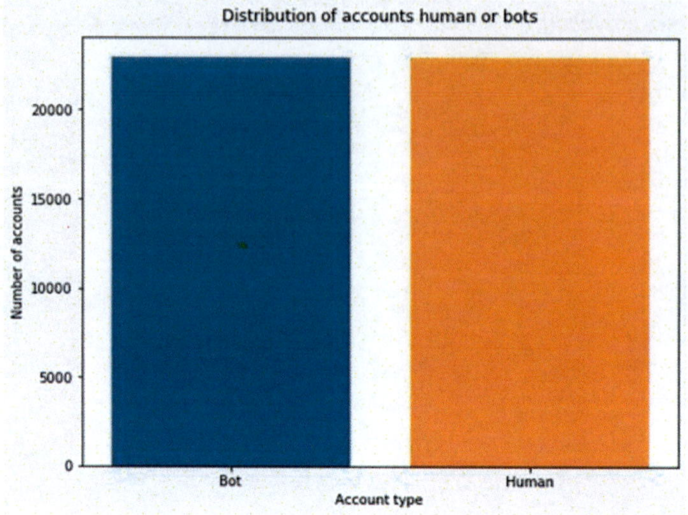

Fig. 4 After applying SMOTE

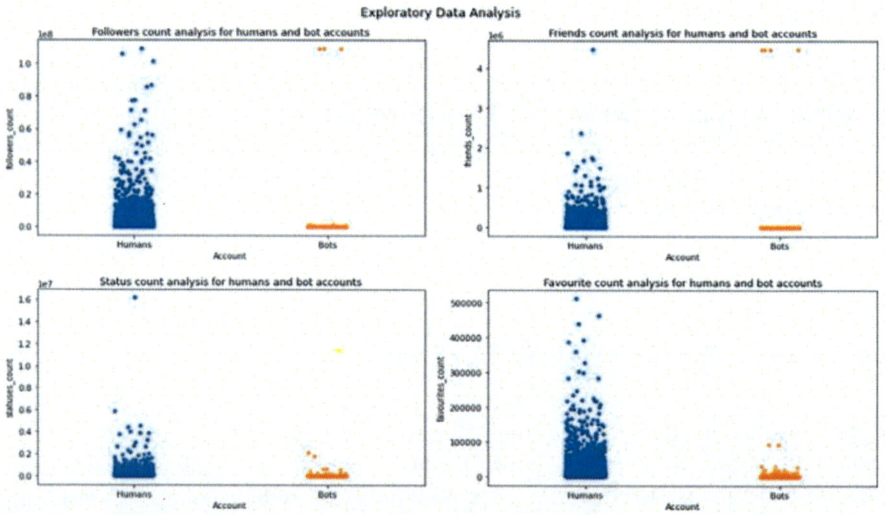

Fig. 5 Analysis of followers_count, friends_count, status_count, and favourites_count with respect to account type (human, bot)

captures maximum variance followed second principal component which is orthogonal to the first principal component and so on [17]. In the work, the 21 features are reduced to 8 features via PCA.

Table 2 List of machine learning algorithms used in the work

Sr. No.	Machine learning algorithm
1	Decision tree
2	Random forest
3	Logistic regression
4	XGBoost
5	Naïve Bayes
6	Support vector machines
7	Multi-layer perceptron

Filter method via correlation

Correlation gives a relationship measure between two variables. For a good machine learning model, feature vectors should be highly correlated with the target and uncorrelated among themselves. In the work after applying correlation, out of 21 features, 16 features are selected (Fig. 5).

Recursive feature addition (RFA)

In RFA, a model is built and an importance score is computed for all the features. In the second phase, it trains the ML model with the most important feature and computes initial performance. It then adds the second important feature and builds a new ML model. Compared to the previous model if there is a performance increase beyond the specified threshold, then that feature is important and it will be retained. This process continues until all the features of the dataset are evaluated. In the work after applying RFA, 21 features are reduced to 11 features.

3.5 Machine Learning Algorithms

After the feature engineering phase, the dataset is split into two parts, training, and validation. The training data is used to design a machine learning model. The validation set is used to evaluate the generalization ability of the model.

The machine learning algorithms used in the work are listed in Table 2.

3.6 Performance Evaluation

For evaluating the performance of classifiers, performance metrics considered are accuracy, precision, recall, $F1$ score, and false positive rate. Table 3 gives the results of all the classifiers in terms of the above-mentioned performance matrix. The best performing models are highlight in bold.

Table 3 Performance analysis table for all the classifiers along with the feature engineering techniques used

Model	Feature selection	Accuracy	Precision	Recall	$F1$-score	False positive rate
Decision tree	All features	98.80	98.88	98.74	98.81	0.012
	PCA	94.63	94.22	95.16	94.69	0.058
	Correlation	98.67	98.66	98.69	98.67	0.013
	Recursive addition	98.74	98.73	98.77	98.75	0.012
Logistic regression	All features	92.29	87.70	98.48	92.78	0.013
	PCA	93.58	92.94	94.41	93.67	0.072
	Correlation	92.40	87.88	98.48	92.88	0.137
	Recursive addition	95.40	92.36	99.04	95.59	0.082
Random forest	**All features**	**99.59**	**99.58**	**99.61**	**99.59**	**0.004**
	PCA	95.89	95.60	96.26	95.93	0.044
	Correlation	**99.61**	**99.62**	**99.62**	**99.62**	**0.003**
	Recursive addition	**99.59**	**99.58**	**99.59**	**99.59**	**0.004**
XgBoost	**All features**	**99.62**	**99.67**	**99.58**	**99.63**	**0.003**
	PCA	96.20	95.73	96.73	96.24	0.043
	Correlation	**99.64**	**99.69**	**99.59**	**99.64**	**0.003**
	Recursive addition	**99.62**	**99.64**	**99.61**	**99.62**	**0.003**
Naïve Bayes	All features	81.43	73.15	99.66	84.37	0.369
	PCA	87.55	82.52	95.46	88.52	0.204
	Correlation	81.67	73.40	99.66	84.54	0.365
	Recursive addition	82.04	73.78	99.70	84.81	0.358
Support vector machine	All features	53.69	52.05	100	68.47	0.931
	PCA	93.52	95.29	91.65	93.44	0.045
	Correlation	53.70	52.06	100	68.47	0.931
	Recursive addition	51.90	51.11	100	67.64	0.967
Multi-layer perceptron	All features	98.36	98.30	98.44	98.37	0.017
	PCA	96.08	95.40	96.88	96.13	0.047
	Correlation	98.47	99.09	97.85	98.46	0.009
	Recursive addition	98.06	99.43	96.69	98.04	0.005

Note The best performing models are highlight in bold

In the work, more importance is given to the accurate detection of bot accounts (Class 1). Hence, accuracy, precision, recall, score, and false positive rate are calculated. For bot detection false negatives, i.e., bot accounts classified as human account should incur more cost or penalty. Low recall indicates that model has lot of undetected bot profiles. Hence, recall is important measure. The focus in bot detection is finding the positive (bot account) cases. However, false alarms cannot be ignored. False alarms are false positives, i.e., human account being classified as bot account. Thus, precision is equally important. Low precision indicates the investigation of accounts that are created by humans. Since precision and recall are equally important for bot detection, the $F1$ score metric is considered for final evaluation. In Table 3, false positive rate is also shown. The focus of the study is also to choose a model which has the lowest false positive rate.

From the performance measure Table 3, it can be seen that many classifiers are giving good accuracy and $F1$ score, but their corresponding false positive rate (FPR) varies. The FPR of support vector machines (SVM) is highest; thus, it is not a good choice of classifier for bot detection. Ensemble techniques like random forest and XGBoost give the highest $F1$ score, and their corresponding FPR is low compared to other classifiers.

It can also be observed that for many classifiers, dimension reduction and feature selection techniques do not significantly alter the performance of the model. Hence, correlation filter technique (15 features) and recursive feature addition can be considered for bot detection. The FPR of most of the classifiers in the study is less than 20%. This indicates that less than 20% of time the false alarm (human accounts classified as bots) will be raised. Lowest FPR is given by XgBoost + correlation, i.e., 0.3%, and its corresponding $F1$-score is 99.64% followed by random forest + correlation gives a good $F1$-score of 99.62% and comparatively low FNR of 0.3%.

The confusion matrix for the models random forest + correlation and XgBoost + correlation is shown (see Fig. 6). Both the models depict very less false negative and false positive values.

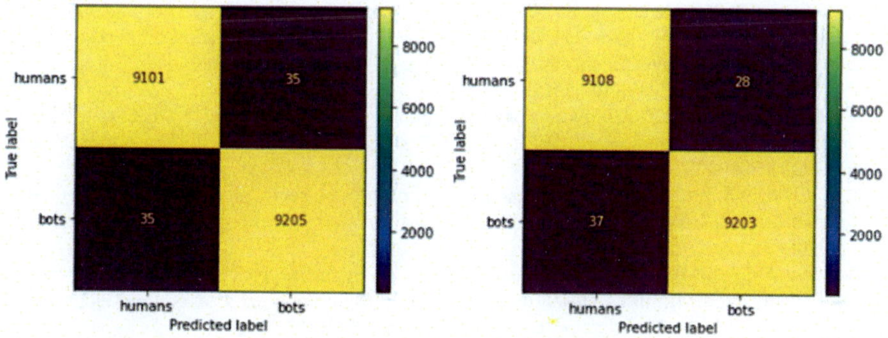

Fig. 6 Left is a confusion matrix for random forest + correlation filter model and right is a confusion matrix for XgBoost + correlation filter model

4 Conclusion

The exponential increase in the use of social media has contributed to the growth of the bot accounts. The objective of this research is to develop an automatic system using machine learning that can precisely detect bot or human accounts on social media platforms. In this work, a new set of attributes are derived from the attributes provided in dataset. We found the derived attributes age_of_account and ratio_of_followers_to_friends to be quite useful. Since dataset has a lot of features, feature engineering is also a critical step. In the work, three featuring techniques are used to remove redundant and correlated features, and seven machine learning models are created using these engineered features. A comparative study of performance metrics is presented in Table 3. Ensemble techniques such as random forest and XgBoost combined with the correlation filter method give the best performance and lowest false positive rate.

In the future, authors would like to extend the work for identifying profiles involved in spreading fake news. The feature set can also be enriched by adding network and graph-related features. Along with bot detection, revealing bot coordination strategies can be a potential research direction ahead.

References

1. Chu Z, Gianvecchio S, Wang H, Jajodia S (2012) Detecting automation of Twitter accounts: are you a human, bot, or cyborg? IEEE Trans Depend Secure Comput 9(6)
2. Kantepe M, Ganiz MC (2017) Preprocessing framework for Twitter bot detection. In: UMBK international conference
3. Dan J, Jieqi T (2017) Study of bot detection on Sina-Weibo based on machine learning. National Key Research and Development Program: The Research on Measurement Method of Service Quality Under the Condition of Network and Information
4. Chen C, Wang Y, Zhang J, Xiang Y, Zhou W, Min G (2017) Statistical features-based real-time detection of drifted Twitter spam. IEEE Trans Inf Forens Secur
5. Khaund T, Kirdemir B, Agarwal N, Liu H, Morstatter F (2022) Social bots and their coordination during online campaigns: a survey. IEEE Trans Comput Soc Syst 9(2):530–545. https://doi.org/10.1109/TCSS.2021.3103515
6. Cardinale Y, Dongo I, Robayo G, Cabeza D, Aguilera A, Medina S (2021) T-CREo: a Twitter credibility analysis framework. IEEE Access 9:32498–32516. https://doi.org/10.1109/ACCESS.2021.3060623
7. Shu K, Wang S, Liu H (2018) Understanding user profiles on social media for fake news detection. In: 2018 IEEE conference on multimedia information processing and retrieval (MIPR), pp 430–435.https://doi.org/10.1109/MIPR.2018.00092
8. Mateen M, Iqbal MA, Aleem M, Islam MA (2017) A hybrid approach for spam detection for Twitter. In: 2017 14th international Bhurban conference on applied sciences and technology (IBCAST), pp 466–471. https://doi.org/10.1109/IBCAST.2017.7868095
9. Fazil M, Abulaish M (2018) A hybrid approach for detecting automated spammers in Twitter. IEEE Trans Inf Forens Secur 13(11):2707–2719. https://doi.org/10.1109/TIFS.2018.2825958
10. Chen C, Wang Y, Zhang J, Xiang Y, Zhou W, Min G (2017) Statistical features-based real-time detection of drifted twitter spam. IEEE Trans Inf Forens Secur 12(4):914–925. https://doi.org/10.1109/TIFS.2016.2621888

11. Gannarapu S, Dawoud A, Ali RS, Alwan A (2020) Bot detection using machine learning algorithms on social media platforms. In: 2020 5th international conference on innovative technologies in intelligent systems and industrial applications (CITISIA), pp 1–8. https://doi.org/10.1109/CITISIA50690.2020.9371778
12. Khaled S, El-Tazi N, Mokhtar HMO (2018) Detecting fake accounts on social media. In: 2018 IEEE international conference on big data (big data), pp 3672–3681. https://doi.org/10.1109/BigData.2018.8621913
13. Derhab A, Alawwad R, Dehwah K, Tariq N, Khan FA, Al-Muhtadi J (2021) Tweet-based bot detection using big data analytics. IEEE Access 9:65988–66005. https://doi.org/10.1109/ACCESS.2021.3074953
14. Van Der Walt E, Eloff J (2018) Using machine learning to detect fake identities: bots vs humans. IEEE Access 6:6540–6549. https://doi.org/10.1109/ACCESS.2018.2796018
15. https://phys.org/news/2016-11-fake-tweets-real-consequences-election.html. Accessed 17 Jan 2023
16. https://www.nyu.edu/about/news-publications/news/2018/november/russian-trolls-relied-on-local-news-more-than-fake-news-in-2016-.html. Accessed 17 Jan 2023
17. Principal Component Analysis (PCA). https://www.geeksforgeeks.org/principal-component-analysis-pca/. Accessed 4 Sept 2023

Artificial Intelligence in Trucking Business Operations—A Systematic Review

Yash Honrao and Shamla Mantri

1 Background

If your company is involved in transporting goods via trucks or boats, you may want to consider using artificial intelligence (AI) to outperform the competition. People can make mistakes, be slow, or have poor planning skills, but AI can solve these issues and complete tasks more efficiently. With AI, you can identify the most efficient method to transport items, such as the quickest route, avoiding traffic, and ensuring timely delivery. AI can also help you understand customer needs and preferences, similar to having an extremely clever friend who supports your efforts. It saves time and money, ensures safety, and facilitates moving goods around. While AI is significantly being used in research for self-driving vehicles, it is also being used in the back-office automation of freight brokerage, which involves middlemen arranging transportation and tracking loads. Machine learning algorithms can detect patterns that are inscrutable to humans, such as allocating loads to drivers. This focuses on crucial issues surrounding safety and logistics. AI is used for both dispatch and in-truck solutions to optimize routes, predict the ideal time of day to schedule a delivery, and match loads with trucks more efficiently than humans.

AI eases the jobs of technicians, managers, and dispatchers by ensuring faster response times, increased fix rates, and better customer satisfaction. It uses GPS tracking and cloud-based computing to determine routes instantly, track inventory and tools in real time, and provide complete clarity to technicians, managers, and the entire business operation. Inaccurate schedules can directly hamper business operations and revenues. AI can help with efficient carrier appointment systems, enabling proper planning and reducing the risk of undelivered goods. AI can also

Y. Honrao (✉) · S. Mantri
Dr. Vishwanath Karad MIT World Peace University, Pune, India
e-mail: yashhonrao2024@gmail.com

S. Mantri
e-mail: shamla.mantri@mitwpu.edu.in

M. S. Kaiser et al. (eds.), *ICT: Smart Systems and Technologies*, Lecture Notes in Networks and Systems 878, https://doi.org/10.1007/978-981-99-9489-2_11

help in vehicle routing and scheduling in supply chain and logistics, which can reduce fuel costs and other vehicle expenses. Transmetrics is an application of predictive analysis and artificial intelligence in transportation that enables automated predictive demand planning and historical reporting for managing workflow more effectively with less manual effort. By taking into consideration all external factors, AI can help companies plan around unnecessary roadblocks and remain efficient. Implementing AI can help trucking companies stand apart from their competitors. AI leverages shipment data to optimize fleets and allowing companies to derive insights. By using AI, companies can increase productivity, efficiency, and customer satisfaction, while also reducing costs and improving sustainability.

Figure 1 shows how with the help of predictive analysis in AI, trucking companies can optimize routes and delivery schedules and also help in scheduling operations. AI can also help with automating things like scheduling and dispatching, so drivers and trucks can be assigned to the best routes based on real-time info. It can also be used in making trucks fully autonomous.

Fig. 1 Brief overview of artificial intelligence in trucking business operations

2 Prior Art

Increasing reliability of logistics on autonomous systems has put forward new challenges for human–machine interaction concepts. Because of the unavailability of competent labor in some regions and to meet the objectives of sustainability improvement and efficiency, it is a need for logistic operators to pursue automation in the process. Comprehensive case studies are performed regarding automated truck driving to test the practical implementation concerns [1]. Logistic nodes, such as terminals, container depots, and loading resources suffer from heavy workloads and simultaneously the companies also face augmented waiting times. The systems predict truck arrival rates and truck waiting times at the warehouses based on previous data and other miscellaneous factors. Based on predicted workloads, the resources can be well-planned and utilized. The companies can smartly carry out their planning of routes to optimize truck waiting times. This also results in reducing traffic and pollution [2].

Teletruck system facilitates dispatches and route planning. It enables the optimization of orders and dynamic planning on the basis of integration with telecommunication facilities. It uses agents along with its subagents that work co-operatively to achieve a flexible structure resource [3]. Drone–truck combined operations are a relatively new concept. It faces various optimization issues including mathematical models, synchronization between drones and trucks, solution methods, and also barriers in implementing DTCO [4].

With continuously increasing demand in modern logistics and urban mobility, vehicle traffic has been significantly rising over the past several decades. Most of the major metropolitan cities are suffering from heavy traffic congestion. This also affects economic development [5]. In Fig. 2, it is shown that AI is a prominent tool when it comes to applying in supply chain management. Some of its major contributions are operational procurement with the help of chatbots and intelligent data, supply chain planning, managing warehouses for optimizing stock, efficient and speedy shipping in turn reducing transportation expenses, and supplier selection by using real-time data.

3 Methodology

In the global corporate, sustainability has become a mandate with two factors impacting the implementation. The first is the need to "green" (make sustainable and environment friendly) the entire supply chain. The second factor is using AI and Big data that is technology digitization. These technologies facilitate the organization and management of their supply chains and in turn improve sustainability [6]. AI promises growth in economy however there is concern that it could replace the job of workers, reshape organizations, and change industry trajectories. Currently, there are limited resources to study whether and how AI will affect various occupations. To

Fig. 2 Prior art of artificial intelligence in trucking business operations

study it better we have a new measure called AI Occupational Exposure. It describes measures for studying the effect of AI on different occupations and markets [7].

We can use neural networks and artificial intelligence methodology to calculate the volume of global logistics expenditure at the country level. Major contributions are former estimates of global expenditures of logistics and the continuous enhancement of the methodology used for estimation. This includes the refinement of the mathematical model along with information availability and quality [8]. Technological development has impacted the road freight transport system. The stable business model had truck at its pivot. The appearing business model also contains technological development of connectivity and automation by looking at the future. Managerial implications suggest that keeping up with emerging businesses is crucial but there is great precariousness in how to react and which resources should be invested and also the coordination of activities [9].

Autonomous electric trucks are expected to transform logistics in the next couple of years. It will also significantly reduce the cost of labor and time required for transportation. Automatic Electric Vehicles (AEVs) will set free up to 30 billion trucking hours every year in the USA alone which is currently spent in traffic or searching for parking space [10] (Fig. 3).

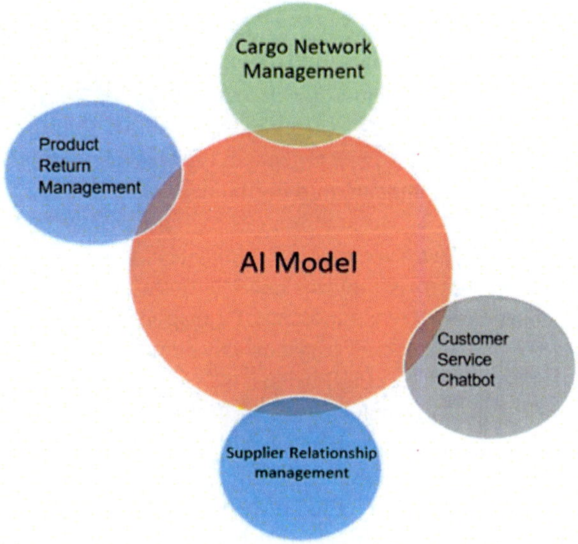

Fig. 3 Methodology of artificial intelligence in trucking business operations

4 Applications

Reliable AI-Technology is powering AI-augmented Business Process Management Systems (ABPMSs) which is a novel concept of process-aware information systems. An ABPMS improves the way business processes are carried out by making them more flexible, context-sensitive, explicable, and proactive. This manifesto outlines an ideal state for ABPMSs and describes the research obstacles that must be overcome to achieve this state. In order to do this, we define the term "ABPMS," describe the lifecycle of a process inside an ABPMS, talk about its key attributes, and produce a list of difficulties in realizing systems with these attributes [11]. In the last two decades, we have seen unprecedented development in artificial intelligence (AI) and machine learning (ML) applications. Studies have been made to examine the current uses of AI in operations management (OM) and management of the supply chain. As industries account for the majority of business-related AI advancements as well as emerging issue areas, studies focus specifically on innovations in health care, manufacturing, and operations of the retail. They go over the primary obstacles and potential uses of AI in certain sectors and also talk about fads [12].

The implementation of the smart warehouse concept has been made possible by breakthroughs in technology, a revolution in business traditions and practices, the necessity to change operations of warehouses as a result of mounting orders and the problems they imply, and a deficiency of managerial talent housing. Moreover, since storage is so non-trivial in the logistics and supply chain, smart warehousing is the need of the hour to upgrade organizational management and efficiency. The use of AI in warehouse operations increases the probability of success in terms of logistics, coordination, and management. Using artificial intelligence in warehouses will improve efficiency. The use of AI in warehousing to provide an intelligent system for automated logistics is suggested by various studies. In order to avail the merchandise that is now accessible in the warehouse at any time, studies have proposed automated storage and retrieval by utilizing the power IOT, AI, and cloud computing [13]. Advancements in digital technologies are making it possible to turn business environments into digital ecosystems. This shift enables new opportunities [14].

Technological advances in AI have led to the development of the transport sector. Although self-driving trucks and autonomous trucks are far from the real world, AI has become a fixed component of transport vehicles. Many studies primarily focus on the technical aspect of assistance systems [15]. Figure 4 shows AI has many implementations in trucking business scheduling operations. For route optimization, AI analyzes factors such as traffic patterns, weather conditions, and road construction. Load matching is handled by AI-powered load-matching platforms that can match suitable shipments with available trucks further optimizing capacity utilization and reducing empty miles. Predictive maintenance with the help of AI can identify potential equipment failures. Driver behavior monitoring can be done with AI to detect and analyze driver behavior.

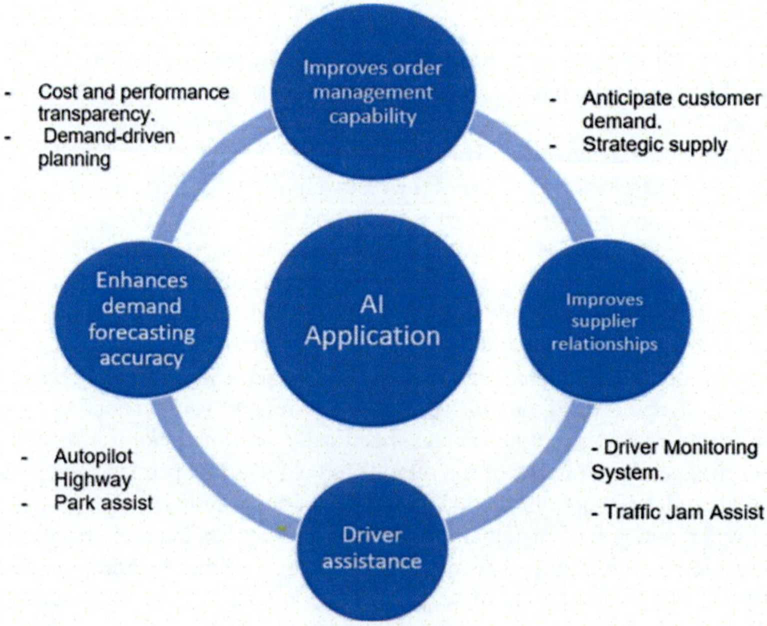

Fig. 4 Applications of artificial intelligence in trucking business operations

5 Current Status

The performance metric of Container Terminal node consists of two indicators: the vessel loading time and the turn-around time of the truck. These are affected by truck arrival and departure. The scheduling of truck appointments is designed to allot a time slot to every container to load in the delivery goods, thereby reducing the truck wait times at the warehouses. This is considered as an explanation to manage truck arrivals and thereby improve the efficiency [16]. There exist different strategies for dispatching problems and analyze the merits and demerits of these strategies [17].

In the trucking business, one of the major drawbacks is the waiting time for trucks at the gates of warehouses or in yards, followed by emissions that are harmful to the environment and result in low productivity of the Container Terminals. The truck appointment problems solve this to achieve more balance in Container Terminals. More efforts need to be taken in order to collaborate with different trucking companies when scheduling external trucks [18]. Mars (modeling autonomous cooperating shipping) models efficient and cooperative truck and order scheduling within partner shipping companies. Distributed AI (DAI) provides tools to eliminate challenges in this domain. The three significant aspects of DAI are cooperation between the agents, decentralized planning, task decomposition, and task allocation. To obtain better solutions for the allocation of resources, an extension of the contract net protocol for the allocation and decomposition of tasks can be used. This solution can be improved significantly and can also be used for implementing dynamic replanning [19].

Another new emerging idea is using drones for last-mile delivery. After the drone drops off, it can either hover back to the warehouse or take a fleet directly to another location for pick-up. Integrating a fleet of trucks and drones can deliver operational excellence. The vehicle travels closer to the boundary to deliver to customers who live closer to the neighboring depot. This ensures us a unique infrastructure to effectively tackle challenges associated with operations [20]. Figure 5 shows how AI is being used more and more in the trucking industry to help plan operations. Basically, AI can analyze a lot of data to figure out the best routes, schedules, and overall planning. Some people are even working on making trucks drive themselves. All of this helps with better planning and decision-making by using a smart computer program that can predict things. This program looks at old data to figure out what might happen in the future. Another important thing AI does is called "predictive maintenance." This just means that computers can tell when a truck might need fixing before it actually breaks down.

6 Future Scope

Due to continuous increasing movements of freight, sustainability in road freight transportation has gained importance. It is difficult for transport managers to manage the risks and mitigate them through economically sustainable strategies. Very little

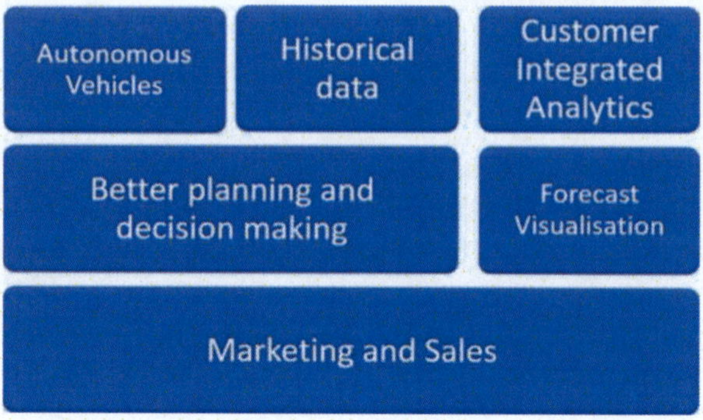

Fig. 5 Current status of artificial intelligence in trucking business operations

research has been done on sustainable risk management practices of road freight transportation. Effect analysis and integrated fuzzy failure mode approach in the selection of risk management planning in the logistics business is used. These are selected considering the cruciality of threats along with a finite budget. Considering evaluations of specialists' judgments and cost–benefit justification gives convincing outcomes [21]. Various scheduling protocols influence resource usage and turn time of trucks in grounded operations. This can compel terminal managers and appointment service providers to alter their existing truck scheduling routine. Simulation facilitates a more practical visualization of the complex operations at warehouses. Simulation also overcomes the limiting assumptions of analytical methods. Such a system keeps the yard train occupied [22].

The value of researching project management in contexts with several projects has been shown in earlier research. When resource capacity is fully committed and project management is schedule-driven, sharing competent resources across concurrent projects presents a difficulty. We employ a simulation on system dynamics built on in-depth groundwork with a highly efficient vehicle manufacturer to delve deeper into this issue. We model what happens when resources initially allotted to one project are taken over to further a later-started product development initiative. Moreover, when staff individuals shift between projects quite often, their productivity overall declines. All these consequences result in the subsequent projects being deferred, permanently impairing the organization's capability to accomplish tasks before the deadline [23]. In recent years, supply networks have become increasingly time-sensitive. Operations in the supply chain that are delayed may have large negative externalities, such as lost sales and clients. Several distribution organizations began using the cross-docking approach to speed up the process of goods distribution within business logistics, decrease related setbacks, and enhance the performance of the operations in the supply chain. Effective truck arrival scheduling is one of the difficult issues in cross-docking facility management. In contrast to the Evolutionary Algorithms, the novel Diploid Evolutionary Algorithm proposed in a study accumulates the genetic data from the parent chromosomes after performing a crossover operation. It aims to solve the problem of scheduling trucks at a cross-docking facility. A developed mathematical model is intended to reduce the overall expenditure of the trucking business. The executed numerical experiments show that the developed algorithm's optimality margin over the instances of small-size problems taken into consideration does not surpass 0.18%. In comparison with a standard Evolutionary Algorithm, the study of the realistic volume issue instances shows that using the created solution method significantly decreases the overall handling time of a truck, total waiting time of a truck, and overall truck delayed departure time by, respectively, 6.14%, 32.61%, and 34.01%. Also, the overall cost of truck service is reduced by 18.17% with the diploid idea [24].

A novel approach to designing and engineering control solutions-based decentralization of control over distributed structures is provided by industrial agent's technology, which takes advantage of distributed computing, multiagent systems, AI tactics, and semantics in the field of infrastructure, manufacturing, and services. The advantages of agent-based industrial systems, particularly with respect to resilience, reconfigurability, productivity, and scalability, all of which sum to a larger competitive edge, are the main drivers for this application. This article tracks the progression of multiagent and holonic system research and development for commercial use. From the 1990s until today, it shows a thorough review of the techniques, designs, and applications of agents in the industrial realm [25]. Figure 6 shows that AI can integrate different road carrier companies for efficient operations through collaboration and partnership. It can fully automate warehouse operations and delivery/shipment notifications. AI can satisfy customers with 24/7 virtual assistance for customers through chatbots in vernacular languages. It can enable real-time shipment tracking. Autonomous vehicles are the biggest challenge that AI can solve.

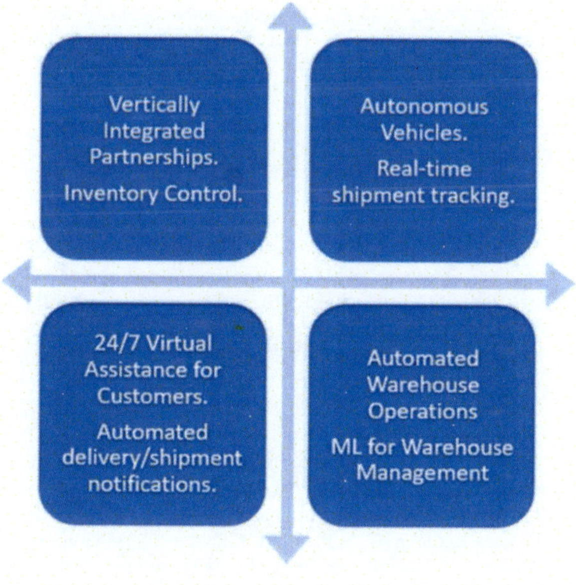

Fig. 6 Future scope of artificial intelligence in trucking business operations

7 Conclusion

To sum up, the integration of AI into the supply chain and logistics industry has led to significant advancements with respect to time-saving, cost efficiency, and operational optimization for businesses. This technology provides companies with the ability to leverage shipment data to derive valuable insights and optimize their fleets, ultimately setting them apart from their competitors. The benefits of AI in the industry include chatbots and intelligent data for operational procurement, supply chain planning, warehouse management, and speedy shipping, as well as real-time information for optimal supplier selection. Automatic Electric Vehicles (AEVs) will set free up to 30 billion trucking hours every year in the US alone which is currently spent in traffic or searching for parking space and also reduce traffic and optimize space. Despite the potential of self-driving trucks and drones for last-mile delivery, more collaboration is needed with different trucking companies when scheduling external trucks. Additionally, research is required to develop sustainable risk management practices for road freight transportation. Looking forward, AI is expected to fully automate warehouse operations and delivery/shipment notifications, and continue to drive innovation in the industry.

References

1. Klumpp M (2018) Automation and artificial intelligence in business logistics systems: human reactions and collaboration requirements. Int J Log Res Appl 21:224–242
2. Hill A, Böse JW (2017) A decision support system for improved resource planning and truck routing at logistic nodes. Inf Technol Manag 241–251
3. Vierke G, Fischer K, Bürckert H-J (2010) Holonic transport scheduling with teletruck. Appl Artif Intell 14:697–725
4. Sah B, Lee J, Chung SH (2020) Optimization for drone and drone-truck combined operations: a review of the state of the art and future directions. Comput Oper Res 123
5. Li J, Cheng H, Guo H, Qiu S (2018) Survey on artificial intelligence for vehicles. Autom Innov 1:2–14
6. Li J, Cheng H, Guo H, Qiu S (2019) Sustainable supply chains in the age of AI and digitization: research challenges and opportunities. J Bus Logist 40:229–240
7. Felten E, Raj M, Seamans R (2021) Occupational, industry, and geographic exposure to artificial intelligence: a novel dataset and its potential uses. Strateg Manag J 42:2195–2217
8. Bowersox DJ, Calantone RJ, Rodrigues AM (2011) Estimation of global logistics expenditures using neural networks. J Bus Logist 24:21–36
9. Lind F, Melander L (2021) Networked business models for current and future road freight transport: taking a truck manufacturer's perspective. Technol Anal Strateg Manag 35:167–178
10. Blitz A, Kazi K (2019) Mapping technology roadblocks and opportunities in the transportation revolution. Strategy Leadersh 47:43–46
11. Dumas M, Fournier F, Limonad L, Marrella A, Montali M, Rehse JR, Accorsi R, Calvanese D, De Giacomo G, Fahland D, Gal A (2023) AI-augmented business process management systems: a research manifesto. Assoc Comput Mach 14:19
12. Dogru AK, Keskin BB (2020) AI in operations management: applications, challenges and opportunities. J Data Inf Manag 2:67–74

13. Pandian DAP (2019) Artificial intelligence application in smart warehousing environment for automated logistics. J Artif Intell Capsule Netw 1:63–72
14. Kopalle PK, Kumar V, Subramaniam M (2020) How legacy firms can embrace the digital ecosystem via digital customer orientation. J Acad Mark Sci 48:114–131
15. Loske D, Klumpp M (2021) Intelligent and efficient? An empirical analysis of human–AI collaboration for truck drivers in retail logistics. Int J Logist Manag 32:1356–1383
16. Abdelmagid AM, Gheith MS, Eltawil AB (2022) A comprehensive review of the truck appointment scheduling models and directions for future research. Transp Rev 42:102–126
17. Alarie S, Gamache M (2002) Overview of solution strategies used in truck dispatching systems for open pit mines. Int J Surf Min Reclam Environ 16:59–76
18. Azab A, Karam A, Eltawil A (2020) A simulation-based optimization approach for external trucks appointment scheduling in container terminals. Int J Model Simul 40:321–338
19. Fischer K, Müller JP, Pischel M (1996) Cooperative transportation scheduling: an application domain for DAI. Appl Artif Intell 10:1–34
20. Ham AM (2018) Integrated scheduling of m-truck, m-drone, and m-depot constrained by time-window, drop-pickup, and m-visit using constraint programming. Transp Res Part C: Emerg Technol 91:1–14
21. Kumar Dadsena K, Sarmah SP, Naikan VNA (2019) Risk evaluation and mitigation of sustainable road freight transport operation. Int J Prod Res
22. Huynh N (2009) Reducing truck turn times at marine terminals with appointment scheduling. Transp Res Rec 100(1):47–57
23. Yaghootkar NGK (2012) The effects of schedule-driven project management in multi-project environments. Int J Project Manag 30(1):127–140
24. Dulebenets MA (2018) A diploid evolutionary algorithm for sustainable truck scheduling at a cross-docking facility. Sustainability 10(5)
25. Leitão P, Mařík V, Vrba P (2013) Past, present, and future of industrial agent applications. IEEE Trans Ind Inform 9:2360–2372

Deep Learning Approach for Early Diagnosis of Alzheimer's Disease

Vaishnav Chaudhari, Shreeya Patil, Yash Honrao, and Shamla Mantri

1 Introduction

Alzheimer's disease is characterized as a neurological disorder leading to the loss of brain cells and the reduction of brain tissue volume. It stands as the primary factor contributing to the onset of dementia. It has been observed that it causes about 60–70% of dementia issues. The current study has shown that there are about 55 million people worldwide who are suffering from dementia. And it is observed from the trend that the number of patients will double the strength. The countries which are mostly affected by it are China, India, and south Asian countries.

Dementia is termed as the impaired ability to remember, decline in thinking, and behavior, and it reduces the person's function independence. It might not be considered seriously by the majority of people, but studies have shown that it is the seventh leading disease which causes death globally. Other than health, it had impacts on physical, social well-being of the patient as well his family. Due to lack of awareness among the people, its diagnosis at initial stages is not possible.

We can classify the Alzheimer in two broad categories as follows:

- Early onset: It happens when the age group of the people diagnosed is below 65. They usually lie in the age group of 40s to 50s. Studies have shown that only 5% of the people diagnosed with Alzheimer's have early onset.

V. Chaudhari · S. Patil · Y. Honrao (✉) · S. Mantri
Dr. Vishwanath Karad MIT World Peace University, Pune, India
e-mail: falseyashhonrao2024@gmail.com

S. Mantri
e-mail: shamla.mantri@mitwpu.edu.in

M. S. Kaiser et al. (eds.), *ICT: Smart Systems and Technologies*, Lecture Notes in Networks and Systems 878, https://doi.org/10.1007/978-981-99-9489-2_12

- Late onset: It happens when the people diagnosed with it are aged 65 and older. Studies have shown that about 95% of the people diagnosed with Alzheimer's have early onset. So, it is shared by the majority of the people.

The diagnosis of the Alzheimer's is carried in the following ways:

If the person is experiencing memory loss, cognitive decline or behavioral change that affects his body to work properly. These are some of the symptoms based on this doctor may prescribe the following test:

- Cognitive and memory test
- Neurological function test
- Blood test
- CT scan or MRI.

On the basis of examining the result the doctor confirms whether the person is suffering from Alzheimer or not and instructs the patient about what measures he/she needs to take. The symptoms of Alzheimer can be observed and identified by the patient or by its family members.

The symptoms of the Alzheimer can be identified based on its stages:

- Early symptoms
  - The patient might experience a loss of memory in the form of forgetting about conversation/events.
  - Tendency to ask questions frequently.
  - Ability to make decisions drop drastically.
- Mild Stage Symptoms
  - There is a high possibility of the symptoms observed in the early stages to scale up.
  - Aphasia: Means the problem with speech/language.
  - They suffer from hallucinations.
  - Their behavior tends to be impulsive.
- Later-Stage Symptoms
  - It is observed that they suffer from significant weight loss.
  - They suffer from dysphagia (swallowing disorder).
  - They gradually lose their body balance and become dependent on personal assistance.

2 Literature Review

Major neurocognitive disorder, which is also referred to as dementia, encompasses a range of symptoms stemming from various diseases rather than a singular condition. These symptoms significantly impact memory, behavior, reasoning abilities, and social skills to the extent that they hinder daily functioning and personal independence. It impacts memory, behavior, thinking, and social abilities severely enough to interfere with one's activities of daily living and social autonomy. Today almost 44 million people worldwide are suffering from a special type of dementia known as Alzheimer's. Subsequently, an integration between machine learning and big data analytics has proven to be a promising solution for its prevention. Using artificial neural networks, there were several approaches developed, which will be evaluated in the following text.

The system proposed by Albright et al. was able to analyze the future cognitive state of the individual [1]. Consequently, the data can also be used to correlate the machine learning model to predict Alzheimer's disease for other patients. Supervised classification techniques were utilized to identify the patient's exposure to Alzheimer's dementia. The framework has proven to be a successful method for practical application. C. Kavitha et al. have generalized using a number of machine learning parameters [2]. In her pursuit of identifying the optimal parameters for Alzheimer's disease prediction, she has integrated a diverse set of machine learning algorithms. These include decision trees, Random Forests, Support Vector Machines, Gradient Boosting, and Voting classifiers. The prediction can further be analyzed using big data, to make it more accurate.

To uncover early diagnostic biomarkers for Alzheimer's dementia, Ankita Sharma and her colleagues have devised an integrated solution utilizing a Hadoop-based big data platform. This platform incorporates non-invasive techniques such as Magnetic Resonance imaging (MRI), MR spectroscopy (MRS), and data from neuropsychological tests [3]. They employ advanced data mining, machine learning, and statistical modeling algorithms in this state-of-the-art, holistic approach. In order, the Hadoop environment was used for data management and storage. Feature extraction was implemented using machine learning algorithms.

Mathew Harper et al. have used a unique approach in its prediction framework [4]. The diagnosis and prognosis of Alzheimer's disease (AD) could experience substantial improvements through the implementation of comprehensive data science technology. This technology would seamlessly integrate various patient data, including brain imaging, blood chemistry, lifestyle information, and other relevant data sources. Furthermore, big data analytics is implemented to analyze the current factors including Genetic data, Inflammation, Infections, Homocysteine, Fasting Insulin Level, etc. The data is then processed, to propose a first-hand treatment intended to slow and stop AD's progression.

Taeho Jo et al. introduced several aspects of deep learning using multiple algorithms, which integrated with big data analytics would certainly prove to be a progressive framework for AD identification [5]. In the following research, the authors have discussed and implemented a number of deep learning architectures including gradient computation, the Deep Belief Network (DBN), and Deep Boltzmann Machine (DBM).

3 Methodology

3.1 Research Method

The primary objective of this study is to leverage deep learning techniques in the analysis of neuroimages to predict the early indicators of Alzheimer's disease. Magnetic Resonance Imaging (MRI) stands out as one of the most advanced and sophisticated approaches for neuroimaging [2].

Magnetic Resonance Imaging (MRI) is a powerful and non-invasive imaging technique capable of producing intricate three-dimensional anatomical images. It is commonly utilized for purposes such as disease detection, diagnosis, and monitoring of therapy progress. Operating on cutting-edge technology, MRI stimulates and detects alterations in the rotational axis of protons within the water present in living tissues [2, 3]. MRI has proven to be a promising technique for Alzheimer's prediction. For the diagnosis of Alzheimer's disease, several frequently employed datasets include the Alzheimer's disease Neuroimaging Initiative (ADNI) dataset, the Harvard Aging Brain Study (HABS) dataset, and the Mayo Clinic Study of Aging (MCSA) dataset. Additionally, DementiaBank, a dataset available through the Open Access Series of Imaging Studies (OASIS), is another valuable resource in this context.

The AD dataset undergoes an initial preprocessing stage that involves several techniques, such as resizing, random rotation, normalization, and brightness adjustment. Spatial normalization is applied to images from diverse subjects to align them with a common template [1–3, 6]. Smoothing is used to improve image quality by reducing noise. Grayscale normalization is employed to map pixel intensity levels to a new and more suitable range. Additionally, the image is segmented into multiple distinct regions using slicing techniques. Random rotation allows you to determine how often each row in a feed will be served. The desired image size is achieved through resizing. Subsequently, the deep learning (DL) model engages in feature extraction and classifies the input data, utilizing the preprocessed data as its input. Finally, the model's performance is evaluated using various metrics, including accuracy, $F1$ score, area under the curve (AUC), and mean square error (MSE).

Feature extractions are then executed upon these images, using certain models of Convolutional Neural Networks (CNN) such as XceptionNet, ResNet-18 [3]. Convolutional Neural Networks (CNN) represent a deep learning system capable of identifying objects within an input image, assigning importance to them through learnable weights and biases, and distinguishing between them [2]. CNNs exhibit significantly lower preprocessing demands compared to traditional classification algorithms. ConvNets have the ability to learn these filters and characteristics autonomously, whereas in earlier approaches, filters were manually designed [2].

Classifiers such as the management and processing of such large data require a special big data analytics (BDA) framework, called Hadoop [3, 7]. This framework allows for the incorporation of a substantial volume of diverse data, which is then subjected to data-specific preprocessing. Following this preprocessing step, the processed data is analyzed, and the results of diagnostic tests are inferred from the findings. There are a number of frameworks used in Hadoop such as Apache Zookeeper, YARN, Oozie, HDFS, and HBase. They are used for data management, preprocessing, and storage, respectively [3].

3.2 Datasets

For accurate prediction of the disease and for deep analysis of the disease we had collected aur dataset from Kaggle. Our dataset contains both tabular and image data for better understanding. For tabular data, we had used the data set provided by publisher Baris Dincer. It characterizes the data into groups such as nondemented, demented, and others. The framework incorporates a range of parameters for data analysis, encompassing factors such as socioeconomic status, the Mini Mental State Examination, the Clinical Dementia Rating, and numerous others.

These factors help in visualization of data and to process it to generate the result. Image-based dataset (i.e., data set obtained by MRI of the brain) has been collected from the research gate publication on deep learning-based prediction on Alzheimer disease [2]. It consists of about 5121 images which are been grouped into four classes as follows (Fig. 1):

- Mild demented
- Moderate demented
- Nondemented
- Very mild demented.

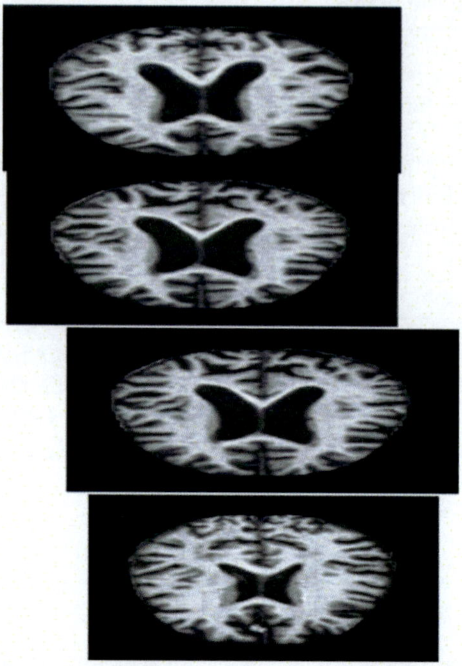

Fig. 1 Type of classes for grouping Alzheimer. **a** Mild demented, **b** moderate demented, **c** nondemented, and **d** very mild demented [2]

3.3 Data Preprocessing

Before assigning an Alzheimer classification, a range of data preprocessing techniques, such as normalization, scaling, rotation, etc., are used. A data preprocessing technique called normalization is used to scale down the picture data's pixel values without changing its shape. It makes sure that the data distribution for each input parameter (in this case, pixel) is consistent. This speeds up convergence while the network is being trained. When normalizing data, each pixel is first subtracted from its mean before the result is divided by the standard deviation. Such data would have a distribution that resembles a zero-centered Gaussian curve. Since we require positive pixel values for picture inputs, we may decide to scale the normalized data in the [0, 1] or [0, 255] range. In the field of computer vision, one important step in the initial data processing is resizing images. When it comes to deep learning models, they tend to train faster when working with smaller images. This is because with a larger input image, the neural network has to learn from a much larger number of pixels, which significantly extends the time it takes to train the model.

4 Architecture

Data collection
↓
Data preprocessing
↓
Data handling
↓
Applying decision tree classifier
↓
Fitting the data
↓
Training the model
↓
Analyzing the result
↓
Displaying the result

4.1 Algorithms

Decision Tree Classifier

When you're working on building a machine learning model, the most crucial consideration is choosing the right approach for the specific dataset and problem you're tackling. Machine learning provides various algorithms, each serving different purposes.

One of these approaches is called "decision tree learning." It's a technique that involves creating decision trees from data to perform tasks like classification and regression. In our research, we employed decision trees to aid in making decisions. The advantage of using decision trees is that they mimic the way humans make decisions, making them easier to grasp. They present their reasoning in a tree-like structure, which simplifies understanding.

5 Result

The model used for the textual data by using discrete tree classification gives an accuracy of 99% and was able to successfully predict the patient report into the specific class and helps to treat the patient (Fig. 2).

Data Visualization

See Figs. 3 and 4.

	precision	recall	f1-score	support
0	0.29	0.17	0.21	12
1	0.85	0.95	0.90	41
2	0.93	0.93	0.93	59
accuracy			0.86	112
macro avg	0.69	0.68	0.68	112
weighted avg	0.83	0.86	0.84	112

Fig. 2 Results

```
data["Age"].hist()

<matplotlib.axes._subplots.AxesSubplot at 0x7f0da138aa50>
```

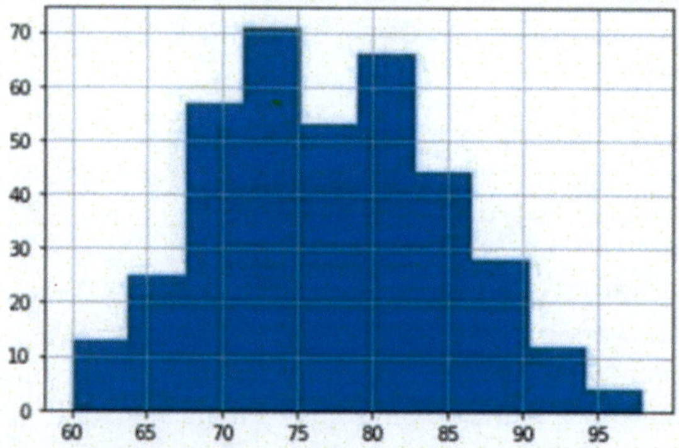

Fig. 3 Plot of Alzheimer's patients at different ages

```
data["SES"].hist()

<matplotlib.axes._subplots.AxesSubplot at 0x7fce4656da10>
```

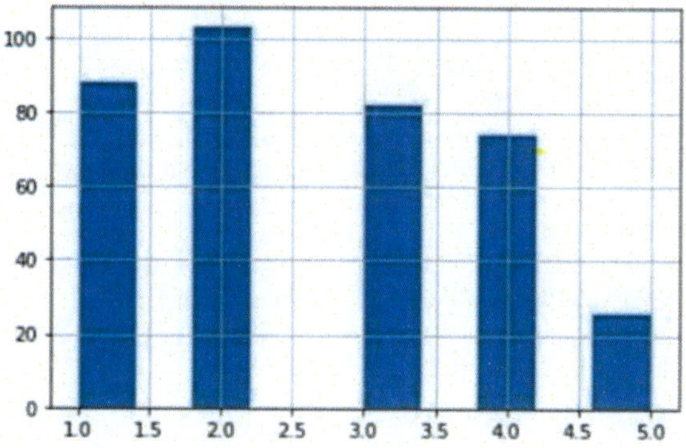

Fig. 4 Plot of patient's symptomatic

6 Conclusion

To develop a robust framework for big data analytics (BDA) in the field of medical research, it's essential to focus on three key aspects: brain imaging, metabolic data, and cognitive scores. These elements play a crucial role in understanding the progression of illnesses and identifying potential early indicators or biomarkers for diagnosis. Collaborative efforts between medical physicists, clinicians, and engineers will be instrumental in creating a practical tool for predicting or diagnosing Alzheimer's disease (AD) in its early stages.

As a result, our model, which is capable of analyzing both text and visual data, has shown promising results. It can accurately categorize different types of Alzheimer's disease from images. This technology has the potential to be applied on a large scale, offering a cost-effective and accessible solution for remote use.

References

1. Digital Economy Papers, No. 233. OECD Publishing, Paris. https://doi.org/10.1787/5jz73kvmvbwb-en
2. Helaly HA, Badawy M, Haikal AY (2022) Deep learning approach for early detection of Alzheimer's disease. Cogn Comput 14:1711–1727. https://doi.org/10.1007/s12559-021-09946-2
3. Shinde S, Satav S, Shirole U, Oak S (2022) Comprehensive analysis of Parkinson disease prediction using vocal parameters. In: 2022 international conference on machine learning, big data, cloud and parallel computing (COM-IT-CON), pp 369–373. https://doi.org/10.1109/COM-IT-CON54601.2022.9850857

4. Kavitha C, Mani V, Srividhya SR, Khalaf OI, Tavera Romero CA (2022) Early-stage Alzheimer's disease prediction using machine learning models. Front Public Health 10:853294. https://doi.org/10.3389/fpubh.2022.853294. PMID: 35309200; PMCID: PMC892771
5. Albright J (2019) Forecasting the progression of Alzheimer's disease using neural networks and a novel preprocessing algorithm. Alzheimer's Dement (N Y) 25(5):483–491. https://doi.org/10.1016/j.trci.2019.07.001. PMID: 31650004; PMCID: PMC6804703
6. OECD (2014) Unleashing the power of big data for Alzheimer's disease and dementia research: main points of the OECD expert consultation on unlocking global collaboration to accelerate innovation for Alzheimer's disease and dementia. OECD
7. Sharma A, Shukla D, Goel T, Mandal PK (2019) BHARAT: an integrated big data analytic model for early diagnostic biomarker of Alzheimer's disease. Front Neurol 8(10):9. https://doi.org/10.3389/fneur.2019.00009. PMID: 30800093; PMCID: PMC6375828

Study of Key Agreement Protocol Implementation in Constraint Environment

Chandrashekhar Goswami, Amit K. Gaikwad, Ansar Sheikh, Swapnil Deshmukh, Jayant Mehare, and Shraddha Utane

1 Introduction

Key establishment is the transaction where creating a common shared secret amongst all legitimate users (parties). Basic approaches for this can be categorized hooked on the key establishment as presented in Fig. 1. Of taxonomy of key establishment schemes. In key transport protocol procedure where one legitimate party contributes in key generation and securely handovers a private (secret scalar) value. In the establishment process, protocols like key agreement protocol procedure where two or more legitimate users derive the common shared secret value where altogether legitimate parties mutually shared the secret. Preferably, nothing can control the final joint value by the users [1–4].

Protocol for key exchange/establishment is always accompanying the process of identification. For example, any unauthorized clients may think of various types of attacks that can happen which are caused by unauthorized users that can create a false connection to the key establishment protocol to conceal as a legitimate user to create a common shared secret key using another communicating party. To avoid active and passive attacks, each party must be assured of the individuality of another

C. Goswami (✉)
MIT School of Computing, MIT ADT University, Pune, India
e-mail: shekhar.goswami358@gmail.com

A. K. Gaikwad · J. Mehare
G H Raisoni University Amravati, Amravati, India

A. Sheikh
St.Vincent Palloti College of Engineering & Technology Nagpur, Nagpur, India

S. Deshmukh
School of Engineering, Ajeenkya DY Patil University Pune, Pune, India

S. Utane
H V P M's COET, Amravati, India

M. S. Kaiser et al. (eds.), *ICT: Smart Systems and Technologies*, Lecture Notes in Networks and Systems 878, https://doi.org/10.1007/978-981-99-9489-2_13

Fig. 1 Taxonomy of key exchange/establishment systems

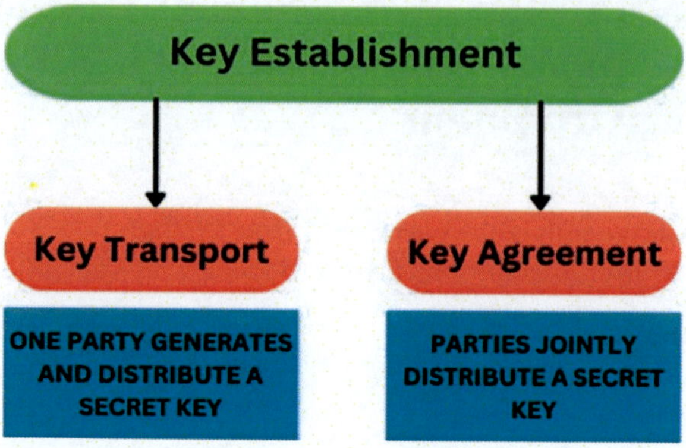

person. A major one is that there is less damage if the key (secret) is revealed. Also, an observer (e.g. attacker) has a smaller amount of ciphertext accessible that was produced through one key, which can make cryptographic attacks much more difficult. Additionally, an attacker is enforced to recognize several keys, if the intended party is concerned in deciphering most parts of the original text. In real-world illustrations, wherever session keys are repeatedly produced which contain a choice of enciphering process in GSM-based devices; in both cases, new keys are produced within a matter of minutes or sometimes even seconds [4–7].

The stimulus for transient keys comprises the succeeding section [8, 9]:

- To bound offered ciphertext (under a fixed key) for cryptanalytic attack;
- Bound acquaintance, with detail to both aspects of effectively termed time and extent of data;
- Circumvent long-term storage of an enormous amount of different secret keys, by producing keys individual when it is necessary;
- To generate liberation across transportations like conferences or presentations. Exercise must elude the prerequisite of keeping formal information transversely required by sessions.

Conventionally, key establishment protocols have challenging issues to plan the protocol. Almost there are numerous experiments regarding key exchange/key agreement. These are as follows [10, 11]:

- Safeguarding the keys is swapped so that the intended dispatcher and receiver can accomplish the encoding (encipher) and decoding (decipher) process.
- Precluding the observer from accomplishment to recognize about the key.
- Propositions the receiver approximate evidence that communication was enciphered by the legitimate user entitled to have directed message.

2 Literature Review

Related to the key agreement and asymmetric key cryptography such as ECC, RSA, EC-DH, DSA is deliberated in this segment. It also offers an overview of the literature relating to this paper. This also includes an introduction to secure and authenticated key agreement protocol which is using arithmetic of Elliptic Curve Cryptography and best usage in cryptographic key agreement protocols. Moreover, it investigates and inspects different widely deliberated and executed identity-based key agreement protocols, as well as the pairing free, certificate-less scheme. Such protocols are presented in the form of security investigation and effectiveness and examination of the projected protocol.

The analysis of several schemes for design methods, such as ElGamal, D–H scheme of key exchange, ECC, and the MQV protocol. Likewise, represents the general approach to cause an attack on key agreement protocols. In conclusion, considering what kind of features resolves to focus on the exploration of the protocols phases of this hypothesis. Collective key establishment and key agreement protocol available in the literature are as follows:

- Mutual (two way) validation with key establishment protocols.
- Mutually agreed on authentication with key agreement protocols.
- Certificate-based authenticated key established/agreement protocols.
- Conference (Group CKA) key agreement protocols.
- Authenticated Diffie–Hellman key agreement protocols.
- Multiple-key key agreement (exchange) protocols.
- Certificate-less (one way) legitimate two-party key agreement protocols.
- Two-party attribute-based key agreement protocols.
- Password-based key agreement protocols.
- Authenticated group key agreement protocols.
- Identity-based key agreement protocols.
- Cluster-based group key agreement protocols.
- ID-based authentication (two pass) with a phase of key agreement protocol.

The following section will emphasize various security and performance issues related to authenticated key agreement (AKA) protocol implementation.

3 Key Agreement Protocol

In cryptography, secure authentication with key agreement protocol is a procedure where many legitimate collaborative events approve on a secrete key in such a way that both encourage the consequence. Key agreement procedures are deliberately the most challenging procedures to project, and further most significant parts of a scheme when it arises to reliability and concealment of data. Numerous authenticated and secure key agreement protocols have been suggested as well as projects, but several of them are deprived of sanctuary evidence. Level specification through the security evidence, the protocol may comprehend flaws that may be oppressed with a fangled compassionate of attack. As of this, it must endlessly inspect protocols to mark certain that they are comprehensive. Key agreement protocols work communally for two or more collaborating events to accomplish protected communication by forming the ephemeral key to encode files that are being transacted between intended and legitimate parties [12–14].

Common characteristics anticipated in key agreement procedures development are as follows [15, 16]:

- Significant endorsement (process of authentication): Mutually agreed secret key value should be acknowledged only through acknowledged parties;
- Perfect (Forward) secrecy: An agreed-upon secret key should persist undisclosed, even if both communicating parties used a longstanding key substantial is conceded.

An authenticated and secure key exchange/agreement procedure follows the procedure of demonstrating confidence consequences in five phases such as scheme of the model, classification of goals within this model, statement of rule books, explanation of protocol, evidence that the protocol encounters its goals within the model.

4 Design Methodology of Key Agreement (Mutual Authenticated) Protocol

The proposed work is likely to be carried out by incorporating techniques and methodologies:

- To assess the current state of the art of elliptic curve cryptosystem and key agreement protocol schemes.
- The results of the analysis phase will be used to create an innovative protocol for key agreement schemes.
- Based on study and analysis, an efficient scheme for key agreement for elliptic curve cryptosystem is proposed and tested against the problems identified.
- The anticipated protocol consists of three major phases of execution: process of system initializing phase, the user action phase with registration, and the phase of mutual authentication with a phase of key agreement (client and server).

5 Design Scheme of Authenticated Key Agreement Protocol

In a wireless communication scenario, legitimate parties and network servers require to authenticate to each other and mutually agree on session key information to be used for encryption initiative in further exchange. Legitimate parties mutually institute a common secret key and contribute with input. Various schemes for an authenticated key agreement which provides authentication process with communicating parties and most appealed for experimental implementation ensures that intended users can be able to establish the session key. Technical families of classification for implementing authentication approaches in which key generation is based on credentials such as password-based, PKI, ID-based, Group Id/conference key. The process of key formation of password-based authentic key exchange strategy permits users to communicate using secret credentials related to passwords in the earlier phase before communication to impart the issue of proper user authentication and exchange of keys. However, such a technique is completely unacceptable for large-scale communication situations because it requires several communicating parties to exchange a password in advance in order to generate the ephemeral key. Public key infrastructure (PKI) is another type of system that makes use of public keys for key exchange.

When it comes to public key-based systems, the most time-consuming activity is the confirmation method of certificate handling; cryptographic approaches are utilized instead for safe communication between parties and to generate a secret for the session key. Keeping public keys secure also takes some extra work. In contrast, in an ID-based (meaning public identity) key exchange method, the user's public identity is used as the user's public key through the usage of relevant information. Communication between the intended parties who wish to create a session key doesn't have to take any extra steps if they use an ID-based authentication technique that is followed by the execution of a protocol for key exchange. As a result, achieving both security and efficiency requires the implementation of ID-based authentication with key exchange protocol [1–7, 12, 13, 15–18].

6 Strategy Schemes of Key Agreement Protocol

Suggested research design procedures are used for planning the different (comparatively better) and proper key agreement. Procedures are constructed on more or less recommended standards for the execution of protocols. In this segment, more focus is on designating the most important of these standards useful for protocol implementation. Also discusses the existing and anticipated issues that are utilized basics of cryptography and schemes discuses as follows [9, 19, 20].

6.1 D–H (Diffie–Hellman) Key Exchange Protocol

To safeguard the transfer of secret values (secret key) between communicating events across an unreliable network communication channel, cryptographic systems like Diffie–Hellman key exchange are used. Even if the legitimate parties have already had some sort of conversation, they can now encrypt their communications over the unprotected network channel using the shared secret key they just generated. The key is not transmitted over the connection, making it impossible for an outsider to learn what it is [9, 19].

6.2 Elliptic Curve-Based Cryptosystems

The Elliptic Curve Cryptography (ECC) is based on number theory which rests on the hardness of resolving the problem of discrete logarithm for the group of an elliptic curve defined on a finite field. Such delinquent is entitled as elliptic curve discrete logarithm problem (EC-DLP) [13, 18, 21].

6.3 MQV Protocol

Sharing a secret using the Menezes-Qu-Vanstone (MQV) method is a secure way for two people to communicate securely. Key pairs consisting of private and public keys are generated and exchanged between users who have mutually shared access. After that, each user calculates an enclosed signature using their private key and the public key of the other party. The signature is used to generate the secret value between parties. Only if their construction follows the public key specifications will the secret values generated by each user be same [20–22].

6.4 Cryptographic Function

There are generally three kinds of numeral-based theoretic problems whose difficultness forms the basis for a usually used public key-based cryptographic protocol like hash function, factorization problem, and discrete logarithm problem. Usually, the trapdoor (one-way property) role function is a mathematical calculation function that can be certainly designed from its route which is always in an onward direction, but difficult to estimate in the reverse direction. Unknown private key based that can be deliberated only in the forward direction. The process of encryption and the digital signature verification is always performed in the forward direction, wherever the inverse direction process is carried out the process of decryption and generating the digital signature. The following section deliberates about examples related to cryptographic functions.

6.5 Hash Functions

The major claim of hash functions used in the concept of cryptography is to achieve message integrity. The generated hash code value transports a cardinal characteristic of the message's contents, which confirms that the message hasn't been changed by an opponent, using a virus. The computed hash function is also termed one-way encryption (Message Digest), without having a key. As an alternative, a fixed-length hash code cost is calculated using the plaintext which makes it difficult to recover the substances or length of the plaintext.

6.6 Factorization Problem

Now with number-theoretic problems of integer factorization (IFP) is nothing but the disintegration of a complex number into the creation of lesser number integers. If digits are supplementarily limited to prime numbers, such the method is also known as prime number factorization. When selected numbers are very enormous, not required no efficient factorization algorithm is acknowledged. The determination carried out by several researchers, the supposed effort are considered at a problem which is a heart of approximately used algorithms in asymmetric key cryptography such as DSA, ElGamal, and RSA. Various extents branch of mathematics, computer science that has been transported to view on the problem, comprising elliptic curves operation, use of algebraic number theory, and quantum computing.

6.7 Discrete Logarithm Problem

There are several theoretic problems on which asymmetric key cryptography is worked. Such cryptographic algorithm is based on the problem of discrete logarithms which is precise concerning groups of multiplicative cyclic properties. Discrete logarithm problem (DLP) is statistically difficult group order can be computed easily and perform fast arithmetic. The rigidity of the expected outcome of discrete logarithms is rested on the concept of cyclic groups.

6.8 EC-DLP (Discrete Logarithm Problem_Elliptic-Curve)

The extent of security of elliptic curve-based cryptography (ECC) is dependent on the hypothesis and inflexibility that uses the concepts of the problem of elliptic curve discrete logarithm problem (EC-DLP). If the groups defined in elliptic curve groups which are used to defined multiplicative notation, at that moment the elliptic curve based on the problem of discrete logarithm difficulty is given for two considered points like $P * k = Q$, but for real-time accomplishment, the value of k would be very large, so it would be infeasible to compute k in this method.

7 ECC Implementation Issues and Consideration

Substantive accords based on key challenges for the constraint environment arise during cryptosystem implementation, such as the potential need to trade off correct display for security or flexibility. To a greater extent than RSA, Elliptic Curve Cryptography (ECC) is amenable to both software and hardware implementations. When compared to traditional hardware, software implementation of Elliptic Curve Cryptography offers faster speeds and higher power consumption. Further, software implementations are notoriously sloppy when it comes to physical security, especially when it comes to the value of key storage [16–18].

By employing ECC arithmetic, specifically the production of domain parameters for implementing elliptic curves. There are computational and communication challenges associated with completing the finite field GF(p) for a very large prime number (p) on a standard computer. Different word sizes are insufficient because different computers need different numbers of bits (words) to characterize the fundamentals of a finite field. The major challenge in demonstrating something is during the time it takes to do the math and carry out the procedure [12, 17, 23].

Around a huge amount of research which is dealing with methods for the achievement of elongated number multi-precision arithmetic proficiently, the most common techniques used in the condition of the modular reduction, whereas arithmetic comes at a cost of time with processors on multi-precision arithmetic, especially with the reduction modulo p operations.

Majorly execution of arithmetic used in an elliptic curve on the finite field like GF(p) or GF(2m) demonstrate a level of confidence of the suggested structure is uncompromising. It has been provided in numerous existing literature reviews that significantly lesser key sizes can be used for ECC rather than other public key cryptosystems like RSA. Numerical calculation is necessary by elliptic curve cryptosystem; therefore, it necessitates to have low scheming power, and therefore, the EC-based cryptosystem is a supplementary suitable for asymmetric key [18]10.

8 Design and Implementation

Graphical and software interface for desktop client and mobile client communication with the server for authentication in the binary field and prime field is implemented using Java and Android. The user interfaces for download with desktop client after verification, id-based authentication, registration phase with IMEI number, and Mac address for desktop and mobile client implementation are carried out in this section. Implementation of underlying finite field and key agreement accomplishment and total time is required for execution finite field and key agreement protocol. This research has attempted to solve a problem relating to the protected, authenticated execution of key agreement protocol. Strategy and development of ID (identity)-based authentication with key agreement protocol which is free bilinear pairing computation, communication overhead, and bandwidth requirement will get reduced. Suggested authenticated key agreement protocol also resisting to probable occurrences of attacks on protocol and also provided desirable security features including perfect forward secrecy, key impersonation, etc.

The most significant and difficult primitives used in cryptography including private and asymmetric key encryption scheme is authenticated key agreement (AKA) phase of the protocol with confirmation, where two or more legitimate events uses traditional values on a key to be used for authorizing the computational privacy and process of authentication amongst communicating parties [23]. Diffie and Hellman in 1976 recommended the first protocol for key exchange that is the most important building block for newly arise protocols [14]. Though, such implementation of the protocol doesn't deal with the verification phase amongst the communicating parties. Consequently, it is inclined to man-in-middle attack. Several such protocols have been designed to solve these issues [17] by proposing proper phase execution of authentication. The phase of authentication can be accomplished by several methods, like the public key infrastructure (PKI), password-based, group-based [6]. This segment is regarded more or less as available authenticated (mutual) key agreement protocols and offered comparative analysis.

9 Conclusion

In order to meet the recommendations of organizations like the American National Standards Institute (ANSI), the National Institute of Standards and Technology (NIST), and the Security Engineering Working Group (SEWG), such cryptosystems employ Elliptic Curve Cryptography (ECC) based on the use of finite field types like GF(p) or GF(2m). This method is effective in a limited space and offers a full safety package. Therefore, identity-based authentication can be implemented via key agreement protocol achieved via ECC, which can give improved security.

The primary advantage of this system is its low computational and communication costs. Anonymity for the user is ensured through a secure process. Identity-based key agreement protocol eliminates the need for a central verification table. In addition, it provides a phase of the key agreement that can be used for mutual authentication. So, this method is able to provide higher security and may be practically implemented in wireless systems for the communication phase.

References

1. Katvickis A, Sakalauskas E, Listopadskis N (2011) Microprocessor implementation of key agreement protocol over the ring of multivariate polynomials. Elektronika ir Elektrotechnika 116(10):1392–1215. ISSN: 1392-1215
2. Abdul-Aziz Gutub A (2010) Remodeling of elliptic curve cryptography scalar multiplication architecture using parallel Jacobian coordinate system. Int J Comput Sci Secur (IJCSS) 4(4):409
3. Chou CH, Tsai KY, Lu CF (2013) Two ID-based authenticated schemes with key agreement for mobile environments. J Supercomput. https://doi.org/10.1007/s11227-013-0962-3
4. Menezes AJ, van Oorschot PC, Vanstone SA (1996) Handbook of applied cryptography, 2nd edn. CRC Press
5. Farouk A, Fouad MM, Abdelhafez AA (2014) Analysis and improvement of pairing-free certificate-less two-party authenticated key agreement protocol for grid computing. Int J Secur Privacy Trust Manag (IJSPTM) 3(1)
6. Weimerskirch A, Stebila D, Shantz S (2003) Generic GF(2m) arithmetic in software and its application to ECC. In: The 8th Australasian conference on information security and privacy (ACISP2003)
7. Baalghusun AO, Abusalem OF, Al Abbas ZA, Kar J (2015) Authenticated key agreement protocols: a comparative study. J Inf Secur 6:51–58. https://doi.org/10.4236/jis.2015.61006
8. Yoon E-J, Yoo K-Y (2009) Robust ID-based remote mutual authentication with key agreement scheme for mobile devices on ECC. In: IEEE international conference on computational science and engineering, vol 2, pp 633–640
9. Yoon E-J, Yoo K-Y (2005) New efficient simple authenticated key agreement protocol. Proceedings of lecture notes in computer science, vol 3595, pp 945–954
10. Chou C-H, Tsai-K-Y, Lu C-F (2013) Two ID-based authenticated schemes with key agreement for mobile environments. J Super Comput. https://doi.org/10.1007/s11227-013-0962-3
11. Bayat M, Aref MR (2013) A secure and efficient elliptic curve based authentication and key agreement protocol suitable for WSN. Elsevier
12. Bansal A, Sharma D, Singh G, Roy T (2012) New approach for wireless communication security protocol by using mutual authentication. Adv Comput: Int J (ACIJ) 3(3). https://doi.org/10.5121/acij.2012.3303
13. Adiga BS, Balamuralidhar P, Rajan MA, Shastry R, Shivraj VL (2012) An identity based encryption using elliptic curve cryptography for secure M2M communication. ACM J IPICS
14. Certicom (2000) Standards for Efficient Cryptography, SEC 1: Elliptic Curve Cryptography, Version 1.0. Available at: http://www.secg.org/download/aid-385/sec1_final.pdf
15. Maheshwari B (2012) Secure key agreement and authentication protocols. Int J Comput Sci Eng Surv (IJCSES) 3(1). https://doi.org/10.5121/ijcses.2012.3111
16. Yu B, Li H (2008) Research and design of one key agreement scheme in Bluetooth. In: 2008 international conference on computer science and software engineering. IEEE. 978-0-7695-3336-0/08. https://doi.org/10.1109/CSSE.2008.1263
17. Yu B (2010) Establishment of elliptic curve cryptosystem. In: IEEE international conference on information theory and information security (ICITIS)
18. Cocks C (2004) An identity based encryption scheme based on quadratic residues, cryptography and coding. In: Institute of Mathematics and Its Applications international conference on cryptography and coding—proceedings of IMA 2001. LNCS, vol 2260. Springer, pp 360–363

19. Mishra D, Das AK, Mukhopadhyay S (2016) A secure and efficient ECC-based user anonymity-preserving session initiation authentication protocol using smart card. Peer-to-Peer Netw Appl 9:171–192. https://doi.org/10.1007/s12083-014-0321-z
20. Le D-P, Nguyen BP (2012) Fast point quadrupling on elliptic curves. SoICT 2012, 23–24 Aug 2012, Ha Long, Vietnam. ACM. 978-1-4503-1232-5
21. Goswami C, Parveen Sultana H (2019) Cross-layer and reliable opportunistic routing with location prediction update vector (CBRT-LPUV) in mobile ad hoc networks (MANET). Int J Recent Technol Eng (IJRTE) 8(1). ISSN: 2277-3878
22. Yoon E-J, Yoo K-Y (2010) A three-factor authenticated key agreement scheme for SIP on elliptic curves. In: 2010 fourth international conference on network and system security. IEEE. ISBN: 978-0-7695-4159-4/10. https://doi.org/10.1109/NSS.2010.101
23. Popescu C (2004) A secure authenticated key agreement protocol. In: Proceedings of the12th IEEE Mediterranean electrotechnical conference, Dubrovnik, Croatia, May 2004, pp 783–786

Covid-19 Disease Prediction System from X-Ray Images Using Convolutional Neural Network

Basam Akshitha and A. Jagan

1 Introduction

Along these lines, it is difficult to differentiate Coronavirus from other lung illnesses. Atomic, nanotechnology, and counteracting agent techniques have been utilized to concoct ways of diagnosing SARS-CoV-2. RT-PCR, a test in light of a subatomic strategy, is typically acknowledged as an ordinary method for diagnosing illnesses from one side of the planet to the other. Yet, a troublesome interaction takes a ton of time and requires a specific staff. The principal objective of this venture is to make programming that will assist specialists with finding Coronavirus illness by utilizing savvy applications. Utilizing the illness-arranged history records will assist with peopling making new connections and seeing groundbreaking thoughts later on. This technique can be utilized to concentrate on the Covid, foresee conduct, analyze sicknesses, and fill prescriptions.

2 Literature Survey

Covid-19 is a very contagious lung disease caused by a new virus called SARS-CoV-2. In December 2019, Covid-19 broke out in Wuhan, China. Since then, it has spread quickly all over the world, causing a pandemic. The virus spreads when a person who has it coughs, sneezes, talks, or even just breathes. It can also spread when someone touches an infected surface and then touches their mouth, nose, or eyes. Covid-19

B. Akshitha · A. Jagan (✉)
B V Raju Institute of Technology, Hyderabad, India
e-mail: jagan.amgoth@bvrit.ac.in

B. Akshitha
e-mail: 21211D5802@bvrit.ac.in

M. S. Kaiser et al. (eds.), *ICT: Smart Systems and Technologies*, Lecture Notes in Networks and Systems 878, https://doi.org/10.1007/978-981-99-9489-2_14

can cause a wide range of symptoms, such as fever, cough, tiredness, body aches, and loss of smell or taste.

The pandemic has had a big effect on people, towns, and businesses all over the world. Governments all over the world have taken different steps to stop the spread of the virus, such as lockdowns, separating people from each other, making people wear masks, and starting vaccine programs. Even with all of these attempts, the virus keeps spreading, and the world is still trying to deal with the pandemic. In this part, the journalists discuss the scientist's latest outcomes. Analysts from everywhere the world have as of late concocted ML and DL-based Coronavirus acknowledgment strategies. In light of Feature Correlated Naive Bayes (FCNB), Mansour, and others thought of a method for sorting out Coronavirus with a most extreme precision of close to 100%. The FCNBP was utilized to bunch patients by involving the weighted Guileless Bayes strategy and different adaptations as the component affiliation. It was additionally contrasted with different strategies. Das and his partners concocted a DCNN model comprised of DenseNet201, Resnet50V2, and Inceptionv3 to distinguish Coronavirus from CX-Beams.

The models were then assembled utilizing another strategy called weighted normal ensembling to get a class esteem. With an exactness pace of 91.62%, they showed improvement over the latest DCNN models. Reshi et al. showed a DCNN model for Coronavirus discovery in light of CX-Beam characterization. The preprocessing 7 ventures for the datasets incorporate adjusting the datasets, having a clinical expert gander at the pictures, and adding more information. This gives the datasets a typical exactness of 99.5%. The CNN model was assessed in two distinct ways.

In the first, the model was tried with 100 X-beam pictures from the preparation dataset, and it was viewed as 100% right. In the subsequent arrangement, the model was tried with an additional arrangement of CX-Beam pictures from Coronavirus, and its exactness was 99.5%. The DCNN model has 68 layers, while the model we are proposing has just 18.

Das et al. offered an elective method for testing for Coronavirus that utilizes CXRaces to break individuals with this condition into three gatherings: positive for Coronavirus, tainted with another sickness, and not contaminated with CNN, VGG16, or ResNet50. At the point when the three learning strategies were scrutinized, VGG16 showed improvement over CNN and ResNet50. The TLCoV model had a $F1$ score of 96.59%, a review of 97.67%, and an exactness of 96.65%. This TLCoV has around 12,410,023 boundaries, which is more accurate than the 6,447,138 boundaries in the model we recommended.

Ayalew [1] and his partners have thought of a technique (DCCNet) for rapidly diagnosing Covid by utilizing a patient's CX-Shaft. CNN and the histogram of oriented gradients (HOG) were referenced as ways of finding things rapidly utilizing CX-Beam pictures from the College of Gondar and online libraries. The DCCNet model did 99.9% of the work to get ready and 98.3% of the work to breeze through the assessment. Crowd did 100% of the work to get ready and 98.5% of the work to finish the assessment. The precision of preparing and testing increased 99.97% and 99.67%, separately, with the blended model [1].

Indumathi [4] and others discussed ways of separating places contacted by Coronavirus into hazardous, safe, and gentle zones. Utilizing the Coronavirus dataset from the Virudhunagar region, they likewise give a DL-based technique for recognizing Coronavirus-impacted zones. The DL accurately predicts the zones with a 98.06% exactness and a 1.94% blunder rate, while the C5.0 calculation just hits the nail on the head 95.92% of the time and commits 4.08% of errors [4].

The essential screening apparatus made by Salau et al. suggested utilizing a support vector machine (SVM) to find and sort Coronavirus. The discrete wavelet transform (DWT) strategy was utilized to get the highlights, and SVM was utilized to sort out what the elements were. This prompted a 98.2% acknowledgment rate. Yadessa et al. told the best way to make a without touch Coronavirus handwashing framework by utilizing an Arduino chip and ultrasonic-based checking gadgets.

Natnael et al. crossed sectional review to figure out the number of taxi drivers who go to the recommended lengths to stop the spread of infections by wearing facial coverings. Narin and his group utilized the DCNN models InceptionV3, ResNet50, and InceptionResNetV2 to sort out what Coronavirus illness was from a CXR picture. They utilized 5-crease cross-validation to make three unique parallel groupings with four classes (Coronavirus, sound, viral pneumonia, and bacterial pneumonia). In view of assessments and execution results, the ResNet50 model has the best order precision: 96.1% for Dataset 1, 99.51% for Dataset 2, and 99.7% for Dataset 3. Contrasted with the model we utilized for our work, the ResNet50 model is much greater and has fifty layers.

Abbas et al. utilized CX-Beams to distinguish Coronavirus illness and thought of the Deteriorate, Move, Form (DeTraC) model, which was 95.12% precise. Up to 95.12% of the time, this mix works impeccably, yet for our situation, it works up to 98.81% of the time. Khan and others presented a DCNN model (CoroNet) for recognizing Coronavirus from CX-Beam in light of the Xception plan. For Coronavirus cases with four classes, CoroNet got a general exactness of 89.6%, with 93% for accuracy and 98.2% for memory. The recommended strategy was 95% exact for ordering things into three gatherings. Contrasted with what we do, this work is less precise.

Maghdid et al. additionally recommended the DL technique (AlexNet) for rapidly and precisely examining Coronavirus cases from CT and CXR checks. The AlexNet organization and altered CNN both got 94.1% precision and the adjusted CNN got 98% exactness. Sethy and his group utilized ResNet50 to extricate highlights from CX-Beam and the SVM classifier to characterize them. They got 95.38% exactness, 95.52% FPR and MCC, and 90.76% Kappa, separately.

Rehman and his group saw how well AlexNet, VGG, SqueezeNet, GoogleNet, MobileNet, ResNet and its varieties, and DenseNet could track down Coronavirus. The best measure of precision was 98.75%. Kumar and co. referenced a concentrate in which CXR pictures were utilized to track down Coronavirus. Coronavirus was related to the assistance of DenseNet and various TL models, like EffcientNet, Xception, GoogleNet, and VGG16. ResNet152V2 showed improvement over the other four models, with 98.15% exactness, while VGG16 and DenseNet both got 99.32% precision. The proposed Troupe technique worked 99.28% of while placing two classes into gatherings. The models worked impeccably, yet contrasted with what we found, their size and number of layers were tremendous. Dadário et al. showed a three-layered indicative rationale approach for Coronavirus. To test how well the proposed approach functions, 4356 chest CT filter pictures were utilized. The outcomes show that the discoveries are both touchy and exact. Ozturk and his partners utilized a DarkNet model as an indicator in the you just look once (consequences be damned) continuous item acknowledgment framework for robotized Coronavirus picture-based recognition. The recommended strategy was intended to give great discoveries to both twofold (no-revelations and Covid) and multiclass requests (no-disclosures, Covid, and pneumonia), and it did as such with an accuracy of 98.08% for matched portrayal and 87.02% for multiclass portrayal. For every one of the 17 convolutional layers in their model, they utilized more than one screening technique.

Yan and his group fabricated a performing multiple tasks man-made intelligence framework that pre-owned CT outputs to accurately distinguish Coronavirus in 89% of patients' lungs.

Singh et al. concocted a method for arranging lung CT pictures to track down Coronavirus. With the assistance of differential development, the CNN's boundaries were found with a 98.24% level of exactness.

G et al. chose to utilize CX-Beam since they figured the outcomes would be great, it was reasonable, and the essential gear was not difficult to get to. Huang, as well as others, in view of what they found all alone, CX-Beam pictures were superior to any remaining approaches to distinguishing Coronavirus.

Zhang [3] and his partners utilized a one-class gathering-based peculiarity finding (CAAD) strategy to differentiate between non-viral pneumonia photographs and viral pneumonia pictures. A portion of these CADD models incorporates taking out features, tracking down botches, and foreseeing modules. Assuming the irregularity score was high or the gauge score was low, popular pneumonia was believed to be the reason. This technique had a region under the bend (AUC) of 83.61% and a responsiveness of 71.70%. The TL-based technique for distinguishing CX-Beam pictures as Covid-19 was made by Apostolopoulos and others.

Two distinct arrangements of information were utilized. The first is comprised of 1427 CX-Beam pictures, of which 224 had Coronavirus diseases. In the second arrangement of information, there are 1442 CXR pictures, and 224 of them are positive for Coronavirus. The outcomes show that DL and CX-Beam imaging might have the option to decide with a precision of 96.78%, a responsiveness of 98.66%, and a particularity of 96.46%.

Farooq and his partners [2] showed Covid-ResNet, which is a CNN model for determining Coronavirus and pneumonia. Utilizing state-of-the-art preparing strategies like discriminative ten learning rates, moderate contracting, and cycle learning rate finding, they make leftover brain networks that are quick and exact. This paper tells the best way to calibrate a ResNet50 structure in three moves toward making models work better and diminishing preparing time.

This is finished by calibrating the organization at each level and continuously lessening the info pictures to $229 \times 229 \times 3$, $224 \times 224 \times 3$, and $128 \times 128 \times 3$. This approach was correct 96.23% of the time in all classes.

Kumar and his group utilized DL to show an interruption identification framework on the haze layer how to differentiate between an assault and customary organization action. V. MK and others gave a full report on the expense viability and DL-based investigation of Coronavirus recognition. They likewise take a gander at customary ways of further developing the DL-based acknowledgment cycle and look into them. Meraihi and his associates gave an inside and out take a gander at how ML-based investigations can be utilized to foresee, analyze, and recognize Coronavirus.

In this review, just 16% of the cases utilized directed learning (RF, SVM, and relapse strategies), while 79% of the cases utilized DL. Alkhodari and others have concocted DL models that utilization Mel-frequency cepstral coefficients (MFCC) and hand-made highlights that come from the first accounts, as well as profound enacted highlights that come from blending CNN and bi-directional long short-term memory units (CNN-BiLSTM). The shallow accounts utilized in the DL strategy.

Asmaa et al. discuss one more work on X-beam pictures called Disintegrate, Move, and Form (DeTraC). DeTraC utilized a class decay strategy to deal with class limits. This gave it a 93.1% precision and a 100% responsiveness.

Bukhari et al. utilized the ResNet50 convolutional brain network thought on lung X-beam pictures that were parted into three gatherings: typical, pneumonia, and Coronavirus

Yujin et al. showed a factual strategy for diagnosing Coronavirus utilizing patches of chest X-beam pictures and few CNN boundaries that can be prepared.

This paper tells the best way to utilize chest X-beam pictures to isolate Coronavirus cases into two gatherings: those that are negative (typical) and those that are positive (tainted). Along these lines, we concocted a profound learning technique for consequently grouping Coronavirus sicknesses that plans to keep an elevated degree of exactness and decrease the quantity of bogus negative cases.

3 Methodology

In this work, X-rays of the chests of Coronavirus and non-Coronavirus patients were utilized. In the coding part, first burden picture assortment. It comes from data about various cases. Individuals who were not on Coronavirus were additionally affected by these photos. Figure 1 illustrates flowchart of Covid-19 disease prediction systems from X-ray images.

3.1 Proposed System

This proposed framework attempts to fix the issues with the Reverse Transcription-Polymerase Chain Reaction (RT-PCR) strategy for diagnosing Coronavirus by adding a programmed Coronavirus recognizing framework in view of chest X-ray (CXR) pictures. For better accuracy and speed, the framework utilizes a Convolutional Neu-

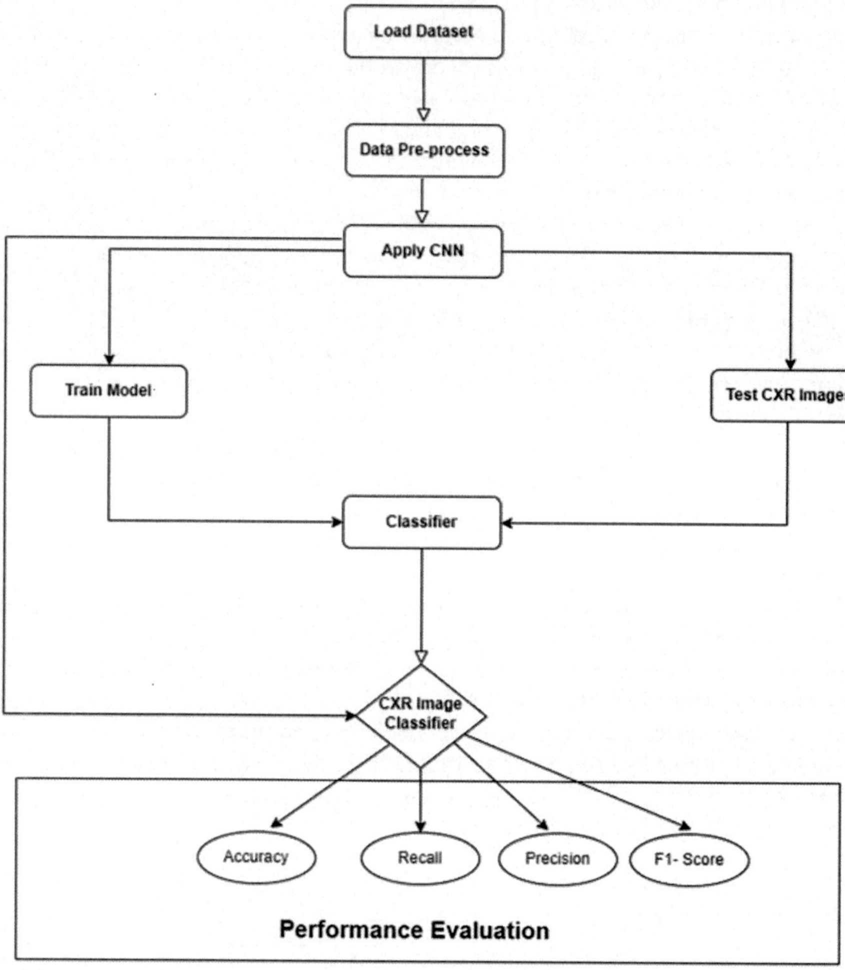

Fig. 1 Flowchart of Covid-19 disease prediction systems from X-ray images

ral Network (CNN) model that is run on a cloud-based stage. The proposed framework is a more exact and successful method for tracking down Coronavirus than the RT-PCR technique. By utilizing cloud-based innovation and CNN-based picture investigation, it intends to make medical services frameworks less occupied and to make finding quicker and simpler for additional individuals.

3.2 Dataset Collection

In this work, X-rays of the chests of Coronavirus and non-Coronavirus patients were utilized. In the coding part, first burden picture assortment. It comes from data about various cases. Individuals who were not on Coronavirus were additionally affected by these photos.

3.3 Dataset Split and Image Resizing

The Coronavirus X-ray picture assortment is comprised of pictures of chest X-rays from Coronavirus cases. These photos have been contracted to $224 * 224$ pixels and given three variety groups (RGB), providing them with the state of $224 * 224$ 3. In ML and PC vision occupations, one of the initial steps is to resize the photos to $224 * 224$ 3. It makes it work with famous apparatuses and models for profound discovering that need normalized input sizes. The width and level of the image, in pixels, are $224 * 224$, and the "3" represents the three variety channels—red, green, and blue—that make up every pixel. By diminishing the Coronavirus CXRay pictures to similar size, the information can be taken care of all the more effectively, put away more productively, and utilized with various neural network plans and picture-handling strategies. This standard size ensures that the photos can be involved well in deep learning models, as convolutional neural networks (CNNs), to do things like arranging, dividing, or tracking down exceptions in Coronavirus examination.

3.4 CNN Model

The grouping of Coronavirus CX utilizing the Convolutional Neural Network (CNN) model. X-ray pictures are made to take a gander at chest X-rays and let know if they are positive or negative for Coronavirus. CNNs are a sort of profound learning model that was made to perceive pictures and interaction with them.

There are many layers in the CNN model, for example, convolutional layers, pooling layers, and completely connected layers. Here is an overall clarification of the parts and what they do:

Convolutional Layers: In the convolutional layers, the crude pictures are gone through a bunch of channels that can be educated. These channels take a gander at the photos, track down the significant parts, and make highlight maps. The channels figure out how to perceive examples, shapes, and designs that are indications of Coronavirus contamination.

Pooling Layers: The pooling layers make the element maps made by the convolutional layers less itemized. Pooling assists with lessening the quantity of focus on the element maps while keeping the main subtleties. Max pooling and normal pooling are two familiar ways of pooling.

Completely Associated Layers: The completely associated layers conclude how the separated highlights are at long last gathered. They take the smoothed element maps from the past levels and guide them to specific result classes, like Coronavirus positive or negative. Actuation capabilities like ReLU or softmax are many times used to add nonlinearity and show the model how to pursue hard decisions.

Dropout: Dropout is a strategy for regularizing CNN models that is frequently utilized. During preparing, it haphazardly eliminates a specific number of neurons. This holds the model back from turning out to be excessively unambiguous and works on its capacity to sum up.

Misfortune Capability and Streamlining: The CNN model purposes a misfortune capability, similar to cross-entropy, to gauge how different the projected names are from the genuine ones. During the preparation cycle, a streamlining strategy like stochastic gradient descent (SGD) or Adam is utilized to limit the misfortune and change the model's boundaries.

Algorithm 1 Covid-19 CXR Images Algorithm

```
Start
Load CXR Image Dataset
Apply Data Pre-process
    Normalize Pixel Values
Apply CNN Model
    Include Convolutional layer to extract spatial feature for CXR Images
Activate ReLU function
Train Model for Random Weights for CXR Images
Evaluate model with Predicted Values
Classify Model Function
Evaluate Performance Metrics
End
```

3.5 *Performance Evaluation*

For the division of Coronavirus CXRay photographs, a Convolutional Neural Network (CNN) model's prosperity is decided by taking a gander at its accuracy, precision, recall, $F1$ score, and other related measures. The objective of this study is to figure out how well the model functions at sorting out the kind of Coronavirus from chest X-rays.

4 Conclusion

In the medical services business, it is elusive and analyze sicknesses in their beginning phases. Customary approaches to finding things take a great deal of time, which shows that it is so vital to create a framework that can find Coronavirus rapidly and in a little space. The fundamental objective is to further develop medical care where there are insufficient specialists. In this review, CX-Beam pictures of both Coronavirus and non-Coronavirus cases were taken a gander at with a Convolutional Neural Network (CNN) model. CNN's settings and hyperparameters were picked with care to get the best speed, size, and figuring productivity. The objective was to make a model that occupied less room and had less CNN levels and highlights. To arrive at these objectives, the CNN model was made with a variety of levels.

References

1. Ayalew AM, Salau AO, Abeje BT, Enyew B (2022) Detection and classification of COVID-19 disease from X-ray images using convolutional neural networks and histogram of oriented gradients. Biomed Signal Process Control 74:103530. https://doi.org/10.1016/j.bspc.2022.103530
2. Farooq M, Hafeez A (2020) COVID-ResNet: a deep learning framework for screening of COVID19 from radiographs. arXiv:eess.IV/2003.14395
3. He K, Zhang X, Ren S, Sun J (2016) Deep residual learning for image recognition. In: 2016 IEEE conference on computer vision and pattern recognition (CVPR), pp 770–778. https://doi.org/10.1109/CVPR.2016.90
4. Indumathi N, Shanmuga Eswari M, Salau AO, Ramalakshmi R, Revathy R (2022) Prediction of COVID-19 outbreak with current substantiation using machine learning algorithms. Springer, Singapore, pp 171–190. https://doi.org/10.1007/978-981-16-6542-4_10

Liquidity Regulation and Bank Performance: The Industry Perspective

Anureet Virk Sidhu, Aman Pushp, and Shailesh Rastogi

1 Introduction

The financial crisis of 2007–08 brought to light an inherent weakness embedded in the banking sector, which had not so far been adequately addressed: liquidity risk. The focal area of the Basel regulations has always been around the capital adequacy of banks. However, during the crisis, it was observed that though the capital levels of banks surpassed the mandatory regulatory requirement, they still struggled as they did not hold sufficient liquidity to meet their obligations, resulting in bank defaults and consequently plunging the entire financial and economic system into one of the worst global meltdowns.

In order to remediate the weaknesses exposed by the crisis, the Basel Committee of Banking Supervision (BCBS) rolled out Basel III: "International Framework for Liquidity Risk Measurement, Standards, and Monitoring" in December 2010. The two fundamental constituents of this new liquidity regime are the Liquidity Coverage Ratio (LCR) standard and the net stable funding ratio (NSFR) standard. The LCR mandates banks to hold sufficient high-quality liquid assets to help them fulfill at least 30 days' obligation in case a stress-like situation arises, whereas the NSFR is designed to achieve long-term resilience by incentivizing banks to finance their activities through more stable funding sources. The two standards aim to achieve different but complementary objectives of stability and resilience. Though these liquidity standards play a pivotal role in ensuring the soundness of the banking industry, many concerns have been raised. The industry practitioners argued that the very design of the new liquidity regulation is such that it will add to banks'

A. V. Sidhu · A. Pushp (✉) · S. Rastogi
Symbiosis Institute of Business Management, Pune Symbiosis International (Deemed University), Pune, India
e-mail: amanpushp4@gmail.com

S. Rastogi
e-mail: shaileshrastogi@sibmpune.edu.in

M. S. Kaiser et al. (eds.), *ICT: Smart Systems and Technologies*, Lecture Notes in Networks and Systems 878, https://doi.org/10.1007/978-981-99-9489-2_15

cost burden and impair performance [1, 2]. Then, there are other studies that try to explain that the net economic benefits of the new regulatory standards would exceed the additional costs by complying with the Basel III requirements [3].

However, the authors observed some gaps, firstly, most of the studies are empirical in nature. The quantitative research illustrates the nature of the relationship that exists between liquidity regulation and bank performance; however, it fails to explain the why aspect of the results. This is further complicated by opposing findings of different studies. It is vital to dig deeper to understand why and how the new liquidity regime affects the different bank performance. Secondly, most of the research centers on developed economies, and we find very few studies conducted for emerging economies.

Finally, the banking industry operates in a dense environment where various bank-specific factors can influence its operations, implying that any bank study done without analyzing these factors' role would essentially remain incomplete. Nevertheless, there are scarce studies that explore the interplay of these elements while examining the influence of liquidity regulation on banks.

The paper attempts to fill these gaps by carrying out an exhaustive qualitative study by conducting in-depth interviews with industry experts to understand how the new liquidity regime would impact the performance of Indian banks. The study adopts a comprehensive performance framework wherein it tries to recognize how LCR and NSFR can impact the profitability measures [Net interest Margins (NIMs) and Return on Assets (ROA)] and the non-performing assets (NPAs) levels of banks. Further, it also seeks to uncover how bank-specific factors like ownership structure (promoters vs. institutional investors holdings), Transparency and Disclosure practices (T&D) followed by banks, and the Information, Communication, and Technology progress of banks (ICT) can vary the interrelationship of liquidity and bank performance.

2 Review of Literature

A dearth of literature explores the relationship between liquidity regulation and bank performance by conducting qualitative research; however, quite a few studies provide empirical evidence on the same, which are discussed in the below sections.

2.1 Liquidity Regulation and Bank Profitability

High-quality liquidity assets have always been in industry demand for both collateralization and investment purposes. The implementation of the LCR standard has further added to the existing demand pressure as banks increase their stock of HQLAs to be regulation-compliant [5, 6]. The augmented demand, on the one hand, drives up the prices of liquid assets, adding to the cost burden of the banks and, on the other hand, impacts the yield of these assets by introducing an HQLA premium [7, 8].

Similar evidence is found for NSFR and bank profitability. When banks adhere to the long-term liquidity standard, they increase their stable funding sources through longer-term liabilities and increased stable deposits [9, 10]. At the same time, on the assets side, banks trade their long-term loan portfolio with short-term investments [11]. The narrower loan spread and a squeezed gap in maturity between bank assets and liabilities result in reduced bank NIMs [12–14].

2.2 *Liquidity Regulation and NPA Levels of Banks*

The Indian banking system has been battling with the problem of high NPA levels for a long time. High NPA levels not only impact the operational efficiency [15] and performance of banks [16, 17] but can disrupt the entire economic system. Therefore, understanding how the NPA levels of banks change in response to the new liquidity regulation is essential. Banks generally hold high-quality assets when complying with the Basel Accords [18]. This behavior has been found to be further strengthened with the advent of a stricter liquidity framework. When banks assume the new liquidity regime, they deliberately construct a more stable and safe investment portfolio, which reduces their overall risk and NPA levels [11, 19].

However, [20] brings forth interesting findings on the subject while studying the interlinkages of the new Basel III regulation and risk-taking for US banks. The author reveals that, in general, banks tend to be less inclined toward risky investments since the new regulatory framework has been stated. However, this risk aversion can alter in the face of competition for banks characterized by undercapitalization and a large stock of risky assets. The author demonstrates that banks with such attributes while operating in an environment of intense competition are often tempted to take on more risk as they conform to the stricter regulatory requirements. The existing literature, thus, highlights that, in general, the new liquidity framework aids in building a robust asset portfolio as banks become more mindful of their investment decisions. However, this can change as banks with certain pre-existing conditions may respond differently to the regulation.

2.3 *Impact of Ownership Structure, T&D, and ICT on the Relationship of Liquidity Regulation and Bank Performance*

Though no studies investigate how ownership structure, T&D, and ICT alter the relationship of liquidity regulation with bank performance, there is evidence documenting the impact of these elements on bank performance, which is elaborated on in the below section. Ownership structure (promoter vs. institutional): There is conflicting evidence on how promoter and institutional holdings affect the bank's performance [21–23]. When turning to institutional holdings, literature provides contrary results for NPAs and bank profits. Studies suggest that institutional investors' inclination toward riskier investments generally leads to high NPA levels [21, 23, 24]. However, the industry expertise of institutional investors enables them to have a more favorable hold on the profits of banks [25].

Transparency and Disclosure Practices (T&D): Better T&D practices lead to improved bank performance [26, 27] through better risk management practices and enhanced public confidence. Since liquidity regulation itself is an integral constituent of T&D practices followed by banks, it is expected to relate positively with T&D to augment bank performance. Information, Communication, and Technology (ICT) has dual effects on banks; on the one hand, it helps in progressing their efficiency, and on the other hand, it help in boosting their business by bringing banking services to customers' doorstep. Thus, various aspects of ICT, ATMs, Internet banking, and debit and credit lead to improved bank performance. Though these advancements are a welcoming change, for a critical industry like banking, they pose new challenges in the form of a more complicated risk and fraud environment, affecting their requirements under the new liquidity regime.

3 Data and Methodology

The objective of the study paper is to determine the real effects of LCR and NSFR on the banking system. The study uses a qualitative approach that aligns with the researcher's preferred subjectivism ontology and interpretivism epistemology. This covers applying the in-depth interview method, a qualitative research instrument. Inductive and descriptive terminology is used to describe this study design. Convenience sampling and snowball sampling are the methods employed in this study to choose participants [28, 29]. Eighteen bankers were approached, and thirteen agreed to the interview. The exploratory investigation conducted included semi-structured interviews. These bankers have expertise working in Indian banks for a minimum of ten years period. In order to allow respondents to freely discuss their opinions and experiences during the interview session, the researchers created twelve questions. Each interview session lasted between 20 and 30 min.

Content analysis is used because it can locate quotes, code the themes, categorize the codes in multiple orders, and map the methodology applied to determine the relationships between the various categories. The study uses NVivo Software to conduct content analysis on the interview data [30]. The interviewer recorded the interviews, and Microsoft 360 software was used to transcribe the transcript. According to the content analysis results, the primary effects of LCR and NSFR on bank operations are mentioned in seventeen themes [30]. The demographic profile of the respondents is tabulated in Table 1. It reflects upon the gender, age, experience, and qualification of the respondents.

Table 1 Respondents profile

S. No.	Designation	Exp. (years)
R1	Vice President, Private Bank	18
R2	Vice President, Private Bank	13
R3	Assistant General, Public Bank	14
R4	General Manager, Public Bank	22
R5	Vice President, Public Bank	12
R6	Assistant Vice President, Private Bank	14
R7	Manager, Public Bank	10
R8	Assistant General, Public Bank	16
R9	General Manager, Public Bank	15
R10	Assistant Vice President, Private Bank	14
R11	Vice President, Private Bank	13
R12	Vice President, Public Bank	12
R13	Deputy General Manager, Public Bank	15

Source Authors Compilation

4 Results and Discussion

The bankers in the industry have highlighted the nuances of the liquidity arrangements in the banking sector. LCR measures the capacity of banks to meet their thirty-day obligation in a crisis situation. NSFR is defined as the amount of the availability of stable funding in relation to the required stable funding of the banks [1]. The themes of this study are divided into primary- and secondary-level themes and are represented in Fig. 1. The major themes of the study are profitability, non-performing assets (NPS), ownership, technology, transparency and disclosure, etc. The first-order themes are 17 in number, and the second-order themes are twelve in number labeled as A to L. The themes and supporting data are displayed in Table 2.

When we study the influence of LCR on profitability, we get to understand that the impact of LCR is not the same in the short and long run. If we consider the short-term scenario, initially, as banks start complying with the new regulation and their liquidity is low, their profits decline. It can be attributed to the additional cost banks must bear to acquire these high-quality assets. So, there is a negative relation between LCR and profitability in the short run. However, as banks become compliant with the regulation and have sufficient stock of liquid assets on their balance sheet, the profits of banks start improving. Industry experts highlight that increased liquidity reduces the risks of banks, which boosts their creditworthiness and enables them to source funds at a relatively lower cost. Further, the HQLA stock of Indian banks is mainly composed of government securities, which give good returns. The combined effect of reduced cost of funds and increased revenue stream facilitates banks in augmenting their profits when banks hold sufficient liquidity.

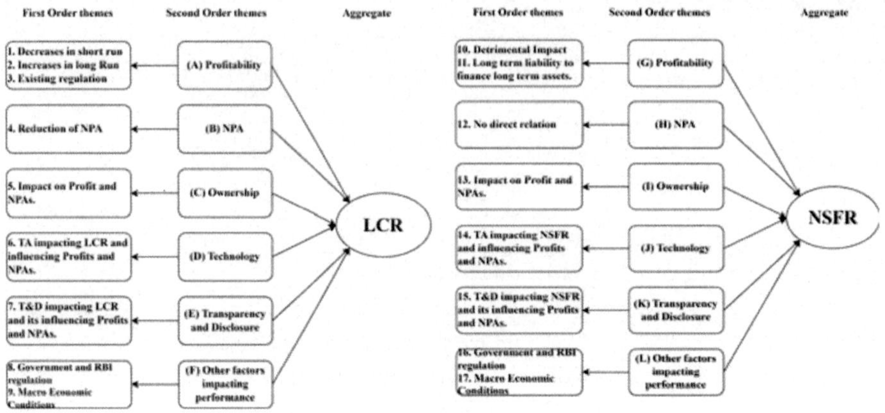

Fig. 1 Theme categorization. *Source* Authors compilation

Similarly, when we study the influence of NSFR on profitability, we observe that NSFR has an adverse impact on the profitability of the banks. The NSFR standard mandates banks to recourse to stable sources of funding. It implies that banks need to increase the maturity of their wholesale funding, issue long-term debt, and hold more stable deposits. All these long-term liabilities on their balance sheet translate into higher interest expenses, resulting in deteriorated NIMs. On the impact of LCR and NSFR on the NPA levels of banks, LCR is found to have a favorable influence on the NPAs, whereas NPAs remain unaffected by NSFR. Industry practitioners contend that the design and incubation of the LCR standard is such that it encourages banks to hold quality assets that are less likely to become NPAs. However, they believe that since NSFR is more focused on the liability side and is silent on the nature of the assets that banks should hold, it will have no bearing on their NPAs.

The findings have vital implications for bankers and regulators. First, the results on the association between LCR and performance call for banks to maintain a balanced liquid asset base that would enable them to meet the regulatory mandate without compromising profits. Second, on the NSFR side, bank strategists must adopt growth-oriented compliance strategies wherein they should use long-term liabilities to finance short-term assets that attract a lower RSF run-off rate. Third, as observed, compliance with long-term liquidity reduces profitability, which may induce the impacted banks to shift their business models and take more risks. National regulators should watch for any such potential changes and take immediate action if needed. Fourth, policymakers/regulators should equip the financial markets in a manner that allows banks to meet their liquidity needs at reasonable funding costs. Finally, banks should explore how ownership structure, T&D, and ICT can be managed to augment the benefits of the new liquidity framework.

Table 2 Representative data for each theme/category

Second-order theme	First-order theme	Representative data
Liquidity coverage ratio		
A. Profitability	1. Decrease in the short run (77%)	"These high liquid assets will decrease the profit. But the important thing is that when banks hold these assets, they will become less risky, and after a point, their profits will improve."
	2. Increases in the long run (69%)	"Initially, banks will have to increase their liquid assets to reach 100% LCR; they will face difficulties. Once banks reach the desired level, the profits of banks can increase."
	3. Existing regulation (69%)	"RBI already requires banks to maintain sufficient liquidity through SLR, CRR, etc. It is a very important aspect of Indian banks, making it strong. The SLR securities usually have a good return."
B. NPA	4. Reduction of NPA (85%)	"This has a healthy impact on the bank's portfolio. The chances of these assets having become NPA chances are less."
C. Ownership	5. Impact on profit and NPAs. (100%)	"Ownership—promoter or institutional, private, or public has a very big role in how the banks function, which then also shows in the profit levels of banks and their NPAs."
D. Technological advancement (TA)	6. TA impacting LCR and influencing profits and NPAs (100%)	"Earlier, you would go and withdraw once and then spend through the month now, with technology, people can withdraw anytime. So, the requirement to keep hard cash has gone down now. But now banks have to manage cash in many places. So, it has become costlier for banks."
E. Transparency and disclosure	7. T&D impacting LCR and its influencing profits and NPAs (92%)	"Corporate governance now, the consumer has become so vigilant because of information available to them. The consumer wants to see that organizations behave responsibly, and LCR and NSFR are its part."
F. Performance	8. Government and RBI regulation (77%)	"Banks are a very important part of the economy, and I think things policies of the government will for sure have a very important impact."

(continued)

Table 2 (continued)

Second-order theme	First-order theme	Representative data
	9. Macro-economic conditions (77%)	"There can be many factors like other macro-economic conditions, market conditions overall economy's performance which can have an impact on banks and then they will also have an effect on LCR."
Net stable funding ratio (NSFR)		
G. Profitability	10. Detrimental impact (100%)	"What it says is that banks should increase the maturity of their liabilities and decrease the maturity of their assets so see this would automatically lead to a decline in the profits of banks."
	11. Long-term liability to finance short-term assets (100%)	"What banks want is to take short-term liability and finance long-term assets. Now, NSFR is forcing banks to finance short-term assets with long-term liability."
H. NPA	12. No direct relation (92%)	"I think NSFR will not have impact on the NPA levels of banks because it is not talking about the quality of assets; it is focusing on maturity."
I. Ownership	13. Impact on profit and NPAs (92%)	"Most of the regulators of the banks in the world do not want financial institutions in the hands of few people because this is a very unique business combination where the leverage is very high, so promoter or institutional holding will definitely impact LCR NSFR and bank profits."
J. Technological advancement (TA)	14. TA impacting NSFR and influencing Profits and NPAs (100%)	"Technology has been the focal point for banks as they are setting up ATMs, promoting internet and mobile banking. All this increases the risk of the banks, so this will show in the LCR and NSFR."
K. Transparency and disclosure	15. T&D impacting LCR and its influencing profits and NPAs (92%)	"There has been much focus on the concept of corporate governance across the industry, as is the case with the Indian banks also. LCR and NSFR need to be published so they become part of the transparency."
L. Performance	16. Government and RBI regulation (77%)	"Regulations of RBI which are made only for Indian banks, will influence the way LCR or NSFR has an impact on banks."
	17. Macro-economic conditions (77%)	"Some of the factors that might play a role for Indian bank's perspective could be the impact of broader economy meltdown—SVB, Credit Suisse"

Note Authors compilation

5 Conclusion

Although the probable impact that liquidity regulation can have on the performance of the banking industry has been investigated in the literature, there are scarce studies that adopt a qualitative approach to explore this relationship. The study results demonstrate that initially, as banks begin to comply with the LCR standard and liquidity is low, the profitability of banks tends to suffer. However, as the LCR of banks increases further and banks hold sufficient liquid assets, the profits of banks start improving. When analyzing the interrelationship of NSFR and bank profitability, it is observed that NSFR adversely impacts bank profits. On the NPA side, experts argue that though NSFR would have no bearing on the NPA levels of banks, LCR has a favorable influence on the same. Further, on the role of bank-specific factors in altering the association of liquidity and bank performance, the industry practitioners are convinced that ownership structure, T&D, and ICT would have a significant pull on the relationship between the two variables. However, the direction of this interaction cannot be determined with certainty.

The findings of the research work would enable the broader set of bank strategists to analyze and modify their compliance strategies by drawing from the views expressed by industry specialists in the study. Most importantly, the study results would assist not only Indian regulators/policymakers but also other comparable organizations of similar emerging economies to understand how liquidity regulation needs to be customized in the context of the respective economies to achieve the desired stability and resilience outcomes.

References

1. Adrian T, Boyarchenko N (2018) Liquidity policies and systemic risk. J Financ Intermediat 35:45–60
2. Du B (2017) How useful is Basel III's liquidity coverage ratio? Evidence US bank holding companies. Eur Financ Manag 23(5):902–919
3. Yan M, Hall MJB, Turner P (2012) A cost-benefit analysis of Basel III: some evidence from the UK. Int Rev Financ Anal 25(5):73–82
4. Taskinsoy J (2018) Effects of Basel III higher capital and liquidity requirements on banking sectors across the main South East Asian nations. Int J Sci Eng Res 9(4):214–237
5. Duijm P, Wierts P, Bank DN (2016) The effects of liquidity regulation on bank assets and liabilities. Int J Cent Bank 12(2):385–411
6. Banerjee RN, Mio H (2018) The impact of liquidity regulation on banks. J Financ Intermediat 35:30–44
7. Roger S, Vlček J (2011) Macroeconomic costs of higher bank capital and liquidity requirements
8. Fuhrer LM, Müller B, Steiner L (2017) The liquidity coverage ratio and security prices. J Bank Financ 75:292–311
9. King MR (2013) The Basel III net stable funding ratio and bank net interest margins. J Bank Financ 37(11):4144–4156
10. Muriithi JG, Waweru KM (2017) Liquidity risk and financial performance of commercial banks in Kenya. Int J Econ Finance 9(3):256

11. Paulet E (2018) Banking liquidity regulation: impact on their business model and on entrepreneurial finance in Europe. Strateg Change 27(4):339–350
12. Pak O (2020) Bank profitability in the Eurasian Economic Union: do funding liquidity and systemic importance matter? N Am J Econ Finance 54
13. Altahtamouni F, Alyousef S (2021) The effect of liquidity according to the requirements of the Basel III Committee on the profitability of banks: evidence from Saudi banks. Int J Econ Bus Admin (IJEBA) 440
14. Khan MS, Scheule H, Wu E (2015) Will Basel III liquidity measures affect banks' funding costs and financial performance? Evidence from U.S. commercial banks
15. Michael JN, Vasanthi G, Selvaraju R (2006) Effect of non-performing assets on operational efficiency of central co-operative banks. Indian Econ Panorama 33–39
16. Stephen Kingu P, Macha DS, Gwahula DR (2018) Impact of non-performing loans on bank's profitability: empirical evidence from commercial banks in Tanzania. Int J Sci Res Manag 6(01)
17. Martiningtiyas CR, Nitinegeri DT (2020) The effect of non-performing loans on profitability in banking sector in Indonesia. In: Proceedings of the international conference on management, accounting, and economy, pp 64–67
18. Podpiera R (2006) Does compliance with Basel core principles bring any measurable benefits?
19. Chalermchatvichien P, Jumreornvong S, Jiraporn P (2014) Basel III, capital stability, risk-taking, ownership: evidence from Asia. J Multinatl Financ Manag 28:28–46
20. Stability B, Schliephake E (2016) Capital regulation and competition as a moderator for banking stability. J Money Credit Bank 48(8):1787–1814
21. Barry TA, Lepetit L, Tarazi A (2011) Ownership structure and risk in publicly held and privately owned banks. J Bank Financ 35(5):1327–1340
22. Iannotta G, Nocera G, Sironi A (2007) Ownership structure, risk and performance in the European banking industry. J Bank Financ 31(7):2127–2149
23. Rastogi S, Gupte R, Meenakshi R (2021) A holistic perspective on bank performance using regulation, profitability, and risk-taking with a view on ownership concentration. J Risk Financ Manag 14(3):111
24. Lim J, Minton BA, Weisbach MS (2014) Syndicated loan spreads and the composition of the syndicate. J Finance Econ 111(1):45–69
25. Saghi-Zedek N (2016) Product diversification and bank performance: does ownership structure matter? J Bank Finance 71
26. Farvaque E, Refait-Alexandre C, Weill L (2012) Are transparent banks more efficient? East Eur Econ 50(4):60–77
27. Zaman R, Arslan M, Siddiqui MA (2014) Corporate governance and firm performance: the role of transparency & disclosure in banking sector of Pakistan. Int Lett Soc Hum Sci 43:152–166
28. Donastorg AD, Renukappa S, Suresh S (2022) Financing renewable energy projects in the Dominican Republic: an empirical study. Int J Energy Sect Manag 16(1):95–111
29. Alam MK (2021) A systematic qualitative case study: questions, data collection, NVivo analysis and saturation. Qual Res Organ Manag: Int J 16(1):1–31
30. Bazeley P, Jackson K (2019) Quantitative data analysis with NVIVO, 2nd edn. Sage

Enhancing Medical Education Through Augmented Reality

Sumit Sawant, Pratham Soni, Ashutosh Somavanshi, and Harsh Namdev Bhor

1 Introduction

The aging population and increased hospitalization rates have resulted in a critical necessity for supplementary medical staff and resources. Efforts have been made to recruit medical and nursing students and provide training materials. However, the high cost of essential training tools and equipment necessitates the search for affordable alternatives. One widely adopted approach for training in various fields is the use of mobile devices and augmented reality (AR) applications [1]. AR is a form of training that enhances or intensifies the current reality by incorporating computer-generated content into specific locations or events. It seamlessly overlays digital content onto our real-world experiences [2]. Technological advancements over the past decade have made AR more accessible for educational purposes, benefiting students from elementary school to professionals. AR in education offers the below features:

- AR gives users an authentic and contextually enriched experience by immersing them in the real-world environment.
- AR enriches the physical surroundings by superimposing interactive and digital virtual information.

Table 1 presents a difference of existing AR systems and proposed work. The HoloLens (Microsoft) offers new features and functions but costs more. The Meta 2, widely used in various applications, provides the widest field of view but still carries a price tag of USD 1495. The Android-based smart eyeglass (Sony) is more affordable but lacks support for users with myopia. Considering these limitations

S. Sawant · P. Soni · A. Somavanshi · H. N. Bhor (✉)
Department of IT, K J Somaiya Institute of Technology, Mumbai, Maharashtra, India
e-mail: hbhor@somaiya.edu

S. Sawant
e-mail: sumit.sawant@somaiya.edu

M. S. Kaiser et al. (eds.), *ICT: Smart Systems and Technologies*, Lecture Notes in Networks and Systems 878, https://doi.org/10.1007/978-981-99-9489-2_16

Table 1 Comparison of AR application devices

	Our system	Microsoft HoloLens	Sony smart eyeglass	Meta 2
OS	Android	Windows	Android	Window
SDK/API	Unity/Vuforia	Unity/unreal/ Vuforia	Smart eyeglass API	Unity SDK
AR features	Static and dynamic	Interactive module	Static and dynamic	Static and dynamic
Propriety hardware	No	Yes (USD 3500)	Yes (USD 699)	Yes (USD 1495)

and the need for a cost-effective solution, we intend to develop our proposed system specifically for the Android platform, eliminating the need for proprietary eye-wear. Users can utilize their own mobile devices with built-in cameras.

2 History of AR

Augmented reality (AR) came into existence in the late 1950s with Morton Heilig, a movie director, who believed that film should be able to immerse viewers in on-screen action. He developed a concept in 1955 called "The Cinema of the Future" and later built a prototype in 1962 called the Sensorama, which predated computerized computing [3]. The Sensorama provided viewers with sensory experiences including sound, visuals, vibration, and smell, although it was not computer-controlled.

In 1968, Ivan Sutherland, a computer scientist from America, developed the concept of head-mounted displays as a means to access worlds of virtual reality. However, the technology was not feasible for mass use at that time. In 1975, Myron Krueger, an American computer developer, created the first "virtual reality" interface known as "Video Place." This interface allowed users to interact with virtual objects at the moment. The expression "virtual reality" was coined by Tom Caudell, a researcher at Boeing, in 1990. The first functional AR technology was devised by Louis Rosenberg at the USAF Lab in 1992.

3 Proposed System

In this project concentrated on augmented reality (AR), we intend to establish a "Smart Learning Software" that offers excellent quality illustrations, a facile-to-use interface, object information visibility, and the ability to bring virtual objects into the true realm. The application is intended to reduce the burden on teachers while augmenting their teaching skills, and it provides students with proper knowledge and an improved grasp of concepts.

The primary objective of this initiative is to establish an immersive learning experience for medical students by leveraging AR technology. By integrating AR into medical training, we intend to encourage the development of fundamental skills such as decision-making, efficient communication, and creative adaptation of global resources to resolve local healthcare priorities. Additionally, AR-based training ensures patient safety by minimizing the occurrence of errors during skills practice.

This initiative stands out for its distinctive approach, providing an extensive and situated educational platform for medical pupils. Through the use of augmented reality (AR), we strive to create a learning environment that not only improves understanding but also enhances practical skills and critical thinking abilities [10]. By leveraging the benefits of AR technology, we aspire to revolutionize medical education and contribute to a generation of competent and compassionate healthcare professionals.

The proposed "Smart Education App" will provide a valuable aid for medical pupils, educators, and medical centers, helping them to fully utilize the potential of augmented reality and create a more engaging and effective learning experience. By promoting active learning and providing real-world context, this initiative seeks to shape the future of medical training and contribute to the growth of care for patients.

4 Implementation

4.1 System Development

The first step in system development entailed establishing a Vuforia Developer Portal, Vuforia engine, and a corresponding target image database. This target image, a predefined non-symmetric pattern, is affixed to a target object or location. Once the application identifies this target image using the device's camera, it triggers AR effects (referred to as Game Objects) and superimposes them onto the target object depicted on the device's screen. This target image, confined to a size limit of 35 MB and stored within the portal, facilitates the computation of the target object's orientation and its distance from the camera [1].

4.2 Marker-Based Softwares

Marker-based applications rely on picture recognition, also known as recognition-based AR. These applications require markers to trigger augmentations. Markers can be distinct shapes or patterns easily recognizable and processed by cameras. They can be paper-based or physical objects existing in the real world. To view the augmented content, the user needs to scan a marker, which can be an object, text, video, or animation, using the device's camera feed. Once the marker has been identified by the gadget, the software overlays digital information onto the marker, permitting the user to see the virtual object.

Building marker-based software requires offering visuals or words in advance to facilitate easier browsing when studying the camera data. In this approach, the objects are hardcoded into the program, making them simpler to recognize. Marker-based AR applications are presently prevalent, particularly in advertising.

4.3 Three-Dimensional Identification and Monitoring

Three-dimensional Identification of images and monitoring are critical aspects of any augmented reality software development kit (SDK). With 3D monitoring, an app can comprehend and improve huge surroundings around the user, such as airports, malls, and resorts. Software that allows 3D identification and monitoring may detect three-dimensional objects including cubes, spheres, toys, and more. This kind of technology is often utilized in online shopping and mobile gaming.

A. **Unity Assistance**

Unity is a widely used cross-platform game technology noted for its popularity and capability. Although mostly utilized for game creation, it may also be leveraged to build AR apps with stunning effects. Unity serves as a flexible platform that enables developers to build cutting-edge experiences or augment old notions with novel ways.

B. **Tools**

Wikitude's SDK7 provides support for simultaneous localization and monitoring. This software offers several capabilities, including real-time augmented reality (AR) application creation, 3D recognition and monitoring, image identification and observation, cloud recognition, services based on location, glasses with sensors integration, and connection with external plugins, such as Unity. The systemic flow of an augmented reality (AR) application entails delineating the sequential steps and decision points navigated by either the user or the application itself. Presented below is a streamlined flow representation for an AR application. It is essential to note that the intricacy of the flow may fluctuate contingent upon the particular features and functionalities incorporated into your AR application (Fig. 1).

5 Result

We have accomplished the effective deployment of augmented reality and its applications by building an app capable of detecting and seeing human bodily components for example the heart, kidney, stomach, and liver in a three-dimensional way. This revolutionary tool has the potential to considerably aid aspiring healthcare professionals in their learning and knowledge of the human body.

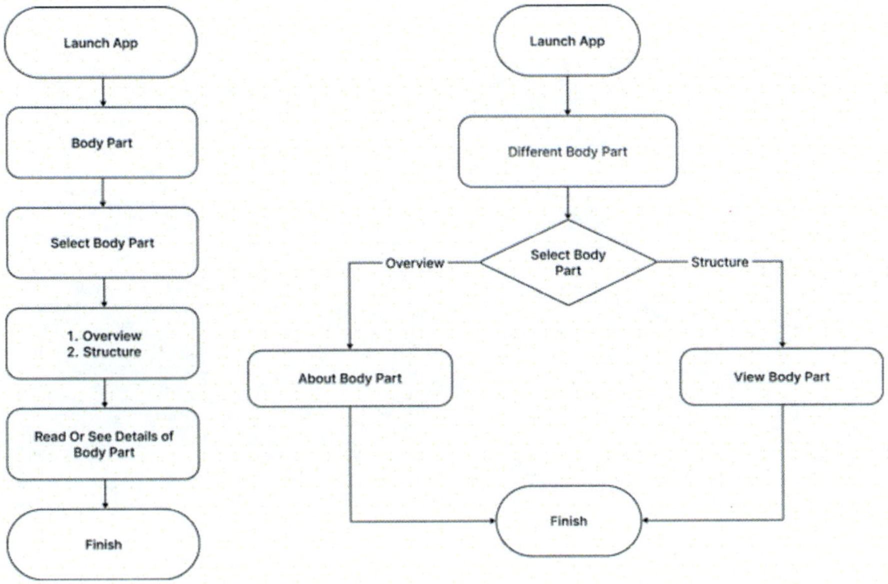

Fig. 1 Systemic flow of an augmented reality (AR) application

Figure 2 showcases the core of our augmented reality (AR)-based Smart Learning Software system for medical education: interactive three-dimensional (3D) models of human organs. These models are a testament to our commitment to providing an immersive and enriching learning experience for medical students. Leveraging AR technology, our system empowers students to engage with and explore intricate anatomical structures in a virtual environment.

In Fig. 3, we showcase a three-dimensional (3D) representation of the lungs as part of our augmented reality (AR)-based Smart Learning Software system for medical education. This figure is a testament to our dedication to providing an immersive learning experience for aspiring medical professionals. Through AR technology, students can interact with and explore intricate anatomical structures like the lungs in a virtual environment. This not only enhances their comprehension of complex concepts but also fosters practical skills and critical thinking.

In Fig. 4, we present a three-dimensional (3D) visualization of the liver, a key element of our augmented reality (AR)-based Smart Learning Software initiative for medical education. This figure illustrates our commitment to creating an immersive learning experience for medical students. By using AR technology, we offer students the opportunity to interact with and explore intricate anatomical structures in a virtual environment.

In Fig. 5, we present a 3D visualization of the kidney, a component of our innovative Smart Learning Software initiative. This initiative leverages augmented reality (AR) technology to provide medical students with an immersive learning experience. The 3D kidney visualization is a representative example of our approach, offering

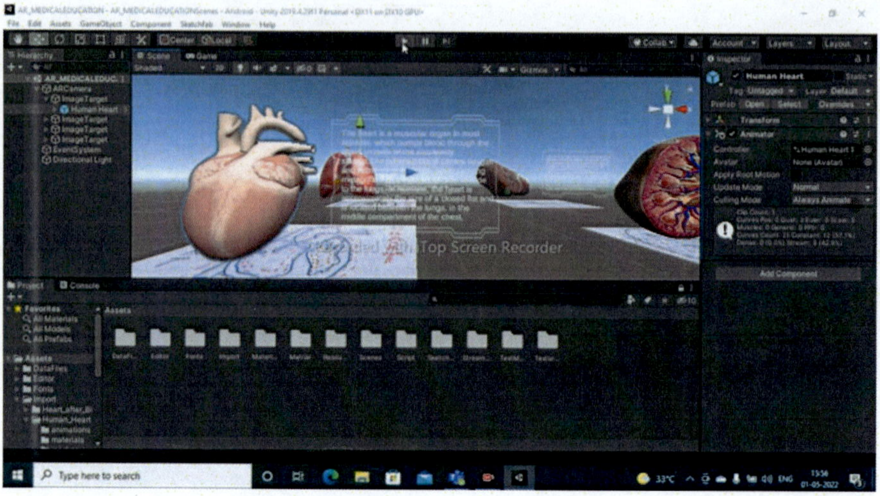

Fig. 2 Interactive 3D models of human organs

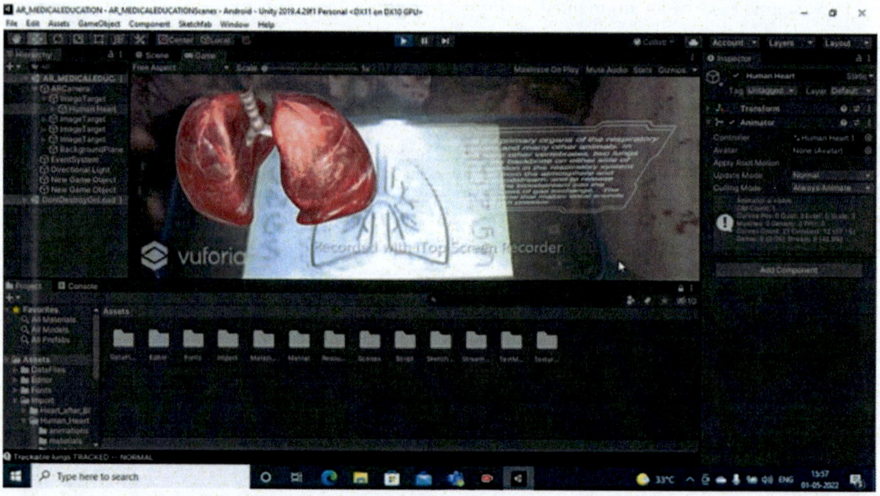

Fig. 3 Three-dimensional representation of lungs in system

students the ability to interact with and understand complex anatomical structures in a virtual environment. This not only enhances their theoretical knowledge but also fosters practical skills and critical thinking [5].

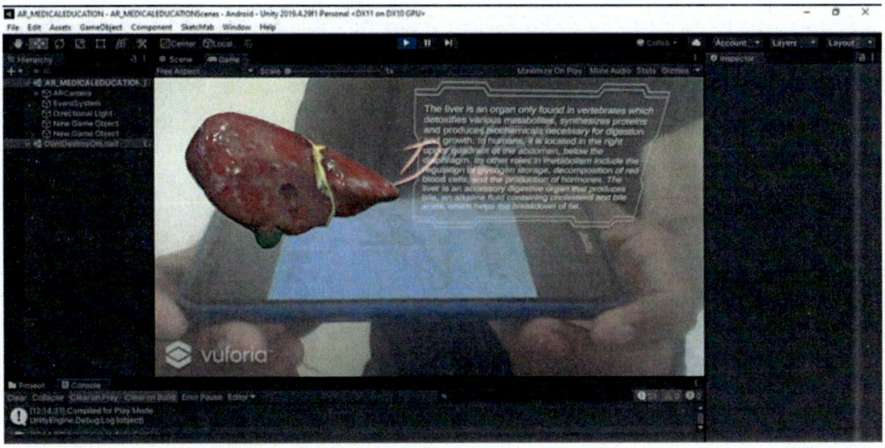

Fig. 4 Three-dimensional visualization of the liver

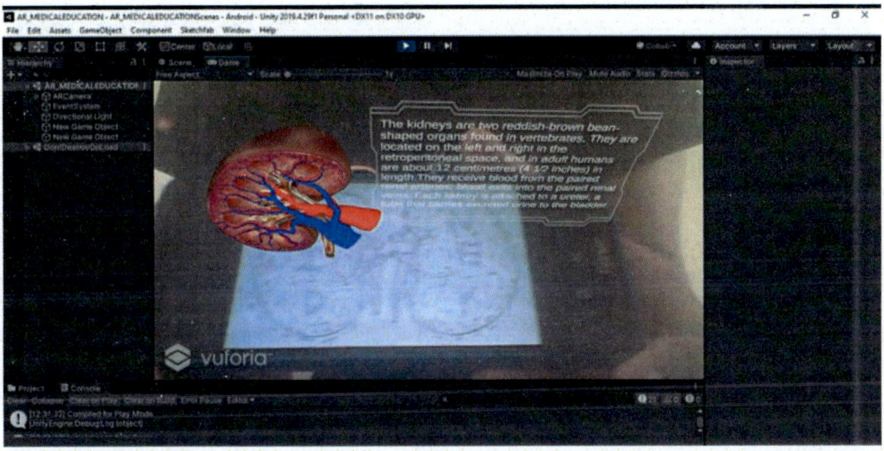

Fig. 5 3D kidney visualization

6 Applications

6.1 Healthcare

The application of augmented reality (AR) has reached substantial acceptability in the medical industry, offering several benefits for physicians and patients. Augmented reality improves patient care by enabling physicians and nurses to visually clarify and demonstrate the patient's actual condition, resulting in enhanced medical outcomes.

Healthcare professionals can leverage AR to improve their knowledge of human structures and physiological processes, facilitating medical care and operations [6].

6.2 Education

The incorporation of augmented reality software has brought about a revolution in the teaching business, bringing up new possibilities and boosting the learning experience. AR is applied to teaching and learning methods, enabling educators to deliver visual effects that enhance understanding and engagement. By merging virtual objects with actual situations, AR helps students experience and grasp complex ideas and concepts that may otherwise be tough in real-world circumstances [9].

6.3 Tourism Industry

The tourism industry, being a large contributor to the GDP of many countries, has embraced the promise of AR technology to improve the holiday experience. Virtual reality applications in tourism provide vivid and instructional experiences for tourists. These programs supply crucial information, such as guides, navigation aids, and translations, making the travelers' trip simpler and increasing their experience of new regions.

6.4 Navigation

AR technology plays a crucial role in navigation systems, both indoors and outdoors. Gerhard Reitmayr developed an outdoor navigation system that helps individuals locate themselves in urban environments. Similarly, Alessandro Mulloni introduced an indoor navigation system. These advancements in AR navigation systems provide users with precise location-based information, enhancing their ability to navigate and explore their surroundings. AR-based navigation is an emerging application with promising potential.

6.5 Surgical Training

Surgeons and surgical trainees can use AR simulations to practice surgical procedures in a risk-free environment. AR can overlay digital surgical instruments and guides onto real patients or mannequins, improving precision and reducing errors [12].

6.6 *Medical Imaging Interpretation*

AR can enhance the interpretation of medical images, such as X-rays, MRIs, and CT scans. It can overlay 3D reconstructions or additional information onto these images, helping students and healthcare professionals make more accurate diagnoses [8].

7 Challenges and Issues with AR

7.1 *Technical Challenges*

The successful functioning of an AR technology depends on effectively managing and handling huge quantities of information. It is crucial to have hardware that is user-friendly, portable, and capable of quickly displaying information [7]. Resolving hardware limitations is essential for the practical use and widespread adoption of AR technology.

7.2 *Lack of Public Awareness*

Despite the potential benefits of AR, many consumers are still unaware of its advantages. There is a lack of public awareness regarding the usability and applications of AR in daily life. Currently, consumers mainly associate AR with trying out virtual wardrobes, glasses, and accessories, but are not fully aware of its broader potential.

7.3 *Lack of Regulation*

Due to the relatively early stage of AR technology, there is a lack of established regulations to support its business applications. The absence of clear standards and guidelines leaves consumers unsure about how to safely and securely use AR in their daily activities. Addressing privacy, security, and data processing concerns is vital for ensuring consumer trust and widespread adoption.

7.4 *Visualization Issues*

In AR systems, registration errors are common and difficult to completely avoid. Rendering objects in a way that accurately displays their position in the user's view is crucial. Improving the quality of virtual objects and ensuring accurate tracking and measurement are key considerations for enhancing the visual experience in AR applications [4].

7.5 *Ethical Considerations*

Using AR in medical education raises ethical questions about informed consent, patient privacy, and the potential desensitization of students to sensitive medical situations [11].

7.6 *Limited Evidence*

While AR holds promise in medical education, there is a need for more research and evidence to demonstrate its long-term benefits and effectiveness.

8 Conclusion

Moving forward, there is great potential for the expansion and development of the app on a larger scale across various fields. While AR is still in its early stages of application in healthcare education, it holds immense promise for enhancing learning experiences in this domain. When contrasted with conventional media like books, videos, or desktop experiences, AR has demonstrated greater effectiveness in educating individuals about intricate spatial structures and functions. This includes areas such as geometric shapes, chemical structures, mechanical machinery, astronomical configurations, and the spatial arrangements of human organs. The implementation of such AR systems in the field has the prospective to significantly enhance the quality of education across all disciplines. By leveraging AR technology, we can facilitate easier comprehension of complex concepts, leading to advancements in various fields.

References

1. Leung K-H, Hung K, Ko C-P, Lo S-F (2022) Design and development of an augmented reality mobile application for medical training. In: 6th IEEE international conference on engineering technologies and applied sciences (ICETAS)
2. Yuen SC-Y, Yaoyuneyong G, Johnson E (2011) Augmented reality: an overview and five directions for AR in education. J Educ Technol Dev Exchange 4(1):119–140
3. Ford J, Höllerer T (2008) Augmented reality and the future of virtual workspaces. In: Handbook of research on virtual workplaces and the new nature of business practices. IGI Global, Santa Barbara, pp 486–502

4. Sirohi P, Agarwal A, Maheshwari P (2020) A survey on augmented virtual reality: applications and future directions. In: Seventh international conference on information technology trends (IIT), 25–26 Nov
5. Billinghurst M, Dunser A (2012) Augmented reality in the classroom. Computer 45(7):56–63
6. Mehta V, Chugh H, Banerjee P (2018) Applications of augmented reality in emerging health diagnostics: a survey. In: International conference on automation and computational engineering (ICACE-2018), Amity University, Greater Noida Campus, U. P., India, 3–4 Oct
7. Terdale JV, Bhole V, Bhor HN, Parati N, Zade N, Pande SP (2023) Machine learning algorithm for early detection and analysis of brain tumors using MRI images. Int J Recent Innov Trends Comput Commun 11(5s):403–415. https://doi.org/10.17762/ijritcc.v11i5s.7057
8. Bhor HN, Kalla M (2022) TRUST-based features for detecting the intruders in the Internet of Things network using deep learning. Comput Intell 38(2):438–462. https://doi.org/10.1111/coin.12473
9. Bhor HN, Kalla M. A survey on DBN for intrusion detection in IoT. In: Zhang YD, Senjyu T, So-In C, Joshi A (eds) Smart trends in computing and communications: proceedings of SmartCom 2020. Smart innovation, systems and technologies, vol 182. Springer, Singapore. https://doi.org/10.1007/978-981-15-5224-3_33
10. Bhor HN, Kalla M (2020) An intrusion detection in internet of things: a systematic study. In: International conference on smart electronics and communication (ICOSEC), Trichy, India, pp 939–944. https://doi.org/10.1109/ICOSEC49089.2020.9215365
11. Bhor HN, Koul T, Malviya R, Mundra K (2018) Digital media marketing using trend analysis on social media. In: 2nd international conference on inventive systems and control (ICISC), Coimbatore, India, pp 1398–1400. https://doi.org/10.1109/ICISC.2018.8399038
12. Bhole V, Bhor HN, Terdale JV, Pinjarkar V, Malvankar R, Zade N (2023) Machine learning approach for intelligent and sustainable smart healthcare in cloud-centric IoT. Int J Intell Syst Appl Eng 11(10s):36–48

A Comprehensive Study on Plant Classification Using Machine Learning Models

A. Karnan and R. Ragupathy

1 Introduction

The identification and categorization of plant species based on their distinctive morphological and physiological properties are a crucial and challenging issue in the science of botany [1]. Traditional methods of plant classification, such as manual observations and identification keys, require significant expertise and can be time-consuming, subjective, and error-prone. There has been growing interest in exercising machine learning models to classify plants automatically [2–5]. Now a days, machine learning focuses on creating models that can use patterns and available information to make predictions. In the context of plant classification, machine learning models can analyze datasets of plant species with large size and extract relevant features from plant species to perform accurate identification [4, 6]. The process of classifying plants might be made more efficient, less expensive, and more accurate with the use of machine learning models. In recent years, there is a significant increase in the number of studies which uses machine learning models for the categorization of plant species [7, 9]. To categorize plant species, these studies used a variety of machine learning algorithms, including decision trees [4], support vector machines [3, 8], random forests [2, 5], and neural networks [18]. Moreover, researchers have also developed various approaches for feature extraction and selection, including image processing techniques and feature engineering [10–13]. Despite the promising results achieved by machine learning models in plant classification, there are still numerous obstacles and constraints to go beyond. For sample, the quality and accessibility of training data may impact the effectiveness and accuracy of machine learning models.

A. Karnan (✉) · R. Ragupathy
Department of Computer Science and Engineering, Faculty of Engineering and Technology, Annamalai University, Annamalainagar, Tamil Nadu 608002, India
e-mail: mekarnan@gmail.com

R. Ragupathy
e-mail: cse_ragu@yahoo.com

M. S. Kaiser et al. (eds.), *ICT: Smart Systems and Technologies*, Lecture Notes in Networks and Systems 878, https://doi.org/10.1007/978-981-99-9489-2_17

As there is a difficult to understand how the model makes its predictions, the interpretability of machine learning models can be an another issue. The literature on this topic covers a range of machine learning algorithms, feature extraction techniques, and evaluation metrics, as well as applications in agriculture, environmental management, and biodiversity conservation. Feature extraction and selection are crucial steps in plant classification while using machine learning models. Several methods have been developed to extract valuable information from plant images, like the scale-invariant feature transform (SIFT), the local binary patterns (LBP), and the histogram of oriented gradients (HOG). Machine learning methods can be made more effective and accurate by using feature selection techniques like mutual information (MI) and principal component analysis (PCA), which can minimize the dimension of the feature space. For the purpose of developing and evaluating machine learning models for classifying plants, publicly accessible datasets are crucial. Various datasets have been developed for different plant species and applications, such as the PlantCLEF dataset [7, 28] for plant identification, Leafsnap dataset [13, 15, 28] for leaf recognition, fruit categorization using the Fruit Recognition dataset [37]. Evaluation metrics are crucial for assessing the performance of machine learning methods for plant classification. Commonly used metrics include accuracy, precision, recall, $F1$-score, and receiver operating characteristic (ROC) curve. The literature on automating and improving plant classification accuracy using machine learning models demonstrates these potentials, with possible applications in agriculture, environmental management, and biodiversity preservation. A thorough review of the body of research on applying machine learning models to classify plants is given in this study. Specifically, the most commonly used machine learning algorithms, feature extraction and selection techniques, publicly available datasets, evaluation metrics, and potential applications are presented. The main obstacles and future objectives for this field of study are also highlighted, emphasizing the potential influence of machine learning models on plant classification and related fields. Overall, this work gives a thorough summary of the state-of-the-art in machine learning methods for classifying plants and identifies key topics for further study. The paper is organized as follows for the remaining paragraphs. The machine learning models for classifying plants are presented in Sect. 2. Section 3 discusses the feature extraction and selection processes used for plant classification. Section 4 describes the datasets and evaluation measures that are available for classifying plants. Section 5 provides challenges and recommendations for the future. Finally, Sect. 6 draws a conclusion.

2 Machine Learning Models for Plant Classification

For classifying plants, a variety of machine learning models have been employed, including more established models like gradient boosting machines (GBM) [43] and Naive Bayes as well as more contemporary models like decision trees [4], support vector machine [3], random forests [5], and K-nearest neighbors [9]. The visual representations of the plant categorization methods are displayed in Figs. 1, 2, 3, 4, and 5.

Naive Bayes is a probabilistic classifier that uses the theorem represented in Eq. (1) to predict the likelihood that it belongs to a certain class based on an instance's attributes.

$$P(\alpha|\beta) = \frac{P(\beta|\alpha)P(\alpha)}{P(\beta)}, \tag{1}$$

where $P(\alpha|\beta)$ is the likelihood of class α given that class β has occurred. $P(\beta|\alpha)$ is the likelihood of class β given that class α has occurred. $P(\alpha)$ is the likelihood of class α occurring independently of class β, and $P(\beta)$ is the likelihood of class β occurring independently of class α.

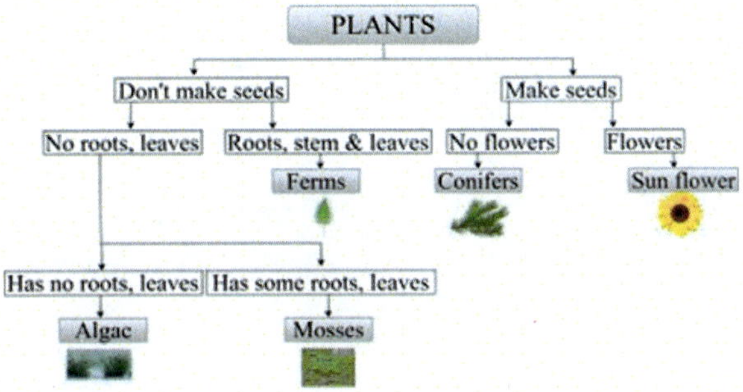

Fig. 1 Illustrates the decision tree method for classifying plants

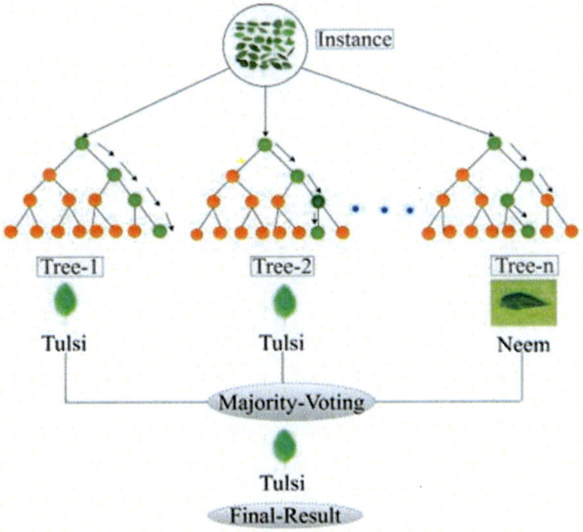

Fig. 2 Illustration of random forest-based plant classification

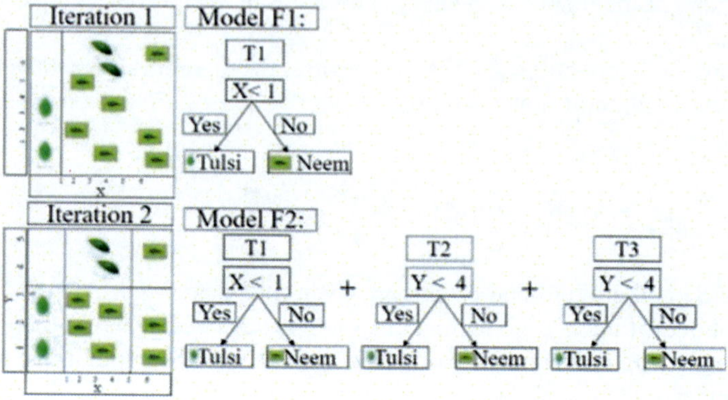

Fig. 3 Plant classification using GBM

Fig. 4 Shows an illustration of plant classification using support vector machine

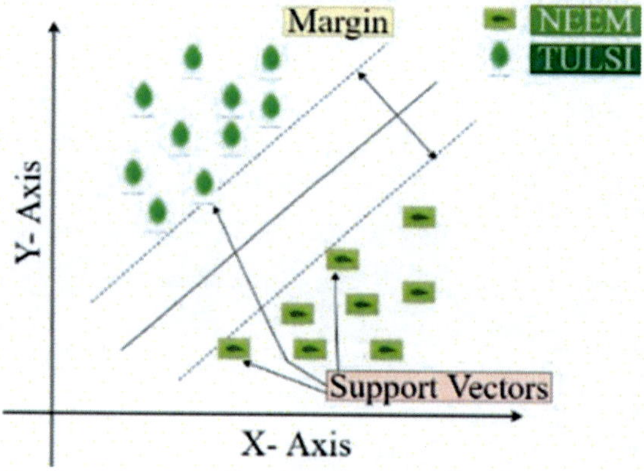

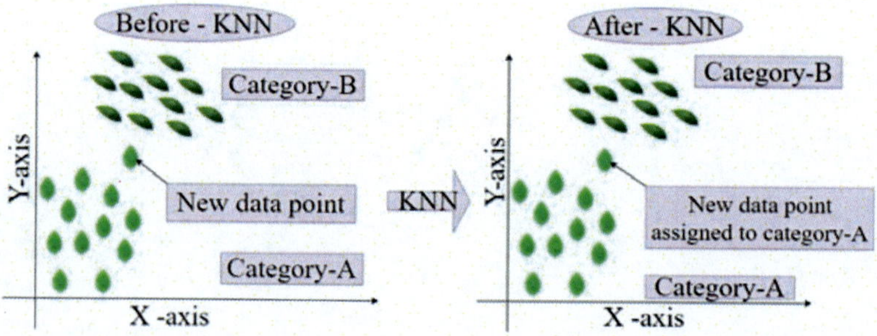

Fig. 5 Shows an illustration of plant classification through the use of KNN

3 Feature Extraction and Selection in Plant Classification

In order to produce precise machine learning models for plant classification, feature extraction and selection become indispensable. The process of extracting features from raw data is the identification of a set of important and informational characteristics that may be utilized to train models. However, feature selection involves choosing a portion of the most important and informational aspects from the set of extracted features. The various feature extraction methods are as follows.

Handcrafted feature extraction techniques. The above methods rely on domain knowledge to design and manually extract features from the data. Handcrafted characteristics include those that are based on shape, texture, color, and other features. Then, using these features, machine learning models are developed.

Hybrid approaches. These approaches involve combining handcrafted features with deep learning features to build more robust machine learning models. Hybrid approaches have shown improvement in the accuracy of plant classification models.

Feature selection techniques. In order to create accurate and effective classification models, feature extraction and selection are essential. Identifying and classifying plants based on their distinctive traits, such as leaf shape, color, texture, or floral characteristics, are known as plant classification. However, high-dimensional and redundant characteristics are frequently found in plant databases, which make it difficult to create efficient models. Feature extraction aims to transform the original raw data into a reduced set of representative features that capture the essential information. Techniques like principal component analysis (PCA) [1] or feature scaling can be employed to reduce dimensionality and remove irrelevant or redundant features. The most significant variation in the data is used by PCA to find principle components which allows for a lower-dimensional representation while preserving important patterns. Additionally, domain-specific feature extraction techniques can be utilized to extract plant-specific characteristics. For instance, in leaf classification, features like contour descriptors [16, 17], vein patterns, or texture measures [14] can be extracted using image processing techniques. Feature reduction methods aim to improve model performance by reducing overfitting and enhancing interpretability through various feature selection methods. Various methods, such as correlation analysis, information gain, and forward/backward feature selection, have been used to identify the most discriminative features [10]. The choice of suitable features relies on the particular traits of the plant dataset and the classification task at hand. Expert knowledge in botany or domain expertise can guide the selection of relevant features. Additionally, machine learning methods like random forests or decision trees can provide insights into feature importance based on their splitting criteria or feature importance. By employing effective feature extraction and selection techniques, the plant classification models focus on the most informative and discriminative features, reducing computational complexity and enhancing classification accuracy. This enables researchers and plant experts to develop robust and interpretable models for plant identification, species classification, and disease detection, supporting various applications in agriculture, conservation, and ecology.

4 Datasets and Evaluation Metrics

To develop and evaluate machine learning models for categorizing plants, high-quality datasets must be accessible. Various datasets have been developed for different plant species and applications, such as leaf classification, fruit classification, and identifying plant diseases. The PlantVillage dataset collection [32], which includes more than 50,000 photos of healthy and diseased leaves from 14 plant species, is one of the most commonly used datasets for classifying plants. Likewise, many more datasets are freely accessible on websites, which are included in Table 1. Some of the widely employed assessment measures used to assess the efficiency of machine learning models for classifying plants are recall, precision, accuracy, $F1$-score, and ROC curve. Table 2 contains a thorough review of evaluation measures used for plant classification.

5 Challenges and Future Directions

Despite recent progress in the categorization of plants using machine learning models, a number of problems need to be addressed/solved. The primary problems along with potential solutions are listed below to use machine learning models for classifying plants.

Availability and quality of training data. One of the key challenges in developing machine learning models for plant categorization is the availability and quality of training data. In many cases, the amount of labeled data available for plant classification is limited, which can affect the accuracy of machine learning models. To address this challenge, researchers can explore alternative sources of training data, such as crowd-sourcing or transfer learning, and develop methods for data augmentation to increase the amount of labeled data available.

Interpretability of machine learning models. The interpretability of machine learning models poses an additional difficulty in classifying plants using these methods. Few machine learning models in particular may be challenging to understand, which may restrict their effectiveness in some applications.. To address this challenge, researchers can develop methods for visualizing and interpreting the learned features and representations of machine learning models and explore the use of explainable AI techniques.

Table 1 Summary of publically available datasets for the plant classification

S. No	Dataset name	No. of classes	No. of images
1	Leafsnap dataset [13, 15, 34]	185	50,000 +
2	Flavia leaf [9, 10, 16, 17, 22, 34]	32	1900
3	Malaya Kew Dataset [19–21]	1500 +	100,000 +
4	Swedish leaf Dataset [22]	50	1800
5	Folio Dataset [12]	32	20
6	Medicinal plant dataset in Bangladesh [23]	10	5000
7	Leaf Dataset [24]	40	7597
8	UCI Dataset [25]	30	116
9	Daucus Carota Dataset [26]	100	300
10	Urban Planter [27]	15	1500
11	Agril Plant Dataset [30]	10	300
12	Foliage Dataset [31]	99	1584
13	PlantVillage Dataset [32]	54	54,306
14	EDEN-ISS Antarctic Dataset [33]	5000	800,000
15	D-Leaf Dataset [22, 34]	22	4502
16	Indian Medicinal Leaves Dataset [39]	40	1500
17	ICL Dataset [18]	220	17,032
18	Plant CLEF 2016 Dataset [40]	1000	113,205
19	PlantCLEF2015 Dataset [41]	1000 +	2.7 million
20	Aarhus University Signal Proce group [42]	12	960
21	Beta Vulgaris Dataset [26]	100	254
22	Oxford102 [29]	102	8189
23	Plant Doc [36]	27	2598
24	Fruits-360 [37]	81	90,380
25	Plants [38]	99	360

Generalization performance. The generalization capabilities of models present a substantial barrier to machine learning-based plant classification. Unknown or fresh data may not be well-received by machine learning models that are overfit to the training set. Researchers might investigate early halting or dropout approaches for regularizing machine learning models to overcome this issue. They can also create ways for assessing the generalization performance of models.

Scalability. As the size of plant classification datasets continues to grow, scalability becomes an important challenge. Large-scale machine learning model training can be expensive computationally and time-consuming, To address this challenge, researchers can explore distributed training methods and develop algorithms that can handle large-scale datasets efficiently. Future studies in this area should concentrate

Table 2 Analysis of evaluation metrics used for the classification of plants

S. No.	Dataset name	Machine learning models	Evaluation metrics	Main findings
1	Plant CLEF	Support vector machines [3, 8]	Precision, recall, accuracy, and $F1$-score	Support vector machines give an accuracy of 88% for the classification of plant species
		Decision tree [4], random forest [2, 5]	Precision, recall, accuracy, and $F1$-score	The decision tree provides an accuracy 83% for the classification plant species
		Gradient boosting machines [43]	Precision, recall, accuracy, and $F1$-score	Gradient boosting machines have an accuracy of 85% in classifying plant species
		K-nearest neighbors [9]	Precision, recall, accuracy, and $F1$-score	The categorization plant species using K-nearest neighbors had an accuracy of 84%
		Ensemble methods (bagging) [6]	Precision, recall, accuracy, and $F1$-score	An accuracy of 86% is obtained by the bagging ensemble while categorizing plant species
		Random forest [2]	Precision, recall, accuracy, and $F1$-score	When diagnosing plant diseases, random forest achieved an accuracy of 93%
2	Leafsnap	Random forest [5]	Precision, recall, accuracy, and $F1$-score	Random forest achieved an accuracy of 82% for identifying plant species
		Support vector machines [8]	Accuracy in Top-1 and Top-5	Support vector machines have a top-1 accuracy 80% for classifying plant species
		AdaBoost [10]	Accuracy ranking one and five	AdaBoost achieved a top-1 accuracy for identifying plant species at 78%

(continued)

Table 2 (continued)

S. No.	Dataset name	Machine learning models	Evaluation metrics	Main findings
3	Flavia	Decision tree [4]	Accuracy, sensitivity, specificity	The decision tree achieved an accuracy of 78% for identifying plant species
		Random forest [5]	Accuracy, sensitivity, specificity	Random forest achieved an accuracy of 83% for identifying plant species
4	PlantVillage	Random forest [2]	Precision, recall, accuracy, and $F1$-score	94% accuracy is grasped by random forest while classifying plant diseases
		Decision tree [4]	Precision, recall, accuracy, and $F1$-score	92% accuracy was reached using the decision tree while classifying plant diseases
5	NWPU-RESISC45	Random forest [5]	Precision, recall, accuracy, and $F1$-score	The classification accuracy 86% was attained by random forest
		Naive Bayes [44]	Precision, recall, accuracy, and $F1$-score	For classifying plant species, Naive Bayes has an accuracy 80%
		Logistic regression [44]	Precision, recall, accuracy, and $F1$-score	For classifying plant species, logistic regression has an accuracy of 82%
6	AGR-FT	K-nearest neighbors [9]	Precision, recall, accuracy, and $F1$-score	K-nearest neighbors classified plant species with a 94% accuracy
7	Swedish Leaf	Logistic regression [44]	Precision, recall, accuracy, and $F1$-score	For classifying plant species, logistic regression has an accuracy of 81%

(continued)

Table 2 (continued)

S. No.	Dataset name	Machine learning models	Evaluation metrics	Main findings
8	Fossil Plant	K-nearest neighbors [9], decision tree [4]	Accuracy, sensitivity, specificity	K-nearest neighbors achieved an accuracy of 74% for identifying fossil plant species
9	Indian Pines	Random forest [5], AdaBoost [10]	Overall accuracy, average $F1$-score	Random forest achieved an overall accuracy of 87% for plant species classification

on overcoming these difficulties and creating machine learning models for plant categorization that are more precise and reliable. The field of plant classification using machine learning models is rapidly evolving, and there are several possible solutions and future guidelines for research in this field. Some of the possible solutions and future guidelines for plant classification using machine learning models are as follows.

Transfer learning. The issue of insufficient training data in plant categorization may be resolved through transfer learning. Transfer learning entails employing pre-trained models that have been polished on a smaller, labeled dataset after being imparted with broad features from a large dataset. Particularly for species with labeled data, this method can significantly increase the accuracy of machine learning models.

Hybrid models. The accuracy and robustness of plant categorization models can be increased by hybrid models that incorporate various machine learning methods. For instance, convolutional neural networks can be used in conjunction with decision trees/support vector machines to boost the accountability of deep learning models while keeping high accuracy.

Explainable AI. Explainable AI techniques can improve the interpretability of machine learning models. These techniques can help researchers to understand how machine learning models make predictions and identify the features and characteristics that are crucial for classifying plants.

Active learning. The effectiveness of plant categorization models can be increased by the use of active learning. Active learning may drastically minimize the number of training examples required while retaining high accuracy by repeatedly choosing the most informative data for labeling.

Integration with additional data sources. Adding additional data sources to machine learning models, such as satellite imagery or environmental data, can increase the precision and sturdiness of models for classifying plants. The precision of models for predicting plant species distributions can be increased, for instance, by including data on soil moisture or temperature.

In summary, there are several potential solutions and future directions for research in plant classification using machine learning models, including transfer learning, hybrid models, explainable AI, active learning, and integration with other data sources. These techniques may enhance the precision, resilience, and effectiveness of plant categorization models and have important applications in agriculture, biodiversity conservation, and environmental management.

6 Conclusion

This article has given a thorough description of how machine learning models may be used to classify plants. Primarily, the different machine learning algorithms and approaches for feature extraction and selection, as well as their advantages and limitations in the context of plant classification have been discussed Additionally, it has been noted how machine learning models could be used in a variety of industries, including agriculture, biodiversity protection, and management of the environment. The key challenges in classifying plants using machine learning models have been determined to be the availability and quality of training data as well as the interpretability of machine learning methods. Furthermore, the potential solutions and future directions for research in this field, including transfer learning, hybrid models, explainable AI, active learning, and integration with other data sources have been discussed. Overall, the potential impact of machine learning models on plant classification and related fields is significant. Additionally, future research directions include improving the interpretability of machine learning models, addressing the challenges of limited training data, and integrating machine learning models with other data sources have been addressed. Finally, there has been a strong emphasis on the necessity for study on the moral and societal effects of applying machine learning algorithms in plant classification and related domains. In conclusion remarks, it may be said that this work offers useful insights into how machine learning methods are currently utilized to classify plants, and it emphasizes the potential effects and prospective paths for further research in the area.

References

1. Bell A, Sejnowski T (1995) An information-maximization approach to blind separation and blind deconvolution. Neural Comput 7(6):1129–1159
2. Breiman L (2001) Random forests. Mach Learn 45(1):5–32
3. Fuentes M, Ritter A, Riquelme JC (2018) Plant species classification using support vector machines with leaf morphology descriptors. Comput Electron Agric 154:196–202. https://doi.org/10.1016/j.compag

4. Ahmad F, Lee HJ, Lee S (2018) Plant species classification using decision tree and ensemble learning methods. J Ambient Intell Hum Comput 9(2):337–349. https://doi.org/10.1007/s12652-016-0457-5
5. Liu Y, Ma F, Cao L, Xu R, Liu X (2020) Identification of tea plant cultivars based on near-infrared spectral data using random forest. Food Anal Methods 13(2):390–397
6. Li Z, Li D, Li D, Li J, Zhang L, Wu Y (2019) Plant species recognition using multi-feature fusion and ensemble learning. J Vis Commun Image Represent 62:33–41
7. Goëau H, Bonnet P, Joly A (2021) Overview of PlantCLEF 2021: cross-domain plant identification. In: CLEF (working notes), pp 1422–1436
8. Kaur J, Singh G, Kumar S, Gupta H (2018) Classification of potato plant diseases using support vector machine. Int J Comput Appl 180(32):17–20
9. Saleem G, Akhtar M, Ahmed N, Qureshi WS (2019) Automated analysis of visual leaf shape features for plant classification. Comput Electron Agric 157:270–280
10. Mahajan S, Raina A, Gao XZ, Pandit AK (2021) Plant recognition using morphological feature extraction and transfer learning over SVM and AdaBoost. Symmetry 13(2), Art. No. 356
11. Wang J, Liu Y, Zhang Y, Yang L, Wang J (2019) Classification of tea plant cultivars based on morphological characteristics using a decision tree. J Agric Sci 11(8):197–203
12. Gajjar VK, Nambisan AK, Kosbar KL (2022) Plant identification in a combined-imbalanced leaf dataset. IEEE Access 10:37882–37891
13. Kumar N, Belhumeur PN, Biswas A, Jacobs DW, Kress WJ, Lopez IC, Soares, Leafsnap JVB (2012) A computer vision system for automatic plant species identification. Int J Comput Vis 98(2):254–272
14. Senthilkumar S, Rajesh R (2010) Plant leaf recognition based on shape and texture features. Int J Comput Theory Eng 2(3):376–380
15. Lalitha B, Raja KB (2014) A comparative study of texture features for leaf classification. Procedia Comput Sci 46:1032–1039
16. Ishikawa T, Hayashi A, Nagamatsu S, Kyutoku Y, Dan I, Wada T, Kochi N (2018) Classification of strawberry fruit shape by machine learning. Int Arch Photogramm Remote Sens Spat Inf Sci 42:463–470
17. Azlah MAF, Chua LS, Rahmad FR, Abdullah FI, Wan Alwi SR (2019) Review on techniques for plant leaf classification and recognition. Computers 8(4):77
18. Amlekar MM, Gaikwad AT (2018) Plant classification using image processing and neural network. Data Manag Anal Innov 839:375–384
19. Ghosh S, Singh A, Jhanjhi NZ, Masud M, Aljahdali S (2022) SVM and KNN based cnn architectures for plant classification. Comput Mater Continua 71(3)
20. Uddin AH (2022) Plant-species recognition from leaf and bark images using neural networks
21. Amlekar MM, Gaikwad AT (2019) Plant classification using image processing and neural network. In: Proceedings of data management, analytics and innovation. Advances in intelligent systems and computing. Springer, Singapore, vol 839
22. Ariyapadath S (2021) Plant leaf classification and comparative analysis of combined feature set using machine learning techniques. Traitement du Signal 38(6):1587–1598
23. Borkatulla B, Ferdous J, Uddin AH, Mahmud P (2023) Bangladeshi medicinal plant dataset. Data Brief 48:109211
24. Panigrahi KP, Das H, Sahoo AK, Moharana MSC (2020) Leaf disease detection and classification using machine learning algorithms. In: Progress in computing, analytics and networking: proceedings of ICCAN, Springer, Singapore, pp 659–669
25. Khatri A, Agrawal S, Chatterjee JM (2022) Wheat seed classification: utilizing ensemble machine learning approach. Sci Program
26. Yan J, Wang X (2022) Unsupervised and semi-supervised learning: the next frontier in machine learning for plant systems biology. Plant J 111(6):1527–1538
27. Rahman A, Abdullah HM, Tanzir MT, Hossain MJ, Khan BM, Miah MG, Islam I (2020) Performance of different machine learning algorithms on satellite image classification in rural and urban setup. Remote Sens Appl Soc Environ 20:100410

28. Chulif S, Lee SH, Chang YL, Chai KC (2023) A machine learning approach for cross-domain plant identification using herbarium specimens. Neural Comput Appl 35(8):5963–5985
29. Patel I, Patel S (2019) Flower identification and classification using computer vision and machine learning techniques. Int J Eng Adv Technol (IJEAT) 8(6):277–285
30. Ghosh S, Singh A (2022) The analysis of plants image classification based on machine learning approaches. In: Emergent converging technologies and biomedical systems: select proceedings of ETBS 2021. Springer, Singapore, pp 133–148
31. Ganatra N, Patel A (2020) A multiclass plant leaf disease detection using image processing and machine learning techniques. IJET 11(2):1082–1086
32. Tan L, Lu J, Jiang H (2021) Tomato leaf diseases classification based on leaf images: a comparison between classical machine learning methods. AgriEngineering 3(3):542–558
33. Nesteruk S, Shadrin D, Pukalchik M, Somov A, Zeidler C, Zabel P, Schubert D (2021) Image compression and plants classification using machine learning in controlled-environment agriculture. IEEE Sens J 21(16):17564–17572
34. Satti V, Satya A, Sharma S (2013) An automatic leaf recognition system for plant identification using machine vision technology. Int J Eng Sci Technol 5(4):874
35. Cleuziou G, Dias G (2015) Learning pretopological spaces for lexical taxonomy acquisition. In: Machine learning and knowledge discovery in databases: PKDD 2015, Porto, Portugal, September 7–11, 2015, proceedings, part II 15. Springer, pp 493–508
36. Singh D, Jain N, Jain P, Kayal P, Kumawat S, Batra N (2020) PlantDoc, a dataset for visual plant disease detection. In: Proceedings of the 7th ACM IKDD CoDS and 25th COMAD, pp 249–253
37. Chopra H, Singh H, Bamrah MS, Mahbubani F, Verma A, Hooda N, Singh AK (2021) Efficient fruit grading system using spectrophotometry and machine learning approaches. IEEE Sens J 21(14):16162–16169
38. Hati AJ, Singh RR (2021) Artificial intelligence in smart farms: plant phenotyping for species recognition and health condition identification using machine learning. AI 2(2):274–289
39. Shailendra R, Jayapalan A, Velayutham S, Baladhandapani A, Srivastava A, Kumar Gupta S, Kumar M (2022) An IoT and machine learning based intelligent system for the classification of therapeutic plants. Neural Process Lett 54(5):4465–4493
40. Chulif S, Lee SH, Chang YL, Chai KC (2023) A machine learning approach for cross-domain plant identification using herbarium specimens. Neural Comput Appl 35(8):5963–5985
41. Chulif, Sophia, Sue Han Lee, Yang Loong Chang, and Kok Chin Chai.: "A machine learning approach for cross-domain plant identification using herbarium specimen. Neural Comput Appl 35(8):5963–5985
42. Leminen Madsen S, Mathiassen SK, Dyrmann M, Laursen MS, Paz LC, Jørgensen RN (2020) Open plant phenotype database of common weeds in Denmark. Remote Sensing 12(8):1246
43. McCarty DA, Kim HW, Lee HK (2020) Evaluation of light gradient boosted machine learning technique in large scale land use and land cover classification. Environments 7(10):84
44. Seka D, Bonny BS, Yoboué AN, Sié SR, Adopo-Gourène BA (2019) Identification of maize progeny genotypes based on two probabilistic approaches. 1:9–13

Critical Analysis of the Utilization of Machine Learning Techniques in the Context of Software Effort Estimation

Chetana Pareta, Rajeev Mathur, and A. K. Sharma

1 Introduction

The software has been the most expensive component of computer system development in recent years. Cost estimate is one of the period and the most difficult aspects of project management. The goal of this tool is to give a true account of estimate assets needed for software development activities. The process of estimating software involves determining the amount of work that will be needed, creating preliminary project timelines, and ultimately estimating the project's total cost. Furthermore, the application of Software Cost Estimation will have an effect on the project's timeframes, team skills, team cohesion, productivity advantage, complexity, dependability, quality, and many other factors. Every stage of the Software Development Life Cycle (SDLC), including implementation, must include a full-time estimating procedure. Its purpose is to stay up with change during the development process. System analysts use software cost estimation to get an idea of the resources needed for a certain software project and their timescales. Size, time, and effort are all essential considerations for estimating costs [1]. Existing methods for evaluating effort encompass include computational and non-algorithmic methods in large numbers. The enthusiasm surrounding effort estimations is increased in software development technologies, particularly in the early stages. There have been numerous estimating models proposed over the years. Expert prediction, mathematical models, such as fuzzy-logical prediction, function point techniques, artificial neural network-based approaches, Bayesian system approaches, decision tree processes, and empirical

C. Pareta · R. Mathur
School of Engineering and Technology, Jaipur National University, Jaipur, India
e-mail: director.soet@jnujaipur.ac.in

A. K. Sharma (✉)
Department of Creative Technologies and Product Design, National Taipei University of Business, Taipei City, Taiwan
e-mail: drarvindkumarsharma@gmail.com

M. S. Kaiser et al. (eds.), *ICT: Smart Systems and Technologies*, Lecture Notes in Networks and Systems 878, https://doi.org/10.1007/978-981-99-9489-2_18

approaches that rely on analogies projects were divided into a number of groups based on the primary elements of their designs. Although it is simple to use and can compute effort in person-months for a project at different phases, Constructive Cost Model (COCOMO) by Boehm is one of the most widely used empirical cost modeling methodologies [2].

The objective of calculating software costs is to:

a. Describe the resources needed to create, validate, and approve the software product and how they will be used to do these activities.
b. A guarantee of the software's quality.
c. Do your best in determining the level of fragility and jeopardy the above measurement presents.
d. Displaying physical development.
e. Promoting environmental support.
f. Assistance with system-level testing of software, including amusement as well as growth use cases.
g. Administrative and support costs.
h. Complete approval and review.
i. Additional audit fees or related costs.

The cost projection is important because:

a. Assist in the organization and emphasis of advancement projects according to a general field-tested approach.
b. Employed to decide which resources should be concentrated on the work and how effectively these resources will be employed.
c. Used to evaluate changes' effects and encourage re-consideration.
d. When assets are better matched to true requirements, activities can be easier to manage and govern [3].

2 Software Effort Estimation Methods

The order in which software effort estimating techniques are used is implied in Fig. 1.

3 Software Effort Estimation Methods

The order in which software effort estimating techniques are used is implied in Fig. 1.

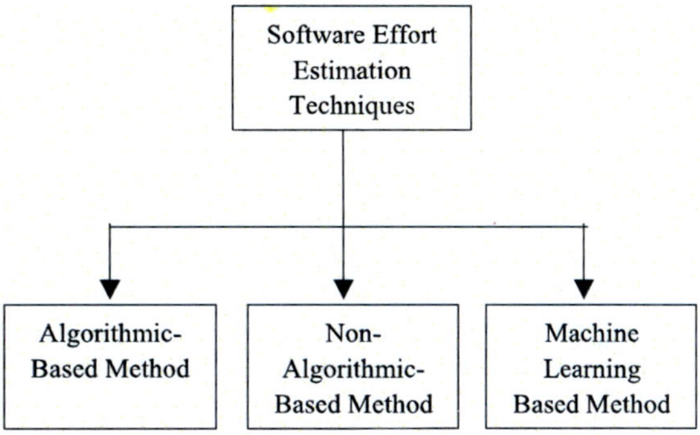

Fig. 1 Software effort estimation techniques

3.1 Algorithmic-Based Methods

The program's cost is calculated using algorithms and an equation. The formula is created by combining models based on cost components. While creating models, the quantifiable technique is employed; the algorithmic approach entails presenting a few mathematical formulae for software estimations. These mathematical estimates, which are based on thorough study and accurate data, take into account the number of functions that must be carried out, the amount of lines of source code (SLOC), as well as additional cost concerns such as speech, development strategy, skill levels, evaluations of risk, and so forth. After a thorough analysis of algorithmic approaches, many models, including COCOMO models, Putnam models, and function point-based models, have been created [4].

(a) **COCOMO Model**

Boehm put forward the Constructive Cost Estimation Model (COCOMO) concept in 1981. A cost estimation approach for LOC-based programming projects. COCOMO is one of the most widely used software estimating models at the moment. COCOMO forecasts the effort and timeline of a software product based on the software's size. COCOMO has also considered factors such as project features, equipment, and production, among others. The business system, the executive's financing system, and the stock administration system are a few instances of programming built on this methodology. COCOMO-II takes into account diverse software development, reuse, and other methodologies. The entire software is partitioned into modules in this paradigm. Accounting sheets and report generators are examples of actions in light of this approach [5].

(b) **Function Point Analysis**

Function-based evaluation provides a different approach for estimating the overall scope and level of difficulty of a piece of software based on the features it offers the user. A function point technique has been used in the cost estimate models ESTI-MACS and SPQR/20. Albrecht originally gave this estimate based on the program's capabilities. By counting certain categories (in terms of organization or handling logic), the overall number of function points is calculated [6].

Objectives of FPA

- It assesses how much utility a client expects and receives.
- It investigates software creation and maintenance without the use of execution innovation.
- It should be a consistent statistic used across projects and organizations.
- It should be simple enough to reduce the overhead associated with the measuring technique.

(c) **Putnam Model (SLIM)**

Putnam's focus acts as a basis for the Rayleigh allocation of project crew level versus time in the Software Life Cycle Model (SLIM). It combines the great majority of well-known size estimate techniques, including ballpark techniques, source directives, capability focuses, and so forth. Estimates are made for the project's work, schedule, and fault rate. In order to normalize the model and create an evaluation model of the order, data from previous initiatives must be collected and examined [7].

3.2 Non-algorithmic-Based Methods

Non-Algorithmic Model unlike Algorithmic approaches; tactics for this gathering rely on in-depth studies and speculation. Non-algorithmic processes need information on earlier projects that are comparable to the proposed venture, and the assessment steps in these techniques are often finished by looking up prior datasets.

(a) **Expert Judgment**

An expert's judgment on a particular matter or unidentified factor is referred to as their expert opinion. When estimating difficult, vague, or poorly understood problems, expert judgment is most useful in (1) circumstances when there are limited empirical facts and (2) these types of scenarios. The fact that expert judgment is so frequently used as a technique for software evaluation can be attributed to both of these qualities [8].

(b) **Top-Down Estimation**

The Macro Model is a top-down estimate technique. After estimating the project's overall cost using the software project's global properties, this approach separates the project into a number of elementary mechanisms or parts. In this technique, the Putnam model is used. When only global characteristics are known, the top-down strategy is better suited to initial estimations of costs. Top-down approach is quite useful in the early phases of software cost evaluation because there isn't much thorough information available [9].

(c) **Bottom-up Estimation**

The bottom-up estimation method is used to estimate the cost of each software element, and the outcomes are then merged to determine the project cost. It makes an effort to estimate a system utilizing the information acquired about the links between the smaller program parts.

(d) **Algorithmic Method**

Several mathematical equations are intended to be produced using the algorithmic technique for software estimates. These numerical calculations account for inputs like the number of required functions, the source lines of code (SLOC) count, as well as other elements that affect costs. They are founded on historical facts and study [10].

(e) **Analogy-Based Estimation**

A subset of case-based reasoning is analogy-based cost estimates (AB). Cases are thought of as incomplete notions of occurrences in space and time. Since AB cost estimates of software projects are based on prior experience from comparable projects, and comparisons are resolved by equating the projects' major features and traits, this is the main reason why they are visible. The dissimilar effect or premium of a project's multiple features, nevertheless, is an essential component of the AB approach that hasn't yet been completely taken into consideration. The cost estimation of software projects incorporates the ethics of real efforts and values. System and component levels can both benefit from the usage of AB estimation methodologies. In some aspects, AB is a suitable sort of EJ because the specialist regularly looks for pertinent circumstances and clarifies the concepts.

The following are the steps for employing estimate using AB:

1. The intended project is currently under wraps.
2. Choosing a precise similar finished project whose characteristics are kept in a historical database [11].

3.3 *Machine Learning Methods*

Machine learning is a method used to develop automated analytical models. The foundation of artificial intelligence (AI) is the notion that machines, including machines and other programmed components, may learn through data. These methods can make predictions about the future based on past data and learn from it. Software cost estimation frequently uses machine learning techniques including support vector regression, regression trees, fuzzy logic, evolutionary algorithms, Bayesian networks, and fuzzy logic. To increase estimation accuracy, researchers suggested numerous machine learning software cost-estimating models [2].

(a) **Artificial Neural Network (ANN)**

ANN stands for artificial neural network, usually referred to as a neural network, is based on the idea of biological neurons. A neuron is a cell-like structure found in the brain. To comprehend a neural network, one must first comprehend how a neuron functions. A neuron is made up of four major components. Dendrites, nuclei, soma, and axon are the four types. These dendrite cells pick up electrical impulses. Soma transforms the electrical signal. The axon transmits the process output to the terminals of the dendrites, which then transmit it to the following neuron. The nucleus is the neuron's beating heart. The neural network is the interconnection of neurons that allows electrical impulses to move across the brain [12].

(b) **Bayesian Networks**

A directed acyclic network that represents a certain domain is known as a Bayesian network. The arcs linking the graph's nodes, which each represent a domain variable, illustrate the connections between those variables. The Bayesian technique is used to assign a weight to each arc [13].

$$\text{Bayes Theorem: } P(A|B) = \frac{p((B|A)).P(A)}{P(B)}$$

A mathematical instance of this is a Bayesian network of a probability distribution over a range of qualities that is used to build a model of the problem area. The Bayesian network representation has the benefit of being a compact representation that may be used for a variety of naturally occurring and complicated issue areas. Additionally, all of the direct relationships between the features of the issue domain are explicitly established before a Bayesian network model is constructed [14].

(c) **Fuzzy Logic**

Fuzzy logic is a sophisticated problem-solving tool that mimics human thinking by encompassing all intermediate options between the numeric values 0 and 1. Furthermore, fuzzy logic provides thinking flexibility by considering all available information and making the best feasible conclusion based on this information [15].

(d) **Support Vector Regression (SVR)**

Regression using support vectors (SVR), the name given to the regression idea in support vector machine (SVM), was introduced in 1996. Since the cost function excludes training points beyond the border when building the model, the SVR focused on a subset of training data. SVRs are more accurate in executing the computation, which is shown to be used into pieces, minima from the surroundings, tiny circumstances during the arrangement, and limiting the control from numbered support vectors, uppermost of the edge, and so on. To more accurately evaluate how well machine learning algorithms anticipate the future, the dataset is segmented into training/test ratios. The influence of training and testing ratios is measured depending on their performance. The model suffers as a result of outliers [16]. SVMs get a separating hyper plane in the simplest case of binary classification, when the margin is maximized. The margin indicates how different each class's criteria are. According to PAC learning, maximizing the margin produces strong generalization abilities. Additionally, SVMs may divide data nonlinearly with little computing cost thanks to the kernel approach. There are probably non-linear interactions among all the types of data seen in the actual world. SVMs have several uses, one of which is the prediction of micrometeorological data. Support vector regression (SVR), on the other hand, uses the same approach as SVMs with the best generalization capability [17].

(e) **Regression Tree**

Every node gets fitted with a regression model to get anticipated *Y* values. With the exception of the *Y* variable's ability to take on ordered values, regression trees are identical to classification trees. A few years prior to THAID, AID20 presented the first regression tree methodology. With the node impurity equal to the sum of squared deviations from the standard deviation and the node predicting the sample mean of *Y*, both the AID and CART tree of regression techniques use Algorithm 1. The result follows the piecewise constant model. Although these models are straightforward to use, their forecast accuracy frequently falls short of that of smoother models. The difficulty in applying this method to piecewise linear models is due to the necessity of fitting for each projection split, there are two linear models—one for every little node. Software Engineering, production management, fisheries management, criminology, and information technology are all examples of occupations. The decision tree uses a non-parametric technique and is unaffected by probabilistic models. It successfully handles large dimensionality by prioritizing variables. This decision tree or regression tree has a root, leaves, branches, and nodes. These trees are continually growing to facilitate the explanation, understanding, and visualization of outcomes. CART is mostly based on impurity metrics such as Entropy or the Gin index. The CART procedure's essential components are selection and halting rules. Binary recursive partitioning was used to create this regression tree. Recursive partitioning is an iterative process that divides data into discrete divisions or branches. Each dividing division with the lowest mean of squares among all partitions is chosen [19]. The regression tree model in this study is built using CART. Starting with the complete dataset, subsets are continuously split into two descendant subsets based on independent factors to create a binary tree. The objective is to create data subsets that are as response variable homogenous as is practical [20].

(f) **Genetic Algorithm**

Genetic algorithms are founded on the laws of evolution and inheritance (GAs), which are mostly search-based algorithms, are founded. The area of computing known as evolutionary computation, of which GA is a part, is quite vast. For any imaginable situation, we provide a broad choice of GA options [21]. Charles Darwin's theory concerning natural selection serves as the foundation for genetic algorithmic algorithms (GA), a kind of heuristic utilized to solve optimization problems. The following list of the most common questions and answers about the GA process:

Step 1: Establish the population.
Step 2: Estimate each chromosome's fitness role in the population.
Step 3: Reproduce chromosomes according to their fitness.
Step 4: Conduct a study on crossover and mutational processes.
Step 5: Go on to step 2 if the necessary pausing circumstances apply [22].

4 Literature Review

In 2022, Ali et al. [23] proposed a unique approach based on Gaussian Process Regression to determine the cost of adaptable software. Training statistics using all three types of batteries—Lithium-Ion, Nickel-Metal Hydride, and Lithium-Polymer—were collected in a climate-controlled lab environment with IEEE 802.15.4-based driver loads operating at temperatures that vary. Using load profiles based on the IEEE 802.15.4 protocol, he explored the discharging graphs showing the temperature-dependent curves of different commercial battery chemistries. Over a wide range of temperatures, voltage, storage ability, and software cost of the battery discharge aspects were estimated. It was suggested that the battery's usable energy be determined by a free software-based cost-estimating technique.

In 2019 Asheeri et al. [24]; proposed an item cost evaluation model based on AI. To anticipate the item cost before all other steps, several AI predictions were used for two public datasets. The results proved how AI approaches may be used to accurately anticipate programming costs. The datasets might be linked to the survey to get more points and groupings in the information sources, which would be represented in an unequaled evaluation and more exact findings. Furthermore, additional ML calculations might be undertaken and linked with upcoming assessments to cover all feasible AI systems.

In 2021 Shafiq et al. [25]; introduced the use of AI at various phases of the programming progress life cycle. The primary goal of this piece is to look at the connections between the stages of the programming progress life cycle and various AI tools, techniques, and types. They attempted, to a limited extent, a complete investigation to determine whether or if AI tends toward certain stages or perhaps specific methodologies. They also offered an order conspire that includes broad AI applications for programming as far as SDLC stages. The arrangement demonstrated the crucial focus of scientists on uncommon stages. This perspective was one of the review's major commitments. This focuses also revealed that the type of crucial examinations in the field of ML and SE were proof-based concerning.

In 2021 De Carvalho et al. [26]; established important aspects for effort evaluation and the findings showed that the ELM model outperformed the other models in the composition for surveying programming plan effort. Thus, including the machine learning technique in the task assessment cycle can increase the odds of coming out on top in terms of the accuracy of time measures and the cost of the endeavor. It was considered that a critical cycle in Software Engineering is the precise and reliable appraisal of the work required to finish an endeavor. As a result, utilizing the approach of contemplating AI increases the endeavor's possibilities of showing to be the best, decreasing the time and expenses of the endeavor.

In 2020 Zhang et al. [27]; introduced DPS alters the search and double-dealing stages and chooses the best cost depending on the price and job duration relationships. Our learning-based calculation extends the vendor's edge when compared to benchmark evaluation strategies and accomplishes a sub-straight lament with both the time skyline and the total amount of effort. In-depth analyses back up the feasibility of DPS. We showed that our calculation produces a sub-straight lament with the time period and the overall amount of effort. Massive scaled replication efforts using real data also point to superior computer execution compared when compared to the best estimation components.

In 2016 Gabrani et al. [28]; centered on the close evaluation of alternative non-algorithmic procedures utilized for analyzing the product exertion by observational assessment of five distinct transformational learning computations. The correctness of these calculations was determined, and the manner these computations behaved in terms of the amount and kind of information was examined. Each of the five approaches was tested on three distinct datasets. The suggested findings were compared to various AI methodologies such as SVR, ANFIS, and so on. It was discovered that advanced learning calculations produced more precise results than AI calculations. It may be rehashed for other new ways until they find a one of a kind procedure that can provide precise and accurate evaluation for a wide variety of datasets.

In 2016 Moharreri et al. [29]; recommended enhancing the manual Planning Poker in Agile. This real effort time differentiates the auto-estimate utilization traits from adaptable tale cards. The approach's validity was examined against other AI computations for exertion expectation. It was proven that some AI tactics performed better in the finishing stages of the job than Planning Poker gauges. The accuracy, materiality, and utility of this assessment approach were assessed, and the findings were shown in the context of a real-world situation. They recommended employing larger datasets and characteristics (engineers' socioeconomics, narrative criticality, and other framework and system) that were not included in this investigation.

In 2020 Martınez et al. [30]; introduced the use of the automated assessment model generator framework, which employs AI procedures to compare the exactness of these models to traditional assessment strategies utilizing a global dataset and an organization's internal dataset. It demonstrated the examination of the outcomes obtained by utilizing the computerized assessment model generator framework. The first case showed the framework's value by outperforming the traditional assessment method with the maximum precision in terms of quality rules thanks to one of the models (LR). This inquiry was conducted in two situations.

In 2018 Mustafa et al. [31]; designed to verify both the overall legality of information mining and the method used particularly for programming gauges. It may entail looking at how well data extraction algorithms can alter and enhance the drivers and advantages of the COCOMO model, which could help to increase the model's evaluation accuracy whilst lowering the edge of its projected inaccuracy.

In 2016 Singh et al. [32]; recommended method for determining product costs involves several challenging processes that unevenly gather data (ex. examination fees, equipment accessibility fees, and labor costs). Learning issues caused by class inconsistency have been tackled, such as those generated by datasets with an overly high number of minority models and classes with more occurrences than others. The following is a list of the most recent events that have been brought to the attention of the public, as well as the most recent events that have been brought to the public's attention: the upcoming events. We researched and spoke about "dynamic inspecting with "dynamic inspecting with MLP and Cart" and "MLP". On datasets, the suggested framework, DyS-MLP-CART, fared better than the current framework, DyS-MLP.

5 Proposed Methodology

In this section discusses the proposed methodology using the flowchart that has been given below. A few machine learning algorithms are used in building prediction models on the basis of evaluation criteria.

The flowchart of our proposed methodology is shown in Fig. 2.

Fig. 2 Proposed methodology

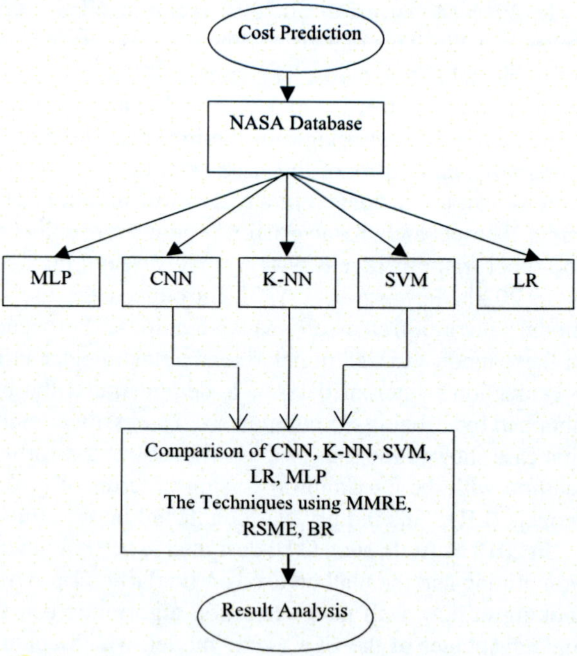

5.1 Techniques Applied

Brief overview of our machine learning algorithms discuss below:

(a) **Multilayer Perceptron (MLP)**

It is the supervised learning neural network that is utilized the most, with input flowing only in one direction and no loops. Learning the optimized bias value () for the optimized function *f* (), which transforms the input into a desired output, is the main goal. When there is a difference between the predicted and actual output, the MLP adjusts the connection weights using a back propagation mechanism. Their primary applications are in the optimization of financial, transportation, fitness, and energy concerns [33].

(b) **Cascade Neural Network (CNN)**

The input–output linkages in CNN are straight and are first trained using the rapid prop approach; there are no hidden units in CNN. Adding a first hidden unit from a set of potential units to the network, it is simultaneously and independently trained using distinct random beginning weights. Until there is no longer any noticeable error reduction, this procedure is repeated. The source of weights for a covert component is initially calculated and then while the weights remain frozen for the unit that results is retrained. A cascaded structure is created as a result of repeating this procedure to create each, which gets input links from the inputs as well as to each successive hidden item that came before it [34].

(c) ***k*-Nearest Neighbors (KNN)**

The method predicts the category of a new sample using samples whose categories were known at the time of classification; hence it is an instance-based reasoning methodology. If *K* is equal to 1, the nearest neighbor category, as established by the NN method is the sample category to be split [35].

(d) **Support Vector Machines (SVMs)**

In both research and practice, support vector machines have been employed successfully. SVMs have been used for a variety of regression analysis and pattern recognition problems, as well as to anticipate, estimate dependencies, and build intelligent robots. The generalization of the Structural Risk Elimination Theory (SRM) concept, i.e., the method's foundation in the ensured risk boundaries of mathematical learning theory, allows SVMs to accommodate very large feature spaces [36].

(e) **Logistic Regression (LR)**

Logistic regression, often known as either logistic modeling or the model of logistic is a popular multidimensional technique for predicting dichotomous outcomes. Because it is perfect for models that deal with decision-making problems. It is commonly used in statistical studies that are published in the literature pertaining to economics and finance [37].

(f) **Random Forest (RF)**

The Random Forest, that Leo Breiman invented, is a group of un-pruned regression or categorization trees produced from a random sample of the training data. Over the induction process, features are randomly selected. The forecast is created by combining the set of recommendations (by majority vote for designation, or by averaged for prediction) [38].

(e) **SMOreg**

When doing regression, SMOreg uses a vector model machine, and when determining parameters, it employs a variety of techniques. The method restores the values that are absent and changes nominal properties into numerical ones [39].

5.2 *Performance Parameters*

To compare the two models that was used:

- **Mean Magnitude of Relative Error (MMRE)**

Its abbreviation is MMRE or Mean Magnitude of the Relative Error, and it provides a unique method for determining the size of all estimations of relative errors. It was developed using the center estimate of more than *N* items from the "Test" set. The relative error is the overall misalignment in the observations assigned by its actual respect [40].

$$\text{MMRE} = \frac{1}{n}\sum_{i=1}^{n}\frac{\text{Predicted Value} - \text{Actual Value}}{\text{Actual Value}}$$

- **Root Mean Square Error (RMSE)**

The term refers to the Roots Mean Square Error. The difference between the attributes of the thing being shown and evaluated that really exist and the predictions provided by an algorithm or estimator is frequently examined using this technique. Simply put, it is the mean square error's square root, as can be seen in the example given below [40].

$$\text{RMSE} = \sqrt{\frac{1}{n}\sum_{i=1}^{n}(\text{Actual Value} - \text{Predicted Value})^2}$$

- **Balanced Relative Error (BRE)**

It is calculated by multiplying the absolute size of the mistake by E, which is equal to the lesser of the anticipated voltage and the actual voltage.

$$\mathrm{BRE} = \left| \frac{C - \hat{C}}{\min\left(c, \hat{C}\right)} \right|$$

where the total number of completed initiatives throughout all metrics is (), the projected cost is (), and the actual price is (C). Additionally, the scale with the lowest value for each scale utilized in this study had the best results [41].

- **R^2 (Correlation Coefficient)**

Regression analysis frequently uses R^2 which is characterized as the proportion understood to overall variance, or variation, or

$$R^2 = \frac{\Sigma E Ę^2}{\Sigma T Ę^2}$$

R^2 values usually signify predicting mistakes relative to a standard, the mean. R^2 varies from one to one, and as this is a ratio, it notifies us of the proportion of the total variation—or mistakes square—explained by the forecasting approach in reference to the mean. The main benefit of R^2 is that it is a straightforward and understandable proportional metric. A drawback of this method is that the reference is the mean, rendering it inappropriate when one observes a substantial pattern in the data and necessitating possibilities, such as Thiel's U-Statistic, which are more suitable for records with a strong trend. Although R^2 is frequently employed in regression analysis, forecasting has not adopted it [42].

- **Mean Absolute Error (MAE)**

MAE is the acronym for the standard deviation across all exact errors.

$$\mathrm{MAE} = \frac{\sum_{i=1}^{n} |y_i - x_i|}{n}$$

y_i	Prediction Value.
x_i	Actual Value.
Prediction error	Actual value-Predicted value.
Absolute error	"Prediction error".
MAE	It is a median of all measurement errors [43].

- **Relative Absolute Error (RAE)**

Absolute relative error a naïve algorithm's remainder or median error is expressed as a ratio to the prediction error using the Relative Absolute Error (RAE) measure. When the proposed approach performs better compared to the naïve method, the calculation outputs a value less than 1. “P” is the expected value in Equation, and “A” denotes the genuine value.

$$\text{RAE} = \frac{\left[\sum_{i=1}^{n}(P_i - A_i)^2\right]^{1/2}}{\left[\sum_{i=1}^{n} A_i\right]^{1/2}}$$

- **Root Relative Squared Error (RRSE)**

This divides overall ZeroR, on the fault through the Mean absolute error is an algorithm that is a classifier that just considers the quantity that occurs most frequently rather than taking into account any predictors. This yields the Root relative error or squared error [44].

$$\text{RRSE} = \sqrt{\frac{\sum_{i=1}^{n}(y_i - \hat{y}_i)^2}{\sum_{i=1}^{n}(y_i - \bar{y})^2}}$$

6 Conclusion

Cost and labor estimation is essential to all forms of project management. Early in the software development life cycle, accurate cost effort estimation is used to draft project plans and budgets. Cost effort estimation remains a challenging topic that compels researchers to investigate and evaluate various solutions. We provided a method for evaluating the efficacy of the machine learning algorithm. After examining the application of machine learning in estimating the cost of any entity, the motivation for estimating the cost of software, in terms of effort put in, in software development initiatives is discussed. Utilizing various algorithms, such as the K-Nearest Neighbors method, the Cascade Neural Network technique, the Support Vector Machine, the Multilayer Perception, and Logical Regression, additional research will be conducted in this domain. This research will aid all software industries in estimating the time required to develop a software project. In the future, more datasets will be incorporated into the research works in order to have a broader perspective and more variety in the dataset inputs, which will reflect better software estimation and a significantly more precise result.

References

1. Hlaing KS et al (2019) A study of tools, methods, and techniques of software cost estimation. IJARIIE 2019:1–4

2. Chirra SMR et al (2019) A survey on software cost estimation techniques. J Softw Eng Appl 2019:226–248
3. Rajkumar G et al (2013) A systematic review of cost estimation models. J Glob Res Comput Sci 2013:1–4
4. Jayapratha K et al (2017) A relative study on cost estimation techniques. Int J Sci Res Mod Educ 2017:1–4
5. Raj M et al (2020) COCOMO and the COCOMO-2 model—a case study. Int Res J Eng Technol 2020:1–4
6. Keim Y et al (2014) Software cost estimation models and techniques: a survey. Int J Eng Res Technol 2014:1–6
7. Khuttan A et al (2012) Proposing a new estimation model based on Putnam model. Int J Sci Res 2012:1–3
8. Han WJ et al (2015) Algorithmic based and non-algorithmic based approaches to estimate the software effort. Int J Multimedia Ubiquitous Eng 2015:1–14
9. Shweta KR et al (2021) Comparative analysis of algorithmic, non-algorithmic, and machine learning models for software cost estimation: a survey. Int Res J Mod Eng Technol Sci 2021:1–11
10. Balaji N et al (2013) Software cost estimation using function point with non-algorithmic approach. Glob J Comput Sci Technol Softw Data Eng 2013:1–7
11. Khan B et al (2020) Software cost estimation: algorithmic and non-algorithmic approaches. Int J Data Sci Adv Anal 2020:1–6
12. Dey A et al (2016) Machine learning algorithms: a review. Int J Comput Sci Inf Technol 2016:1–6
13. Nudelman Z et al (2018) Using Bayesian networks and machine learning to predict computer science success. Int J Comput Intell Inform 2018:1–16
14. Ramaswami M et al (2012) Student performance prediction modeling: a Bayesian network approach. Int J Comput Intell Inform 2012:1–5
15. Khairy G et al (2023) A proposed gasification framework using sentiment analysis and fuzzy logic in higher education. J Theor Appl Inf Technol 2023:1–10
16. Arumugam J et al (2022) A comparative study of bit coin price prediction using SVR and LSTM. Int J Creative Res Thoughts 2022:1–8
17. Kaneda Y et al (2016) Sliding window-based support vector regression for predicting micrometeorological data. Elsevier, 1–9
18. Loh W-Y et al (2011) Classification and regression trees. WIREs Data Mining Knowled Disc 2011:14–23
19. Jadhav P et al (2020) A comparative study of linear regression and regression tree. In: International conference on communication and information processing, 1–10
20. Tran VT et al (2010) A classification and regression trees (CART) model of parallel structure and long-term prediction prognosis of machine condition. Structural Health Monit 2010:1–12
21. Lambora A et al (2019) Genetic algorithm—a literature review. In: International conference on machine learning, big data, cloud and parallel computing, 1–5
22. Carbonne Y et al (2015) Genetic algorithm as machine learning for profiles recognition. In: 7th international conference on evolutionary computation theory and applications, 157–165
23. Ali O et al (2022) On-line WSN SoC estimation using Gaussian process regression: an adaptive machine learning approach. Sci Direct 61:9831–9848
24. Asheeri MMA et al (2019) Machine learning models for software cost estimation. IEEE, 1–6
25. Shafiq S et al (2021) A literature review of using machine learning in software development life cycle stages. IEEE, 140896–140921
26. De Carvalho HDP et al (2021) Extreme learning machine applied to software development effort estimation. IEEE, 92676–92688
27. Zhang X et al (2020) Dynamic pricing and placement for distributed machine learning jobs. IEEE, 1–9
28. Gabrani G et al (2016) Effort estimation models using evolutionary learning algorithms for software development. IEEE, 1–6

29. Moharreri K et al (2016) Cost-effective supervised learning models for software effort estimation in agile environments. IEEE, 1–6
30. Martnez JIS et al (2020) Analysis of automated estimation models using machine learning. IEEE, 1–7
31. Mustafa AB et al (2018) Predicting software effort estimation using machine learning techniques. IEEE, 1–8
32. Singh R et al (2016) Dynamic sampling approach with MLP and cart. IEEE, 1–5; Grid. IEEE, 2014, 1–9
33. Naskath J et al (2022) A study on different deep learning algorithms used in deep neural nets: MLP SOM and DBN. Springer, 1–24
34. Nechyba MC et al (1997) Cascade neural networks with node-decoupled extended Kalman filtering. IEEE, 1–6
35. Wang L et al (2019) Research and implementation of machine learning classifier based on KNN. IMMAEE, 1–6
36. Nayak J et al (2015) A comprehensive survey on support vector machine in data mining tasks: applications & challenges. IJDTA 2015:1–18
37. Strzelecka A et al (2020) Application of logistic regression models to assess household financial decisions regarding debt. Sci Direct 2020:1–10
38. Ali J et al (2012) Random forests and decision trees. IJCSI 2012:1–8
39. Assim M et al. Software defects prediction using machine learning algorithms. ICDABI, 1–6
40. Pareta C et al (2018) Predicting the accuracy of machine learning algorithms for software cost estimation. Springer, 1–9
41. Abdulmajeed AA et al (2021) Predict the required cost to develop software engineering projects by using machine learning. FISCAS 2021:1–12
42. Makridakis S et al (2018) Evaluating accuracy (or error) measures. INSEAD 2018:1–41
43. Dillen A et al (2020) Comparative analysis of machine learning and deep learning algorithms for software effort estimation. ICDIIS 2020:1–12
44. Jayasingh SK et al (2016) Comparison between J48 decision tree, SVM and MLP in weather forecasting. SSRG-IJCSE 2016:1–6

Review of Recent Research and Future Scope of Explainable Artificial Intelligence in Wireless Communication Networks

Vijay, K. Sebasthirani, J. Jeyamani, M. Gokul, S. Arunkumar, and Amal Megha John

1 Introduction

The ability to shed light on the decision-making processes of complex machine learning models makes explainable artificial intelligence (XAI) a hot topic. This article provides a comprehensive overview of recent research efforts in explainable artificial intelligence (XAI) for wireless communication networks. Section 2 discusses the importance of XAI in wireless networks, emphasizing the critical need for transparency and interpretability in AI systems. Section 3 provides a detailed overview of the existing XAI techniques specifically applied to wireless communication networks. The techniques examined include a range of approaches aimed at improving explainability and confidence in AI algorithms. Section 4 discusses the future scope and possible directions of XAI in wireless communication networks while highlighting new challenges and opportunities for further advances. At last, Sect. 5 concludes the article with a summary of the key findings and highlights the importance of XAI in facilitating transparent and reliable decision-making in wireless networks. The AI techniques related to visualization with case studies, Hypothesis Testing, and Didactic Statement hold the most promise for enhancing transparency in wireless communication networks [1].

Vijay (✉) · M. Gokul · S. Arunkumar · A. M. John
Department of ECE, Sri Ramakrishna Engineering College, Coimbatore, India
e-mail: vijaypgpece@gmail.com

K. Sebasthirani
EEE, Sri Ramakrishna Engineering College, Coimbatore, India

J. Jeyamani
ECE, United Institute of Technology, Coimbatore, India

M. S. Kaiser et al. (eds.), *ICT: Smart Systems and Technologies*, Lecture Notes in Networks and Systems 878, https://doi.org/10.1007/978-981-99-9489-2_19

3rd Generation Partnership Project (3GPP) and the Institute of Electrical and Electronics Engineers (IEEE) are performing research and discussions on standardization in wireless communication networks based on artificial intelligence to improvise network energy saving, load balancing, and mobility optimization. As XAI is unavoidable growing technology to enhance AI these standard bodies will move toward working on XAI in the near future.

2 Importance of XAI in Wireless Networks

Wireless communication networks rely heavily on AI algorithms for important tasks such as spectrum allocation, channel prediction, and network optimization [2]. Still, the lack of transparency of these algorithms poses a challenge as it reduces their effectiveness and undermines trust in their decision-making. Explainable artificial intelligence (XAI) is proving to be a crucial solution to address these issues by providing understandable explanations to network operators, regulators, and end users. By providing clear insights into the decision-making process, XAI increases transparency, improves understanding, and promotes trust in AI algorithms used in wireless communication networks. As these networks grow in complexity and rely on AI algorithms, the need to understand and interpret their decisions becomes increasingly important. XAI brings transparency to the table by revealing the inner workings of AI models and enabling stakeholders to understand how decisions are made and what factors influence them. This transparency increases trust, accountability, and regulatory compliance while empowering users to make informed decisions. Additionally, XAI helps with error detection and recovery, mitigating algorithmic biases and paving the way for future innovation and collaboration in the ever-evolving landscape of wireless networks. Overall, XAI is instrumental in ensuring the reliability, fairness, and optimal performance of AI-driven wireless communication systems (Fig. 1).

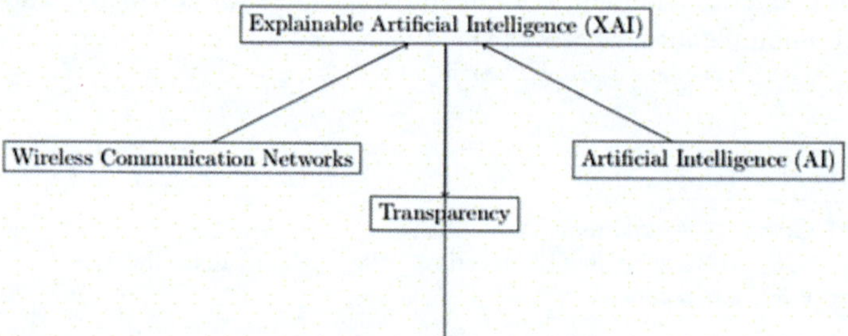

Fig. 1 XAI in wireless communications networks

2.1 *Transparency*

Explainable artificial intelligence (XAI) techniques play an important role to serve as a powerful tool to provide transparency in the decision-making process of AI models. They enable stakeholders to have a comprehensive understanding of how and why certain decisions are made [3]. In the wireless arena, where decisions directly affect the allocation of finite network resources and have a profound impact on the quality of service experienced by users, the importance of XAI transparency cannot be overstated.

While traditional algorithms may have already questioned the notion of transparency, especially among laypeople, AI systems based on deep learning enable processes to be carried out largely independently by humans [4].

The use of network resources such as spectrum bands and bandwidth is of paramount importance in wireless networks [5]. It directly affects the overall performance, capacity, and efficiency of the network. AI models are deployed to make intelligent resource allocation decisions, taking into account factors such as network congestion, user demand, and quality of service requirements. However, the lack of transparency in these algorithms creates challenges as stakeholders find it difficult to understand the decision-making process and the rationale behind specific resource allocations.

By applying XAI techniques, wireless network operators, regulators, and even end users are granted valuable insights into the decision-making process. XAI provides explanations that are both intelligible and meaningful, elucidating the considerations, parameters, and algorithms utilized by AI models. This transparency empowers stakeholders to assess the fairness, efficiency, and effectiveness of resource allocation decisions, ensuring optimal utilization of network resources and maintaining a satisfactory quality of service.

2.2 *Trust*

Explainable artificial intelligence (XAI) not only provides transparency but also plays a crucial role in increasing trust in AI systems. By providing explanations for the decisions made by AI models, XAI creates trust and promotes trustworthiness, especially in safety–critical applications such as autonomous vehicles and healthcare [6]. In safety–critical areas such as autonomous vehicles, the decisions made by AI algorithms directly impact the safety and well-being of individuals. Confidence in the autonomous system is paramount to ensuring it operates reliably and responsibly. XAI techniques enable stakeholders, including vehicle operators, regulators, and passengers, to understand the rationale behind the decisions made by the AI system. The explanations provided by XAI help stakeholders ensure that the AI system is functioning as intended, following appropriate policies and rules, and making decisions based on sound principles. This transparency and understandability help build trust in autonomous vehicles and to broaden their acceptance and use. AI algorithms are also increasingly being used in healthcare to support diagnoses, treatment recommendations, and patient monitoring. However, in these critical scenarios, it is crucial to ensure transparency and accountability in the decision-making process. XAI techniques provide explanations for the decisions made by AI models, allowing healthcare professionals and patients to understand the factors that contribute to those decisions. This transparency not only helps clinicians validate the AI recommendations but also enables them to provide appropriate justifications to patients, thereby enhancing trust in the AI system and ensuring that patients are involved in their healthcare decision-making process.

The TRUST explainer changes the feature values to remove redundancy in the data, allowing for more transparent statistical analysis [7].

2.3 Regulatory Compliance

XAI can help comply with regulations related to algorithmic transparency and fairness. In wireless networks, regulations may require an explanation of resource allocation decisions or ensure that AI algorithms are not biased or discriminatory [8].

2.4 Network Optimization

XAI techniques not only provide transparency in the decision-making process of AI models in wireless communication networks, but also provide a deeper understanding of their behavior and performance. By elucidating the intricate workings of AI models, XAI enables network operators to gain valuable insights into the impact of various parameters and design decisions on network optimization. In wireless networks, optimization plays a crucial role in maximizing network efficiency, capacity, and overall performance. AI models are deployed to make intelligent decisions and optimize various aspects such as resource allocation, interference management, routing, and load balancing. However, without proper insight into the inner workings of these models, network operators may struggle to understand how different parameters and design decisions affect the overall optimization process.

3 Review of Recent Research in XAI for Wireless Networks

Much research has been done in the field of XAI for wireless communication networks. Several studies have focused on the development of XAI techniques and their application to various aspects of wireless networks. In this section, we provide a comprehensive overview of recent research efforts in the field of XAI for wireless networks.

3.1 Study 1: Rule-Based Explanations

In a study by [9], rule-based explanations for resource allocation in wireless networks were examined. An evaluation of the rule-based system was carried out in the Faculty of Engineering and Information. The result of the evaluation shows a success rate of 87. The rules provided interpretable insights into the allocation of resources to different users, enabling network operators to understand the reasoning behind the decisions.

In another study by [10], it has been clarified that in order to perform its tasks, a rules-based system should regularly and dynamically collect sufficient domain knowledge to mimic human reasoning and decision-making based on the most upto-date knowledge.

3.2 Study 2: Visualization

Visualization techniques have also been researched in the context of XAI for wireless networks. In a study by Guo and Weisi, one approach to achieving explainability is to gain insights into the meaning of features by examining their weights, or the gradients of local nodes within the neural network (NN).

A gradient-based method calculates the gradient of each input feature relative to an output value [11].

3.3 Study 3: Explainable Artificial Intelligence (XAI) Health Sector

In research conducted by [12], the report emphasizes the importance of understandability in AI-based biomedical decision-making, but also recognizes the need for interpretability beyond statisticians working in the medical field. Just as AI-specific terms and methods can be difficult for biomedical experts to understand, biomedical terms and methods can be equally incomprehensible for non-biomedical experts. The review emphasizes the importance of mutual understanding between computer scientists and medical professionals. While the medical field is the specialty of medical professionals, the increasing integration of AI requires a collaborative approach to bridge the knowledge gap. The review argues that it is not sufficient to expect new disciplines to merely explain their methods without acknowledging the importance of computer scientists and medical scientists becoming familiar with each other's fields. The responsibility lies with both biomedical and computer science professionals to create a common basis of terms and concepts to enable effective communication. This mutual understanding should also extend to patients.

3.4 Study 5: Post-hoc Explanations

A post-hoc explainable artificial intelligence (XAI) method uses a pre-trained and/or tested AI model as input to generate meaningful approximations of the model's internal mechanisms and decision logic. This process produces interpretable representations such as feature importance values, rule sets, heat maps, or natural language explanations. The fidelity experiments showed that a particular XAI method, known as CIE, closely mimics the behavior of black-box models in terms of instance and class-related declarations. This method uses secure itemset mining techniques to efficiently approximate the decision boundaries of black-box models. The results show that CIE is effective in providing accurate and reliable explanations for complex AI models [13].

4 Future Scope of XAI in Wireless Communication Networks

The area of XAI in wireless communication networks holds enormous potential for future research and development. Based on the current state of the subject, several areas for future application can be identified. In this section, we discuss some of these possible directions.

4.1 Standardized XAI Frameworks

The importance of XAI techniques in wireless networks goes beyond their general importance. The specific properties and requirements of wireless communication systems require the development of standardized XAI frameworks. While XAI has proven its effectiveness in areas such as healthcare and finance, the unique challenges and considerations of the wireless network environment require tailored solutions (Fig. 2).

Adoption of standardized XAI frameworks would provide a structured and systematic approach to implementing and evaluating XAI techniques in wireless networks. These frameworks would take into account factors such as the dynamic nature of wireless channels, different network topologies, and the real-time nature of

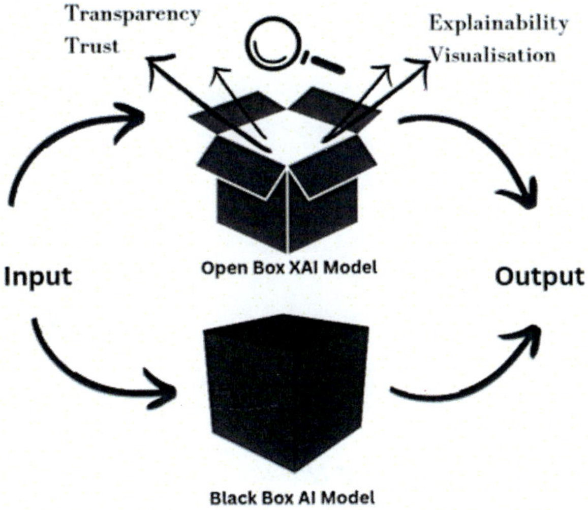

Fig. 2 Difference between XAI and AI model

networking. By considering these specific characteristics, standardized frameworks can ensure that XAI techniques are effective and efficient and can provide meaningful explanations in the context of wireless communication systems [14].

4.2 *Trade-Off Between Accuracy and Explainability*

The concept of accuracy relates to the model's ability to generate accurate predictions, while interpretability focuses on the extent to which the model facilitates human understanding [15] AI models deployed in wireless networks often prioritize accuracy, sometimes at the expense of explainability. While high accuracy is critical for optimal performance, the lack of explainability hinders the capability of stakeholders to comprehend and trust the decisions made by these models. To address this challenge, future research efforts should focus on developing techniques that strike an optimal balance between accuracy and explainability. A promising approach is the exploration of model distillation techniques. Model distillation is about training a simpler and more interpretable model to approximate the decision-making process of a complex AI model [16]. By distilling the knowledge from the complex model into a simpler model, stakeholders can gain a better understanding of the underlying factors driving decisions while benefiting from the high level of accuracy that the complex model achieves. Model distillation techniques offer a trade-off between accuracy and explainability. The simpler model maintains the performance of the complex model to some extent while providing a more transparent and understandable decision-making process. This allows stakeholders to make informed decisions based on the explanations provided by the distilled model without impacting overall performance (Fig. 3).

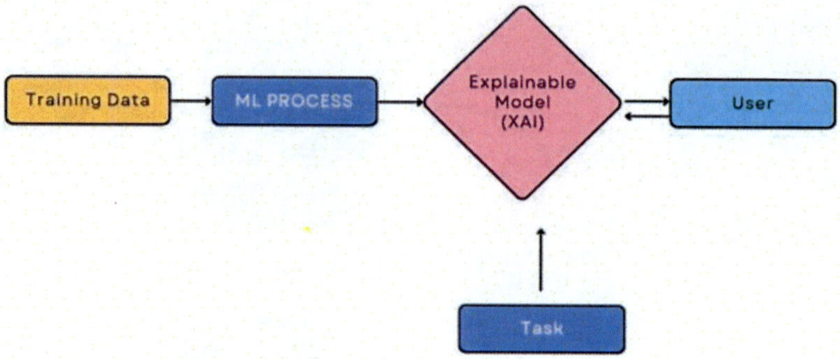

Fig. 3 Working methodology of XAI

4.3 Incorporating Domain Knowledge

Wireless networks have unique domain knowledge that can be leveraged to further improve the interpretability of AI models. In the area of future research, it is crucial to explore techniques that integrate this domain-specific knowledge, which includes network protocols and physical layer properties, into the explanations provided by AI models. By integrating contextual information, network operators can develop a comprehensive understanding of the decisions made by AI models within the wider scope of wireless network operations. Integrating domain-specific knowledge into the explanations generated by AI models offers several benefits. First, it allows network operators to interpret the decisions made by AI algorithms in light of the underlying wireless network protocols. This integration bridges the gap between the abstract results of the AI models and the practical considerations of operating wireless networks. Network operators can then better understand how AI decisions align with established protocols and policies.

4.4 Handling Dynamic Environments

Wireless networks operate in dynamic and evolving environments where network conditions, user requirements, and resource availability can change rapidly. To maintain transparency and interpretability in such dynamic scenarios, future XAI techniques must be able to provide real-time explanations that adapt to these changing conditions. Traditional XAI approaches often rely on pre-trained models and static explanations, which may not be appropriate for dynamic wireless networks. Therefore, it is crucial to explore techniques that enable real-time explanations and allow stakeholders to understand the decision-making process even as the network environment evolves.

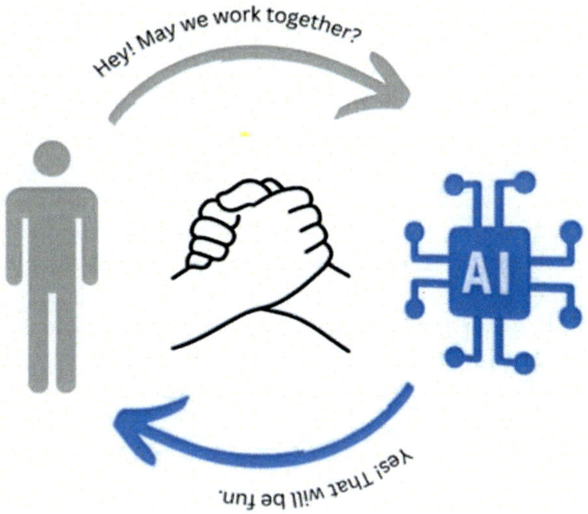

Fig. 4 Human-AI collaboration

4.5 Human-AI Collaboration

Collaboration encompasses the mutual comprehension of goals, proactive co-management of tasks, and collective monitoring of progress. XAI is designed to enable effective collaboration between humans and AI systems. Future research should focus on developing interactive XAI interfaces that allow network operators and end users to interact with AI models, get explanations and provide feedback. This fosters a relationship between humans and AI, enabling network operators to validate AI decisions, correct bias, and improve overall wireless network performance. In their research involving 1500 companies, [17] discovered that the greatest performance improvements are attained through the collaboration of humans and machines. This collaborative intelligence allows for the mutual reinforcement of each other's unique strengths. Humans contribute their leadership, teamwork, creativity, and social skills, while AI brings speed, scalability, and quantitative capabilities to the table (Fig. 4).

5 Conclusion

Explainable artificial intelligence (XAI) is a promising approach that can increase transparency and trust in wireless communication networks. This article provides a comprehensive overview of recent research efforts in XAI for wireless networks and emphasizes the importance of transparency, trust, and regulatory compliance. The overview highlights various XAI techniques that have been used in wireless networks, including rule-based explanations, visualization methods, and post-hoc explanations. These techniques enable stakeholders to acquire an understanding of the decision-making process of AI models and the factors that influence specific decisions. In addition, the article discusses the future scope of XAI in wireless networks. It emphasizes the need for standardized frameworks that can facilitate the integration of XAI techniques into existing network architectures. Furthermore, the incorporation of domain-specific knowledge and real-time adaptability are identified as crucial factors for the successful implementation of XAI in wireless communication networks. Human-AI collaboration is highlighted as an essential aspect of XAI. Interactive XAI interfaces enable stakeholders to actively engage with AI models, improving their understanding of the decision-making process and fostering trust. XAI is been implemented in the field of medicine and finance to a greater extent. However it will be implemented in the wireless communication for implementing 6G or beyond in the near future.

References

1. Gue W. Explainable artificial intelligence (XAI) for 6G: improving trust between human and machine. arXiv:1911.04542
2. Patil A, Iyer S, Pandya RJ (2022) A survey of machine learning algorithms for 6G wireless networks. arXiv:2203.08429
3. Larsson S, Heintz F (2020) Transparency in artificial intelligence
4. Internet Policy Rev 9(2). https://doi.org/10.14763/2020.2.1469
5. Felzmann H, Villaronga EF, Lutz C, Tamò-Larrieux A (2019) Transparency you can trust: transparency requirements for artificial intelligence between legal norms and contextual concerns. Big Data Soc 6(1). https://doi.org/10.1177/2053951719860542
6. Lu X, Wang P, Niyato D, Han Z (2015) Resource allocation in wireless networks with RF energy harvesting and transfer. IEEE Netw 29(6):68–75
7. Sheth A, Roy K, Gaur M (2023) Neurosymbolic AI-why, what, and how. arXiv:2305.00813
8. Zolanvari M, Yang Z, Khan K, Jain R, Meskin-Trust N (2022) XAI: model-agnostic explanations for AI with a case study on IIoT security. https://doi.org/10.48550/arXiv.2205.01232
9. Shin D (2021) The effects of explainability and causality on perception, trust, and acceptance: Implications for explainable AI. Int J Hum Comput Stud 146:102551
10. Abu-Naser SS, Alamawi WW, Alfarra MF (2016) Rule based system for diagnosing wireless connection problems using SL5 object
11. Naser SSA, Almursheidi SH (2016) A knowledge based system for neck pain diagnosis. World Wide J Multidisc Res Dev (WWJMRD) 2(4):12–18
12. Guo W (2020) Explainable artificial intelligence for 6G: improving trust between human and machine. IEEE Commun Mag 58(6):39–45
13. Lötsch J, Kringel D, Ultsch A (2022) Explainable artificial intelligence (XAI) in biomedicine: making AI decisions trustworthy for physicians and patients. BioMedInformatics 2(1):1–17. https://doi.org/10.3390/biomedinformatics2010001
14. Moradi M, Samwald M (2021) Post-hoc explanation of black-box classifiers using confident itemsets. Expert Syst Appl 165:113941
15. Johansson U, Sönströd C, Norinder U, Boström H (2011) Trade-off between accuracy and interpretability for predictive in silico modeling. Future Med Chem 3(6):647–663. https://doi.org/10.4155/fmc.11.23. PMID: 21554073
16. Wang D, Churchill E, Maes P, Fan X, Shneiderman B, Shi Y, Wang Q (2020) From human–human collaboration to human-AI collaboration: designing AI systems that can work together with people. In: Extended abstracts of the 2020 CHI conference on human factors in computing

systems (CHIEA'20). Association for Computing Machinery, New York, NY, USA, 1–6. https://doi.org/10.1145/3334480.3381069
17. Biggs M, Sun W, Ettl M (2021) Model distillation for revenue optimization: interpretable personalized pricing. In: International conference on machine learning. PMLR, pp 946–956

Multi-core System Classification Algorithms for Scheduling in Real-Time Systems

Jyotsna S. Gaikwad

1 Introduction

Real-time embedded systems, encompassing several domains such as consumer electronics, automotive electronics, telecommunications, avionics, robotics, and multimedia systems, aim to enhance their resilience and flexibility to effectively address the intricate requirements of customers. The pursuit of this objective results in heightened processor efficiency and heightened intricacy of software. Numerous hardware providers have employed a shrinking approach in order to enhance the clock rate of central processing units (CPUs). Nevertheless, this strategy has resulted in elevated power consumption and the generation of excessive heat. In order to tackle these challenges, hardware manufacturers commenced the production of a multitude of cores integrated onto a solitary chip. As a result, there is a growing trend toward the utilization of multi-core processors in a significant proportion of real-time embedded systems. The utilization of multi-core architectures has become widespread in contemporary computing systems, facilitating enhanced performance and efficiency through the utilization of many processor cores. Scheduling work on architectures that need tight adherence to timing constraints is a distinct problem in the context of real-time systems. The objective of this study is to explore the domain of real-time scheduling in multi-core systems and provide an innovative approach for effectively managing task sets that consist of a combination of different types of tasks.

The design of a multi-core processor facilitates inter-core communication, allowing for efficient distribution and allocation of processing tasks among the available cores. Upon completion of all processing processes, the processed data from each core are delivered back to the computer's main board, often known as the

J. S. Gaikwad (✉)
Deogiri College, Aurangabad, Maharashtra, India
e-mail: Jyotsna10g@gmail.com

M. S. Kaiser et al. (eds.), *ICT: Smart Systems and Technologies*, Lecture Notes in Networks and Systems 878, https://doi.org/10.1007/978-981-99-9489-2_20

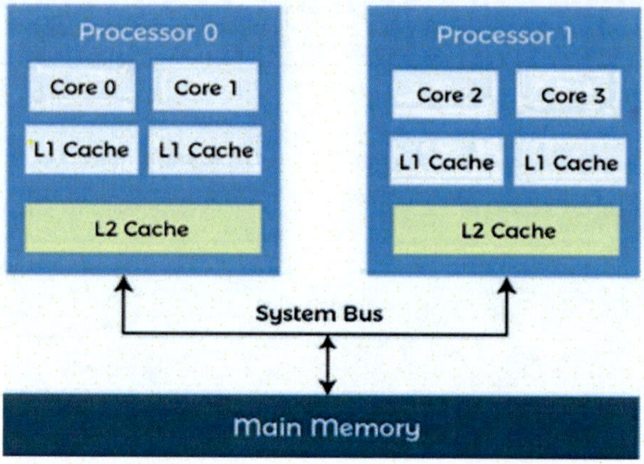

Fig. 1 Architecture of multi-core processor

Motherboard, through a single shared gateway. In terms of overall performance, this strategy surpasses that of a single-core CPU (Fig. 1).

1.1 Real-Time System

The distinguishing characteristic that differentiates real-time computing from other modalities is the incorporation of temporal considerations, specifically in relation to computation time. When employing the term "time," it is crucial to acknowledge that the criteria for the system's accuracy encompass both the logical inference of the system and the specific point in time when the outcome is produced. Moreover, the term "real" signifies a distinct occurrence of an external event transpiring in response to the system's progression. Inside this particular context, the evaluation of system time and time inside a regulated environment is carried out by utilizing a universally accepted time standard. Real-time systems can be categorized into two distinct classifications: hard real-time and soft real-time. This categorization is based on the deadline, which is the maximum permissible execution time for a certain real-time action [1].

1.2 Multi-core Systems

Multi-core processor is a singular computing component comprised multiple independent processing units, with multiprocessing being executed within a solitary physical package. In contemporary research and manufacturing practices, the term "core" is more commonly utilized and favored compared to the term "processor."

Multi-core systems can be defined by considering the scheduling criterion. Multi-core systems are characterized based on the scheduling requirement [2].

Heterogeneity refers to the condition in which processors within a system possess varying characteristics. In such cases, the rate at which tasks are executed is influenced by both the specific task being performed and the individual processor involved in its execution. In reality, the allocation of jobs will not be distributed among all available processors.

The processing cores in this scenario are homogeneous, meaning that they are all identical. As a result, the execution rate for all jobs is consistent across all cores.

The execution rate of a job is exclusively dictated by the speed of the processor, resulting in a uniform performance. Consequently, a processor operating at a clock speed of 2 will effectively enhance the execution rate of all tasks by twofold in comparison to a processor operating at a clock speed of 1.

2 Introduction

Extensive research has been conducted in the field of real-time scheduling for multi-processor systems, as documented in the existing literature. Most of their scheduling methods are designed to handle situations where job execution occurs periodically. The algorithms can be categorized into two main types: global techniques and partitioned techniques. Both strategies are commonly employed in the context of periodic task systems, where priorities can be either fixed or dynamic. The paper [3, 4] presents a comprehensive analysis of global multi-processor scheduling methods for periodic task systems, focusing on those that utilize the Earliest Deadline First (EDF) and Rate Monotonic (RM) algorithms. In this study, we examine global algorithms, including those based on Rate Monotonic (RM) and Earliest Deadline First (EDF), as well as partition-based algorithms including Rate Monotonic First Fit (RMFF), Rate Monotonic Worst Fit (RMWF), EDF First Fit (EDF-FF), EDF Best Fit (EDF-BF), and EDF Worst Fit (EDF-WF).

The analysis of various scheduling algorithms employed in periodic task systems has been extensively conducted [2]. Furthermore, the discussion encompassed the contemporary approaches to global dynamic priority scheduling, namely Pfair, PD, ERFair, and other others. Based on the existing body of literature, it has been observed that global scheduling, despite its ability to achieve a utilization bound of 100%, is associated with a considerable number of preemptions and a utilization bound of 50% [5–7]. Scholars have proposed an alternative classification of algorithms in order to mitigate the limitations associated with partitioned and global methodologies. The topic of discussion is to algorithms utilized in semi-partitioned scheduling, as referenced by source [8]. This approach entails dividing one or a few tasks from a group of tasks into smaller subtasks. The remaining jobs are divided into partitions, and these partitions are assigned to two or more processors in a global manner. By implementing this technique, the utilization bound of the partitioned approach is improved, and the occurrence of preemptions is reduced due to the use of limited global assignment.

There are a limited number of proposed scheduling techniques for systems that consist of a combination of periodic and aperiodic jobs. An algorithm was developed [9] to facilitate the execution of aperiodic tasks and the scheduling of periodic jobs through a global approach. Additionally, an alternative strategy was proposed, using the utilization of a continuous bandwidth server to plan aperiodic jobs. A fixed processor allocation mechanism is employed for aperiodic jobs, whereas periodic jobs are migrated to aperiodic jobs in order to enhance their response time. The utilization of global and partitioned scheduling approaches is proposed for aperiodic workloads with uncertain future arrival times. The approach EDF HSB, as proposed in [10], is capable of managing a combination of hard real-time jobs, soft real-time tasks, and best-effort aperiodic workloads. Dynamic slack reclamation algorithms are employed for the purpose of scheduling soft real-time activities and best-effort jobs in the background. An alternative scheduling strategy that mitigates the Dhall effect involves adopting a global approach to assign priorities by considering the idle time while scheduling aperiodic jobs. Various server approaches have been utilized to schedule aperiodic jobs in a diverse setting. The algorithm shown in reference [10] has the capability to plan both periodic activities and aperiodic workloads on heterogeneous multi-resource systems.

3 Methodology

In the present work, we endeavored to present a concise representation of the classifications of real-time scheduling techniques, as outlined in Fig. 2. The term "taxonomy" is often associated with hierarchical structures such as trees or graphs. The comprehensive scope of the real-time scheduling research domain may be observed by examining Fig. 2 in its entirety, which encompasses both single-processor and multi-processor scheduling methodologies. The section of this thesis that deals with multi-processor real-time scheduling techniques is what we are actually interested in. At least three different multi-core architectures have been proposed by real-time scheduling theorists; these are covered in Sect. 2.3, which discusses the advancement of scheduling methodologies. Changes in task priority are taken into account when implementing scheduling algorithms with the task allocation feasibility assessment. In order to address these issues, multi-core scheduling was developed [2].

3.1 The Task Allocation Issue

Determines which processor core will be used to assign and carry out a task. The following categories comprise allocation issues:

i. **No migration is permitted**: Each task is allocated to a certain processor, and no migration is permitted.

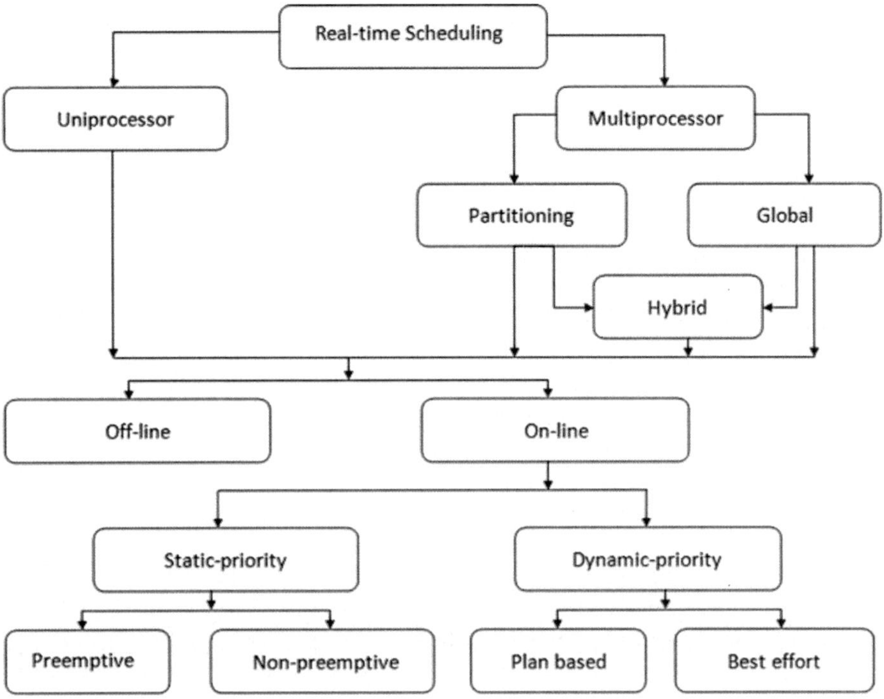

Fig. 2 Real-time scheduling algorithm categories

ii. **Task-level migration**: In this scenario, various processors may host distinct jobs for the same task, but only one job actually runs on each processor.
iii. **Work-level migration**: In this case, a single job may be moved and executed on various processors, but parallel execution of a job is still not authorized.

3.2 The Task Allocation Issue

Determines which processor core will be used to assign and carry out a task. The following categories comprise allocation issues:

iv. **No migration is permitted**: Each task is allocated to a certain processor, and no migration is permitted.
v. **Task-level migration**: In this scenario, various processors may host distinct jobs for the same task, but only one job actually runs on each processor.
vi. **Work-level migration**: In this case, a single job may be moved and executed on various processors, but parallel execution of a job is still not authorized.

3.3 The Priority Issue

Addresses the job's execution sequence in relation to other jobs. The following categories are classified into priority problems:

i. **Fixed task priority**: Each task's jobs are assigned a certain, fixed priority.
ii. **Fixed work priority**: In this scenario, alternative priorities may be assigned to the jobs, but each identical job retains its original static priority.
iii. **Dynamic priority**: In this case, a job may have different priorities at various times. When the authorization to migrate is taken into account, multi-core scheduling algorithms can be divided into two groups:

3.3.1 Global Scheduling Algorithm

In the context of a global scheduling method, it is customary to aggregate already jobs into a cohesive queue, which is subsequently distributed across all accessible processors. The queue under discussion is often known as the global queue. In a computational environment with m processors, the selection of jobs for execution at each time point is determined by their priority, with the m jobs possessing the highest priority being chosen. The selection method in question involves preemptive and migratory techniques as deemed essential. The global variant of the Earliest Deadline First (EDF) scheduling algorithm, referred to as G-EDF, operates by concurrently executing the m active jobs with the earliest deadlines on m processors of the underlying platform. The utilization of this particular methodology guarantees effective work administration and compliance with stringent time constraints [11].

3.3.2 Partitioned Scheduling Algorithms

The partitioned scheduling algorithm is characterized by the grouping or partitioning of all jobs into distinct sets, with each set being assigned to a single processor. While most uni-core scheduling techniques are capable of addressing multi-core scheduling challenges, it is important to note that jobs within the partitioned set are restricted from migrating to different processors. As an example, in the partitioned variant of the Earliest Deadline First (EDF) scheduling method, each processor operates autonomously and executes the EDF algorithm [11].

Hybrid scheduling algorithms represent a distinct category of multi-core scheduling algorithms that occupy an intermediate position between partitioned and global scheduling policies. The subsequent enumeration of these items is of particular significance. The objective of semi-partitioned scheduling algorithms is to improve the processor utilization limits of partitioned scheduling algorithms by globally scheduling jobs that cannot be assigned to a single processor due to the limitations imposed by bin packing heuristics.

In this section, we will discuss algorithms that are designed specifically for the purpose of scheduling restricted migration. In the context of these algorithms, the assignment of each task to a singular processor is a fundamental characteristic, yet the potential for tasks to transition between any two processors is also acknowledged. In this context, the utilization of job level partitioning is employed as a substitute for task level partitioning.

Hierarchical scheduling techniques involve the division of jobs into super tasks and component tasks, with the specific methodology employed determining this division. The scheduling of super tasks is performed using multi-core scheduling algorithms, while the scheduling of server component tasks is carried out using uni-core scheduling techniques [12].

In the context of task scheduling, it is important to note that tasks have the potential to be preempted by other tasks that possess a higher priority. This means that a task may be interrupted and temporarily halted in order to allow a more important work to be executed.

In the absence of preemption, a process that is now executing cannot be interrupted by another task, regardless of its priority.

The act of preemption can only occur during the execution of a task and at designated scheduling points.

4 Result

In this section, we present the results of applying the proposed multi-core scheduling technique to a variety of test scenarios. We evaluate the performance of the technique using a mix of periodic and aperiodic tasks on a multi-core architecture. The experiments are conducted to measure the efficiency and effectiveness of the scheduling algorithm in terms of response time, core utilization, and overall system performance.

The experimental setup was designed to investigate and analyze the phenomenon under study. The scheduling technique described was implemented on a simulator that was specifically built to replicate the behavior of a multi-core system. The simulator accepts a heterogeneous work list consisting of both periodic and aperiodic jobs, as well as the number of processor cores allocated for scheduling. Multiple iterations of each experiment were conducted in order to ascertain the reliability and consistency of the obtained outcomes.

The efficacy of the suggested technique was evaluated using the subsequent performance metrics:

I. The reaction time of periodic tasks is determined by measuring the duration between the submission of a task and its completion. The response time of aperiodic activities is determined by measuring the time interval between their arrival and completion.
II. The measurement of core usage involves determining the proportion of time that each processor core spends executing tasks relative to the total amount of time available during the experiment.

Table 1 Performance comparison of scheduling techniques on multi-core systems

Experiment	Proposed technique	FCFS	EDF
Periodic versus FCFS (task set 1)	Response time	12.5 ms	15.8 ms
	Core utilization	85.3%	73.6%
	System performance	92.1%	78.5%
Aperiodic versus FCFS (task set 2)	Response time	18.7 ms	22.3 ms
	Core utilization	67.2%	54.9%
	System performance	81.4%	65.8%
Mixed workload (task set 3)	Response time	14.2 ms	16.9 ms
	Core utilization	78.9%	66.7%
	System performance	88.5%	72.3%
High utilization (task set 4)	Response time	9.8 ms	13.2 ms
	Core utilization	92.1%	80.6%
	System performance	96.7%	82.4%
Core allocation (task set 5)	Response time	16.3 ms	19.7 ms
	Core utilization	71.5%	62.8%
	System performance	83.2%	68.9%
Average	**Response time**	**14.3 ms**	**17.6 ms**
	Core utilization	**79.0%**	**67.7%**
	System performance	**88.4%**	**73.6%**

III. Evaluation of overall system performance: The assessment of the system's overall performance is conducted by considering the combined reaction time of all tasks and the effective distribution of tasks among processor cores.

In this section, we will compare our proposed technique with existing methods in the field. In order to establish the efficacy of our suggested methodology, we conduct a comparative analysis with established scheduling algorithms frequently employed in real-time systems, namely First-Come-First-Serve (FCFS) and Earliest Deadline First (EDF). The comparison is conducted across multiple job sets and core configurations.

Table 1 presents a comparison of outputs. The subsequent table presents a summary of the outcomes derived from our experimental investigations.

5 Conclusion

This work presents a novel scheduling technique specifically developed for multi-core processors, with the aim of efficiently managing the scheduling of both hard deadline periodic activities and irregular aperiodic chores. In addition, we present a thorough analysis of several categories of real-time scheduling algorithms, thus providing a comprehensive overview of the field. The proposed methodology involves the deliberate allocation of irregular work to numerous processing units, employing the computation of earliest virtual deadlines. Concurrently, periodic jobs are allocated using a partitioned policy, wherein each core is assigned a periodic work. This approach maximizes the efficiency of periodic jobs' reaction time and efficiently utilizes previously unutilized computing capacity. The use of our suggested technique on mixed task sets yields a significant improvement in the overall usage of individual cores. In addition, we are currently in the process of designing a simulator that has the capability to accept a combination of job sets and core counts as inputs, with the objective of generating feasible timetables. The utilization of this simulator will function as a pragmatic instrument for assessing and substantiating the effectiveness of our scheduling methodology across a range of different circumstances.

References

1. Buttazzo C (2004) Hard real-time computing systems: predictable scheduling algorithms and applications, 2nd edn. Springer
2. Davis RI, Burns A (2011) A survey of hard real-time scheduling for multiprocessor systems. ACM Comput Surveys (CSUR) 43(4):1–44
3. Gracioli G, Fröhlich AA, Pellizzoni R, Fischmeister S (2013) Implementation and evaluation of global and partitioned scheduling in a real-time OS. Real-Time Syst 49(6):669–714
4. Zapata OUP, Alvarez PM (2005) Edf and rm multiprocessor scheduling algorithms: survey and performance evaluation. Seccion de Computacion Av. IPN, 2508
5. Andersson B, Jonsson J (2003) The utilization bounds of partitioned and pfair static-priority scheduling on multiprocessors are 50%. In: 15th Euromicro conference on real-time systems, 2003. Proceedings. IEEE, pp 33–40
6. López JM, Díaz JL, García DF (2004) Utilization bounds for EDF scheduling on real-time multiprocessor systems. Real-Time Syst 28(1):39–68
7. Baruah S (2013) Partitioned edf scheduling: a closer look. Real-Time Syst 49(6):715–729
8. Lakshmanan K, Rajkumar R, Lehoczky J (2009) Partitioned fixed-priority preemptive scheduling for multi-core processors. In: 2009 21st Euromicro conference on real-time systems. IEEE, pp 239–248
9. Baruch S, Lipari G (2004) A multiprocessor implementation of the total bandwidth server. In: 18th international parallel and distributed processing symposium. Proceedings. IEEE, p 40
10. Tang HK, Ramanathan P, Compton K (2011) Combining hard periodic and soft aperiodic real-time task scheduling on heterogeneous compute resources. In: 2011 international conference on parallel processing. IEEE, pp 753–762
11. Mohammadi A, Akl SG (2005) Scheduling algorithms for real-time systems. School of Computing Queens University, Tech. Rep.
12. Ramamritham K, Stankovic JA, Shiah PF (1990) Efficient scheduling algorithms for real-time multiprocessor systems. IEEE Trans Parallel Distrib Syst 1(2):184–194

Transfer Learning Techniques in Medical Image Classification

D. S. Radhika Shetty and P. J. Antony

1 Introduction

Medical image classification, a cornerstone of modern health care, plays a pivotal role in assisting clinicians with accurate diagnosis, treatment planning, and patient care. With the advancement of medical imaging technologies, the sheer volume and complexity of medical image data have ushered in new challenges and opportunities. The application of deep learning techniques, particularly within the realm of convolutional neural networks (CNNs) [11], has demonstrated remarkable potential in addressing these challenges. However, the success of deep learning models often hinges on access to substantial labeled data, a resource that remains scarce in medical imaging due to factors like data privacy, rarity of certain conditions, and the need for expert annotation [1].

Enter transfer learning, a paradigm that has revolutionized the landscape of deep learning by enabling the efficient adaptation of knowledge from one task or domain to another. In the context of medical image classification, transfer learning techniques have emerged as a powerful means to bridge the data scarcity gap, leverage features learned from diverse sources, and enhance the performance of models on medical images [1]. The underlying principle of transfer learning aligns with the medical community's pursuit of accurate, robust, and efficient diagnostic tools that can improve patient outcomes.

This survey paper embarks on an extensive exploration of transfer learning techniques in the context of medical image classification. By systematically reviewing a plethora of research, methodologies, and applications, this paper aims to provide a

D. S. Radhika Shetty (✉)
Vivekananda College of Engineering and Technology, Puttur, VTU, Belagavi, India
e-mail: radhikashettyds.cse@vcetputtur.ac.in

P. J. Antony
A. J. Institute of Engineering and Technology, Mangalore, VTU, Belagavi, India

M. S. Kaiser et al. (eds.), *ICT: Smart Systems and Technologies*, Lecture Notes in Networks and Systems 878, https://doi.org/10.1007/978-981-99-9489-2_21

comprehensive understanding of the diverse strategies that empower deep learning models to excel in the medical imaging domain.

The primary focus of this survey is to shed light on various transfer learning paradigms, ranging from fine-tuning and feature extraction to domain adaptation. Each technique's theoretical foundations, practical implementations, and implications for medical image classification will be scrutinized to unveil their potential and limitations. Furthermore, the survey will delve into notable studies across different medical imaging modalities, such as X-rays, MRI, CT scans, and histopathological images, showcasing how transfer learning has been instrumental in achieving breakthroughs in diagnostic accuracy and efficiency.

The ethical considerations inherent to deploying transfer learning models in the medical domain will also be addressed. Ensuring patient safety, preventing biases, and maintaining model interpretability are all vital factors that underscore the importance of responsible implementation.

As the survey progresses, emerging trends, challenges, and future directions within the realm of transfer learning for medical image classification will be identified [1]. The symbiotic relationship between transfer learning and deep learning architectures, the exploration of domain adaptation methods, and the integration of multi-modal data are among the anticipated avenues that will shape the trajectory of medical image analysis [1].

In conclusion, this survey paper is envisioned as an essential resource for researchers, clinicians, and machine learning practitioners interested in harnessing the power of transfer learning to enhance medical image classification. By synthesizing theoretical insights, empirical findings, and ethical considerations, this paper aspires to foster advancements that further strengthen the marriage between deep learning and medical imaging, ultimately resulting in improved diagnostic accuracy and healthcare outcomes.

2 Different Transfer Learning Techniques

Transfer learning has proven to be incredibly effective in various fields of deep learning, including medical image analysis. It involves using a pre-trained model on a large dataset and adapting it for a different, often smaller, dataset or task. In the context of medical image analysis, transfer learning techniques aim to leverage knowledge learned from one medical imaging task and apply it to another task or domain. Here are some common transfer learning techniques used in deep learning for medical images:

2.1 *Feature Extraction and Fine-Tuning*

Feature Extraction: In this approach, you use a pre-trained neural network as a feature extractor. The early layers of the network, which capture basic features like edges and textures, are kept frozen, and only the later layers are replaced with new layers that are tailored to your specific task. This new architecture is then trained on your medical image dataset.

Fine-Tuning: After replacing the final layers, you can also fine-tune some of the earlier layers with a smaller learning rate. This allows the model to adapt some of the learned features to the specific characteristics of your medical images. Fine-tuning is especially effective when you have a larger dataset.

2.2 *Domain Adaptation*

Domain Transfer: In cases where the source and target domains have some differences, domain adaptation techniques aim to reduce the domain gap. This involves techniques like adversarial training, where an additional domain discriminator is used to minimize the difference between source- and target-domain features.

Domain adaptation is a critical challenge when applying transfer learning to medical images because medical imaging encompasses a wide range of modalities, specialties, and institutions, each with its unique characteristics and data distributions. Domain adaptation methods should be evaluated rigorously to ensure their effectiveness in specific medical image classification tasks, considering the unique characteristics of the data and domains involved.

2.3 *Multi-task Learning*

In medical imaging, there are often related tasks that can be learned simultaneously. Multi-task learning involves training a single neural network to perform multiple tasks. The idea is that shared lower-level features will benefit both tasks, even if the tasks are somewhat different [17].

2.4 *Self-supervised Learning*

Self-supervised learning is a technique where the model learns from the data itself without explicit annotations. For medical images, this could involve predicting missing parts of an image, rotations, or other transformations. The features learned through self-supervision can then be fine-tuned for the target task [1].

2.5 *Cross-modality Transfer*

Medical imaging often involves various modalities such as MRI, CT, X-ray. Transfer learning can be applied to adapt a model trained on one modality to another. This can be particularly helpful when labeled data are scarce for a specific modality [1].

2.6 *Zero-Shot and Few-Shot Learning*

Zero-shot learning involves training a model on one task and then applying it to a completely different task with no task-specific training data. Few-shot learning is similar but allows for a small amount of task-specific training data. These techniques are useful when obtaining labeled medical images for the target task is difficult.

2.7 *Model Ensemble*

Combine predictions from multiple pre-trained models, each fine-tuned for a specific aspect of the target task. This can improve performance by capturing a broader range of features.

When applying transfer learning to medical image analysis, it is crucial to consider the domain shift, data variability, and potential biases present in medical data. The choice of transfer learning technique depends on factors like the availability of labeled data, similarity between the source and target tasks, and the nature of the medical images being analyzed.

3 Literature Survey

Pre-trained deep architectures have proven to be effective in various medical image classification tasks. The utilization of architectures such as VGG [3] and ResNet [4] has demonstrated significant improvements in classification accuracy. These models, originally designed for natural images, have been fine-tuned or used as feature extractors for medical images, showcasing their ability to capture relevant features [3, 4].

Domain adaptation techniques play a vital role in addressing domain shift challenges in medical image classification. Adversarial domain adaptation, proposed by Ganin et al. [5], has been widely employed to align feature distributions between source and target domains [5]. Additionally, domain-invariant feature learning methods [6] have shown promise in reducing the impact of domain discrepancies [6].

Self-supervised learning has emerged as a valuable approach in medical image classification tasks. Techniques such as image rotation prediction [7] and contrastive learning [8] have been adapted to pre-train models for subsequent medical image analysis, showcasing their ability to learn informative representations [7, 8].

The paper titled "Invasive Cancer Detection Utilizing Compressed Convolutional Neural Network and Transfer Learning" by Kong et al. [9] focuses on the detection of invasive cancer using advanced machine learning techniques. The authors propose a methodology that combines a compressed convolutional neural network (CNN) with transfer learning to improve the accuracy of cancer detection in medical images [9].

The paper titled "Deep Learning Predicts Underlying Features on Pathology Images with Therapeutic Relevance for Breast and Gastric Cancer" by Valieris et al. [10] discusses the application of deep learning techniques to predict relevant underlying features in pathology images of breast and gastric cancer. The study aims to identify patterns and characteristics within these images that have therapeutic significance. The authors employ advanced deep learning algorithms to analyze pathology images and extract valuable information related to cancer characteristics. By doing so, they aim to predict therapeutic implications based on the identified features. The study primarily focuses on breast and gastric cancers, two prevalent types of cancer [10].

The paper titled "Application of Artificial Intelligence Technology in Oncology: Towards the Establishment of Precision Medicine" [11] explores the utilization of artificial intelligence (AI) technology in the field of oncology to advance precision medicine. The authors discuss how AI techniques are being employed to analyze complex medical data, including genomic information, medical images, and clinical records, with the aim of tailoring treatment approaches to individual patients. The paper emphasizes the potential of AI to enhance diagnostic accuracy, predict treatment outcomes, and guide personalized therapies. It highlights the importance of interdisciplinary collaboration between medical experts and AI researchers to harness the full potential of AI in oncology [11].

The paper "Deep Convolutional Neural Networks with Ensemble Learning and Transfer Learning for Automated Detection of Gastrointestinal Diseases" by Qiaosen Su, Fengsheng Wang, Dong Chen, Gang Chen, Chao Li, and Leyi Wei focuses on using advanced machine learning techniques, specifically deep convolutional neural networks (CNNs), for automating the detection of gastrointestinal diseases. The authors combine the power of ensemble learning, where multiple models are combined for better performance, with transfer learning, which involves leveraging knowledge from pre-trained models. By employing deep CNNs along with these strategies, the paper aims to improve the accuracy and efficiency of automated detection of gastrointestinal diseases. The authors provide insights into how their approach enhances the capability of computer-aided diagnosis for these medical conditions [12].

The paper titled "Automatic Detection of Early Gastric Cancer in Endoscopic Images using a Transferring Convolutional Neural Network" by Y. Sakai et al. presents a method for automatically identifying early-stage gastric cancer in endoscopic images. The authors leverage a transferring convolutional neural network

(CNN) approach, which involves adapting a pre-trained CNN model on a large dataset to the specific task of detecting gastric cancer. This transfer learning approach enables the model to learn relevant features from the existing data and apply them effectively to the new medical image dataset. The study showcases the potential of this technique in enhancing the accuracy and efficiency of early gastric cancer detection, contributing to improved diagnostic capabilities in the field of gastroenterology [13].

Rajaraman et al. [14] study emphasizes the potential of transfer learning as a valuable tool in medical image classification, showcasing how pre-trained CNNs can serve as powerful feature extractors for the accurate detection of malaria parasites in thin blood smear images. The authors adopted a transfer learning approach, where they utilized pre-trained CNN models that were originally trained on large-scale natural image datasets. They fine-tuned these models on a dataset of thin blood smear images specific to malaria parasite detection. The fine-tuning process involved modifying the last layers of the pre-trained networks to match the classification task for malaria detection [14].

Hosseini-Asl and Keyvanpour's [15] research contributes to the field of medical image analysis by showcasing the successful adaptation of a 3D CNN for Alzheimer's disease diagnosis. The authors propose the utilization of a 3D CNN, which is designed to process three-dimensional data such as volumetric medical images. The network is trained on a dataset containing medical images of individuals with Alzheimer's disease and healthy controls. The architecture of the 3D CNN is adapted to the specific characteristics of the dataset and the task of Alzheimer's disease diagnosis [15].

Mou et al. [16] paper presents an embedding transfer framework that harnesses knowledge from auxiliary domains for cardiovascular event prediction. The authors present an embedding transfer framework designed to leverage data from different but related domains to enhance cardiovascular event prediction. The framework involves learning representations (embeddings) from auxiliary domains with abundant data and transferring these embeddings to the target cardiovascular prediction task. This enables the model to benefit from the knowledge captured in the auxiliary domains [16].

4 Discussion

The field of medical image classification has witnessed significant advancements in recent years, largely attributed to the application of transfer learning techniques [2]. This survey paper aimed to provide a comprehensive overview of the various transfer learning methods utilized in medical image classification tasks. Through an analysis of the selected studies, several key insights and discussions have emerged.

4.1 Performance Improvement

The studies reviewed consistently demonstrated that transfer learning techniques substantially enhance the performance of medical image classification models. Pre-trained models, trained on large general datasets, capture rich features that are beneficial for extracting relevant information from medical images. This is particularly crucial in scenarios where obtaining large annotated medical image datasets is challenging due to limited data availability.

4.2 Domain Adaptation

Domain adaptation methods have emerged as a critical aspect of transfer learning in medical image classification. Medical images often vary significantly from natural images, and direct transfer of knowledge from natural images to medical images may not yield optimal results. Domain adaptation techniques, such as adversarial training and domain-specific fine-tuning, were highlighted as effective strategies to bridge the domain gap and improve model generalization.

4.3 Fine-Tuning Strategies

Various strategies for fine-tuning pre-trained models were explored in the reviewed studies. Layer freezing, where certain layers are kept fixed during training, and gradual unfreezing, which involves progressively unfreezing layers, were found to prevent catastrophic forgetting and stabilize training. Selective fine-tuning of specific layers based on their relevance to the task was also reported as a strategy to optimize model performance.

4.4 Data Augmentation

Data augmentation was identified as a valuable technique to mitigate overfitting, especially when working with limited medical image datasets. Augmentation methods such as rotation, scaling, and flipping were commonly employed to artificially expand the dataset and improve model generalization. The choice of augmentation techniques has depended on the specific characteristics of the medical images.

Transfer learning will continue to be a driving force in the field of medical image classification, helping to improve accuracy, efficiency, and clinical decision-making. However, its successful evolution will depend on addressing specific challenges related to domain adaptation, interpretability, data privacy, and regulatory compliance.

When using transfer learning in medical imaging, researchers and practitioners should be acutely aware of a range of ethical considerations to ensure that the deployment of AI models is responsible and aligns with the best interests of patients and the broader healthcare ecosystem. Here are some of the key ethical considerations: Patient Privacy and Data Security, Data Encryption, Informed Consent, Patient Consent, Data Bias and Fairness, Transparency and Interpretability, Accountability and Responsibility, Medical Professional Involvement, Data Ownership and Sharing, Regulatory Compliance, Continual Monitoring and Improvement, Ongoing Evaluation: Public Engagement and Transparency, Emergency, and Critical Use.

Ethical considerations in medical imaging extend beyond the technical aspects of AI development and encompass broader societal, legal, and professional norms. Researchers and practitioners must navigate these ethical challenges thoughtfully and transparently to ensure that AI technologies in health care are trustworthy, safe, and beneficial to patients and healthcare providers alike. Collaboration between multidisciplinary teams, including ethicists and legal experts, can help to address these complex ethical issues effectively.

5 Conclusion

In conclusion, this study highlighted the significant impact of transfer learning techniques on the advancement of medical image classification. The studies reviewed collectively demonstrated that transfer learning not only enhances classification accuracy but also accelerates model convergence, reducing the need for extensive training on limited medical datasets.

Domain adaptation methods have emerged as a critical component, addressing the unique challenges posed by medical images and their distinct characteristics compared to natural images. Fine-tuning strategies, along with thoughtful layer selection, have proved to be pivotal in achieving optimal results while avoiding overfitting.

The adoption of transfer learning in medical image classification holds immense promise for improving diagnostic accuracy, aiding medical professionals in making informed decisions, and ultimately enhancing patient care. As the field continues to evolve, it is expected that further research will explore innovative transfer learning architectures, domain adaptation techniques, and data augmentation strategies tailored specifically to the nuances of medical image analysis.

References

1. Alzubaidi L, Al-Amidie M, Al-Asadi A, Humaidi AJ, Al-Shamma O, Fadhel MA, Zhang J, Santamaría J, Duan Y (2021) Novel transfer learning approach for medical imaging with limited labeled data. Cancers 13:1590. https://doi.org/10.3390/cancers13071590
2. Siddalingaswamy CKPC, Pathan S, D'souza N (2021) A multiclass skin lesion classification approach using transfer learning based convolutional neural network. In: 2021 seventh international conference on bio signals, images, and instrumentation (ICBSII), Chennai, India, pp. 1–6. https://doi.org/10.1109/ICBSII51839.2021.9445175
3. Simonyan K, Zisserman A, Visual Geometry Group (2015) Very deep convolutional networks for large-scale image recognition. Department of Engineering Science, University of Oxford, ICLR
4. He K, Zhang X, Ren S, Sun J (2015) Deep residual learning for image recognition. Microsoft Research
5. Ganin A, Massaro E, Gutfraind A et al (2016) Operational resilience: concepts, design and analysis. Sci Rep 6:19540. https://doi.org/10.1038/srep19540
6. Ching T, Himmelstein DS, Beaulieu-Jones BK, Kalinin AA, Do BT, Way GP, Ferrero E et al (2018) Opportunities and obstacles for deep learning in biology and medicine. J R Soc Interface 15(141):20170387
7. Gidaris S, Singh P, Komodakis N (2018) Unsupervised representation learning by predicting image rotations. arXiv:1803.07728
8. Chen T, Kornblith S, Norouzi M, Hinton G (2020) A simple framework for contrastive learning of visual representations. In: Proceedings of the 37th international conference on machine learning, PMLR, 119, 1597–1607
9. Kong B, Sun S, Wang X, Song Q, Zhang S (2018) Invasive cancer detection utilizing compressed convolutional neural network and transfer learning. In: 21st international conference, Granada, Spain, 16–20 Sept 2018, proceedings, part II. https://doi.org/10.1007/978-3-030-00934-2_18
10. Valieris R, Amaro L, Osório CABT, Bueno AP, Rosales Mitrowsky RA, Carraro DM, Nunes DN, Dias-Neto E, Silva ITD (2020) Deep learning predicts underlying features on pathology images with therapeutic relevance for breast and gastric cancer. Cancers 12(12):3687. https://doi.org/10.3390/cancers12123687. PMID: 33316873; PMCID: PMC7763049
11. Hamamoto R, Suvarna K, Yamada M, Kobayashi K, Shinkai N, Miyake M, Takahashi M, Jinnai S, Shimoyama R, Sakai A et al (2020) Application of artificial intelligence technology in oncology: towards the establishment of precision medicine. Cancers 12(12):3532. https://doi.org/10.3390/cancers12123532
12. Su Q, Wang F, Chen D, Chen G, Li C, Wei L (2022) Deep convolutional neural networks with ensemble learning and transfer learning for automated detection of gastrointestinal diseases. Comput Biol Med 150:106054. https://doi.org/10.1016/j.compbiomed.2022.106054. Epub ahead of print. PMID: 36244302
13. Sakai Y, Takemoto S, Hori K, Nishimura M, Ikematsu H, Yano T, Yokota H (2018) Automatic detection of early gastric cancer in endoscopic images using a transferring convolutional neural network. In: Annual international conference on IEEE engineering medical biology society, July 2018, 4138–4141. https://doi.org/10.1109/EMBC.2018.8513274. PMID: 30441266
14. Rajaraman S, Antani SK, Poostchi M, Silamut K, Hossain MA, Maude RJ, Jaeger S, Thoma GR (2018) Pre-trained convolutional neural networks as feature extractors toward improved malaria parasite detection in thin blood smear images. PeerJ 6:e4568. https://doi.org/10.7717/peerj.4568
15. Hosseini-Asl E, Ghazal M, Mahmoud A, Aslantas A, Shalaby AM, Casanova MF, Barnes GN, Gimel'farb G, Keynton R, El-Baz A (2018) Alzheimer's disease diagnostics by a 3D deeply supervised adaptable convolutional network. Front Biosci (Landmark Ed) 23(3):584–596. https://doi.org/10.2741/4606. PMID: 28930562
16. Mou L, Ghasemzadeh N, Dey D (2018) Learning from others: an embedding transfer framework for cardiovascular event prediction. In: Proceedings of the 24th ACM SIGKDD international conference on knowledge discovery & data mining, pp 1050–1059

17. Tu Z, Bai X (2010) Auto-context and its application to high-level vision tasks and 3D brain image segmentation. IEEE Trans Pattern Anal Mach Intell 32(10):1744–1757. https://doi.org/10.1109/TPAMI.2009.186

Integrating AI Tools into HRM to Promote Green HRM Practices

Jasno Elizabeth John and S. Pramila

1 Introduction

Sustainability has emerged as one of the most important aspects of any business organization. The central objective of corporate governance is to attain an elevated level of performance and profitability while safeguarding against management prioritizing their interests at the expense of stakeholders [1]. However, this goal is pursued without considering the environmental damage that it can create. As stated by the Chartered Institute of Personnel and Development in London [2], the core of organizational sustainability revolves around the idea of reinforcing the environmental, societal, and economic aspects of business operations. It highlights how the concept of sustainability enables a business to thrive while ensuring the well-being of future generations' needs is not overlooked [3].

Green HRM manifests an organization's commitment to sustainability, where HR strategies, policies, and practices are designed to not only enhance employee well-being but also to reduce the environmental footprint of its operations. To fulfill their sustainability objectives and to meet the expectations of the organizational stakeholders, both internally and externally, companies are increasingly turning to AI [4]. At the nexus of AI and Green HRM lies the prospect of reshaping the conventional HR functions, and revamping the traditional process of HRM like recruitment, development, engagement, and performance management to be eco-friendly and resource-optimized.

Incorporating AI tools into HRM practices presents a novel dimension of optimization and innovation, enabling organizations to harness data-driven insights, predictive analytics, and automation to achieve their sustainability objectives. The

J. E. John (✉) · S. Pramila (✉)
Christ (CHRIST (Deemed to be University), Delhi NCR, India
e-mail: jasno.john@res.christuniversity.in

S. Pramila
e-mail: pramila.s@christuniversity.in

M. S. Kaiser et al. (eds.), *ICT: Smart Systems and Technologies*, Lecture Notes in Networks and Systems 878, https://doi.org/10.1007/978-981-99-9489-2_22

ingenious employment of AI can contribute to reducing energy consumption, minimizing waste, and enhancing operational efficiency, all of which are central to the themes of Green HRM.

2 Literature Review

The organizations have developed strategies to deal with the ecological footprint's reduction, in addition to addressing economic concerns. In today's corporate landscape, to thrive in the corporate arena and to facilitate shareholder profit, organizations must concentrate on a holistic range of factors, encompassing not only economic and financial aspects but also social and environmental factors [5]. The Agenda 21 action plans for sustainable development also highlight the need for interdisciplinary coordination to integrate environmental and social issues into business practices [6].

The term 'Green HRM' was coined in 1996 by Wehrmeyer in his book 'Greening People,' marking the inception of its evolution. The author discussed improving sustainability through the most vital resource—Human Resources. It elaborated the idea of raising awareness about the significance of these practices, introducing policies for their practical implementation, and incentivizing individuals who adopt environmentally sustainable behavior through rewards to encourage their continued adoption. This also resulted in a substantial shift in work attitudes and personal motivation concerning their jobs and the organization [7]. Considering the current heightened emphasis on corporate environmental sustainability, contemporary HR managers have been tasked with the added duty of integrating Green HR principles into the corporate mission statements and HR policies. Green HRM fosters an environmentally conscious behavior and culture, both at the organizational and individual levels [8].

HRM as a critical function that directs the workforce, has been entrusted with the responsibility to drive initiatives that align with sustainability goals. The integration of AI tools into HRM processes presents an exceptional scope to achieve this target, underlining the tenets of Green HRM. A varied array of AI applications centered on Human Resource Management (HRM) is employed within organizations. These applications enhance the cost-efficiency of HR processes and consequently enhance the overall employee experience. This, in turn, leads to heightened levels of employee commitment and satisfaction while reducing instances of employee turnover [9]. We anticipate technology to not only streamline back-office operations but also progressively assume the more human-centric aspects of HR responsibilities. While these technologies undoubtedly enhance efficiency, reduce bias, and enhance the HR function's value within organizations, it is essential to recognize that the human touch will remain indispensable for achieving success [10]. Maintaining a harmonious equilibrium between technology and the human element is essential to ensure that employees' distinct requirements and emotions continue to be efficiently addressed. These tools should be integrated and utilized in such a way that they minimize human efforts and contribute to the efficient utilization of resources, thus fostering organizational sustainability.

3 Objectives

The primary purpose of this study is to:

1. To understand how AI tools support various factors of Green HRM.
2. To explore some real-world examples of AI-powered tools in HRM.
3. To list the challenges faced by organizations in implementing AI-powered tools in HRM.

4 Methodology

The research methodology for this study involves utilizing secondary data sources to investigate the integration of AI tools into HR practices to promote Green HRM practices. Secondary data, collected from various academic databases, industry reports, and organizational publications, will be analyzed to examine how AI tools are being employed to foster environmentally sustainable HRM strategies.

5 Exploring AI Tools for Enriching Green HRM Practices

The Human Resources Department within an organization plays a pivotal role in upholding the company's culture of sustainability [11]. Hence, it is imperative for all organizations emphasizing long-term success to enhance their Human Resource Management practices. This entails developing an effective framework encompassing elements such as performance appraisal and reward systems that incorporate environmental considerations into the evaluation process [12]. Additionally, organizations should implement comprehensive training and empowerment programs to facilitate the assimilation of new skills and competencies among employees, particularly in those aiming to be environmentally responsible and achieve long-term success. By leveraging AI's analytical ability and automation possibilities, organizations can position their HRM strategies with the concept of sustainability, ultimately leading to the creation of a greener, more socially responsible workforce.

5.1 Green Recruitment

Green Recruitment centers its efforts on implementing a paperless recruitment process to minimize its environmental footprint by adopting eco-friendly practices aimed at mitigating environmental [13]. Organizations are using Internet facilities to drive major green practices, which include online interviews, company websites, job portals, video conferencing methods, etc. Resumes are submitted electronically, and communication takes place via email and telephone, enabling organizations to minimize waste associated with printing and mailing hard copies of resumes. Adopting eco-friendly recruitment practices contributes to the company's reduced carbon footprint by diminishing paper usage and unnecessary travel. Green Recruitment facilitates a comprehensive evaluation of candidates' alignment with the organization's environmental culture and values [14] by assessing candidates' environmental knowledge, values, and beliefs [15]. The utilization of AI in HRM would also help to mitigate the influence of subjective factors like nepotism and favoritism in employee recruitment and selection [16].

5.2 Green Selection

Effective selection techniques are essential components of any organization's HRM function. Green selection means hiring employees who are well-informed about environmental practices and are more behaviorally inclined toward the protection of the environment. These days companies even consider candidates' environmental concerns and interests as selection criteria. They also try to include environment-related questions while evaluating the applicants during the screening process. Through the deliberate selection and retention of like-minded employees, the Green Recruitment process underscores an organization's commitment to collaboration for the betterment of environmental performance [17]. By deploying AI's analytical expertise, organizations can easily identify individuals who align with their environmental goals and possess the necessary skills and eco-conscious values. AI-driven algorithms sift through resumes and online profiles, identifying candidates with relevant experience in environmentally friendly sectors. This action not only ensures a talent pool committed to sustainability but also optimizes resource allocation and reduces operational carbon footprints.

5.3 Green Performance Management

Green performance management appraises the employee's environment-friendly behaviors concerning the organizational policies and strategies. This addresses the organizational policy challenges and the imperative of environmental conservation by incorporating environmental management within the framework of performance management. Numerous organizations have integrated environmental sustainability objectives into their performance appraisal system. This enables managers to engage in ongoing discussions with employees regarding their performance and provide feedback throughout the year, rather than solely during the annual evaluation process. Through AI-powered analytics, companies can monitor and assess ecological indicators, resource consumption, and carbon emissions in real time. This helps organizations track and reward employees who contribute to environmentally conscious initiatives. AI-driven analytics highlights trends and identifies areas for improvement, guiding strategic decisions to enhance sustainability efforts.

5.4 Green Training and Development

Green training and development initiatives center on educating employees about eco-conscious behaviors and environmentally sustainable practices that contribute to the organization's achievement of its sustainability objectives. This approach enhances employees' awareness of the organization's environmental goals and fosters a deeper understanding of various facets and principles of environmental management. Leveraging AI, organizations can tailor learning pathways that equip employees with skills aligned with sustainable practices and green technologies. Through personalized learning algorithms, AI adapts content delivery to individual needs, ensuring optimal comprehension and engagement. Companies use AI tools to enhance learning and development among their employees like free digital libraries to explore content [18].

5.5 Green Compensation Management

Green Compensation and Rewards represent a comprehensive incentive system, encompassing both financial and non-financial elements, with the overarching goal of attracting, retaining, and motivating employees to actively support and contribute to environmentally sustainable objectives [19]. Often, employees may not strongly align with the organization's environmental objectives. Nevertheless, this approach can significantly enhance employee engagement and foster their commitment to exert maximum effort in pursuit of organizational goals. The facets comprising Green Compensation and Rewards encompass Competency-Based Bonuses, Behavioral and Technical Incentives, Recognition for Green Environmental Performance, and Incentives for the adoption of environmentally responsible behavior [20]. AI's analytical capabilities allow organizations to quantitatively assess employees' various eco-actions such as their efforts in reducing carbon footprints, resource utilization, and embracing eco-friendly practices. By objectively measuring these contributions, AI facilitates fair and transparent compensation structures that align with environmental goals.

5.6 Green Employee Discipline Management

Employee discipline is one of the vital functions of any organization. Employee discipline also contributes to the success of any organization. Acting against the employees for their undesirable behavior is as important as rewarding them for any desirable behavior. Failure of group members to adhere to established rules can lead to the potential collapse of the organization, giving rise to chaos, confusion, disobedience, disloyalty, as well as anti-social and anti-organizational behaviors, ultimately harming the interests of all stakeholders [21]. By analyzing data patterns, AI can identify potential violations and deviations, prompting timely corrective actions. This integration ensures that environmentally conscious behaviors are consistently reinforced and maintained, fostering a culture of accountability and positive environmental impact.

5.7 Green Employee Retention

Organizations allocate substantial resources toward the recruitment of highly qualified employees and dedicate significant efforts to their training and development to foster long-term retention. Employee retention encompasses organizational initiatives aimed at cultivating a motivating work environment through sound practices and thoughtfully crafted policies that prioritize employee needs [22]. AI's data-driven insights enable organizations to personalize retention strategies by recognizing employees who actively contribute to sustainability efforts. This enables tailored retention plans that encompass eco-friendly perks, skill development, and recognition for sustainability contributions.

6 Examples of AI-Driven HR Tools

Pymetrics: This automated screening system is designed to identify candidates who exhibit attributes such as systemic thinking, resilience, and business acumen through the analysis of behavioral cues [23].

User: Unilever.

Contribution to Green HRM: Pymetrics' AI-driven assessments help match candidates to roles based on their cognitive and emotional traits, reducing the likelihood of mismatches. By placing the right people in the right roles, organizations can lower turnover rates, minimize resource waste associated with training and onboarding, and improve overall efficiency.

HireVue: HireVue offers a seamless integration within the Microsoft Teams platform, facilitating virtual interviews by providing structured questions and ratings directly within the Teams experience [24].

User: Microsoft.

Contribution to Green HRM: HireVue's video interview platform helps the candidates to attend the interviews virtually eliminating the need for travel. This contributes to lower carbon emissions by minimizing travel-related energy consumption and aligns with sustainable practices by utilizing digital communication tools.

Coursera for Business: Coursera's platform offers a wide range of online courses that allow employees to learn remotely [25].

User: IBM.

Contribution to Green HRM: The platform enables employees to access various training materials remotely. By reducing the need for employees to travel to training locations, organizations can significantly lower carbon emissions associated with commuting.

Degreed: Degreed's AI-driven platform offers personalized learning paths for employees [26].

User: Airbnb.

Contribution to Green HRM: It helps to reduce the need for printing training materials and minimizing waste. It also enables remote learning, reducing the need for physical resources and transportation.

Glint: Glint serves as the People Success Platform, harnessing real-time personnel insights to assist global organizations in enhancing employee engagement, fostering employee development, and driving improvements in business outcomes [27].

User: LinkedIn.

Contribution to Green HRM: Glint's AI-powered platform gathers employee feedback and analyzes engagement levels. By digitizing the survey and analysis process, organizations avoid the need for printed survey forms and associated administrative tasks.

Peakon (acquired by Workday): Peakon's platform uses AI to collect and analyze employee feedback to improve engagement and performance [28].

User: Capgemini.

Contribution to Green HRM: By utilizing a digital platform for feedback, companies reduce the need for paper-based surveys and feedback forms.

PayScale: PayScale empowers organizations to accurately determine job pricing, shape compensation strategies, and furnish executives with precise information swiftly, mitigating the risk of errors [29].

User: Adobe.

Contribution to Green HRM: PayScale's AI-driven compensation management platform provides real-time salary data and market insights. This eliminates the need for extensive research and paper-based benchmarking, reducing resource consumption and promoting data-driven decision-making.

7 Challenges of Using AI-Driven HR Tools

AI-driven HR tools offer significant benefits to streamline processes including improved efficiency, better decision-making, and enhanced candidate and employee experiences. However, they also come with their fair share of challenges when it comes to the implementation and usage of such tools. Below listed are a few instances:

- *Resistance to Change*: Any implementation of new tools disrupts the existing process leading to resistance from the employee side as they may be sometimes required to change the way they function. Sometimes, they are so accustomed to traditional methods that they find it difficult to adapt to the new requirements [30]. Employees may also feel uncomfortable about AI tools making decisions that may impact their careers directly or indirectly.
- *Loss of Human Touch*: Excessive dependence on AI-driven HR tools can lead to a disconnection from the human aspects of the HR process. While AI-based systems excel in talent identification, there remain certain tasks that require human intervention, including negotiations, assessing cultural compatibility, and fostering interpersonal relationships [31]. Also, core human values like relationship management, conflict resolution, understanding employees' requirements, and providing empathetic support might be bargained if we rely completely on AI for HR processes.
- *Data Availability and Quality*: AI-driven decisions largely rely on the dataset that they are trained on. Any outdated, inaccurate, or incomplete data can adversely affect the performance of these tools. AI models also lack an understanding of complex human relations, emotions, and values. This may lead to misinterpretation of data. AI tools also require continuous updates and improvements to remain effective.
- *Data Privacy and Security*: AI-powered HR tools generally require access to sensitive employee data, including health records, personal information, etc. This could raise serious concerns about the security and confidentiality of the data. AI tools can also become targets for hackers who seek to misuse personal information. A robust cybersecurity system, access controls, and data encryption are essential for the protection of such data.
- *High Cost and Lack of Skills*: Developing, implementing, and maintaining AI-driven tools are an irreversible decision that involves a huge investment. Organizations must perform a cost-benefit analysis to understand the potential benefits that they would derive from implementing the tools. However, a major challenge

would be a highly skilled and trained workforce who can handle these tools [32]. If the employees resist learning to use the tools, the actual benefits will be lost.

8 Conclusion

The future potential of Green Human Resource Management (HRM) appears promising for all HRM stakeholders, including organizations, individuals, employees, practitioners, and academics. The current generation of HR managers bears the responsibility of raising awareness among younger generations and organizational members regarding Green HRM, the green movement, sustainable utilization of natural resources, and aiding corporations in environmental stewardship to preserve these resources for future generations, thereby promoting sustainable development [33]. The integration of AI tools into HRM presents a driving force in promoting Green HRM practices. As organizations these days are more inclined toward sustainable growth, AI-driven insights equip HR professionals to make informed decisions that are in line with environmental objectives. The combination of technology and sustainability enables efficient utilization and allocation of resources, minimizing wastage, thereby streamlining the processes. AI endorses the adoption of eco-friendly policies, which in turn helps the employees to enhance their green knowledge and actively contribute toward green objectives. This will not only benefit the organization in terms of cost saving but will also contribute to the culture of sustainability, positioning them as diligent leaders in a rapidly changing world.

References

1. Zhelev C (2020) Corporate governance and marketing management issues and their reflection on customer satisfaction. In: 5th international conference on engineering and formal sciences, Brussels, 24–25 Jan 2020
2. CIPD (2012) Responsible and sustainable business: HR leading the way—a collection of 'ThoughtPieces.' Chartered Institute of Personnel and Development, London
3. Boudreau JW, Ramstad PM (2005) Talentship, talent segmentation, and sustainability: a new HR decision science paradigm for a new strategy definition. Hum Resour Manage 44(2):129–136
4. Garg V, Srivastav S, Gupta A (2018) Application of artificial intelligence for sustaining green human resource management. In: 2018 international conference on automation and computational engineering (ICACE). IEEE, Noida, India, pp 113–116
5. Daily BF, Bishop J, Steiner R (2007) The mediating role of EMS teamwork as it pertains to HR factors and perceived environmental performance. J Appl Bus Res 23:95–109
6. International Institute for Sustainable Development (2012) SD timetable 2012 [online]. https://www.iisd.org/sites/default/files/publications/sd_timeline_2012.pdf, 16 Oct 2021
7. Mishra RK, Sarkar S, Kiranmai J (2014) Green HRM: innovative approach in Indian public enterprises. World Rev Sci Technol Sustain Dev 11(1):26–42
8. Cherian JP, Jacob J (2012) A study of green HR practices and its effective implementation in the organization: a review. Int J Bus Manage 7(21):1–15. https://doi.org/10.5539/ijbm.v7n21p25

9. Malik A, Budhwar P, Patel C, Srikanth N (2020) May the bots be with you! Delivering HR cost-effectiveness and individualised employee experiences in an MNE. Int J Human Resour Manage 33:1148–1178. https://doi.org/10.1080/09585192.2020.1859582
10. Gikopoulos J (2019) Alongside, not against: balancing man with machine in the HR function. Strateg HR Rev. https://doi.org/10.1108/SHR-12-2018-0103
11. Harmon J, Fairfield KD, Wirtenberg J (2010) Missing an opportunity: HR leadership and sustainability. People Strategy 33:16–21
12. Jabbour CJ, Jabbour LS, Govindan K, Teixeira AA, Freitas WR (2013) Environmental management and operational performance in automotive companies in Brazil: the role of human resource management and lean manufacturing. J Clean Prod 47:129–214
13. Aranganathan PA (2018) Green recruitment, a new-fangled approach to attract and retain talent. Int J Bus Manage Res 8(2):69–76. https://doi.org/10.24247/ijbmrapr20189
14. Jackson SE, Renwick DWS, Jabbour CJC, Muller-Camen M (2011) State-of-the-art and future directions for green human resource management: introduction to the special issue. German J Human Resour Manage: Zeitschrift Für Personal forschung 25(2):99–116. https://doi.org/10.1177/239700221102500203
15. Renwick DWS, Redman T, Maguire S (2013) Green human resource management: a review and research agenda. Int J Manag Rev 15(1):1–14
16. Kshetri N (2021) Evolving uses of artificial intelligence in human resource management in emerging economies in the global South: some preliminary evidence. Manag Res Rev. https://doi.org/10.1108/mrr-03-2020-0168
17. Masri HA, Jaaron AA (2017) Assessing green human resources management practices in Palestinian manufacturing context: an empirical study. J Clean Prod 143:474–489. https://doi.org/10.1016/j.jclepro.2016.12.087
18. Gupta M (2021) The role of artificial intelligence in adopting green HRM practices. In: Reinventing manufacturing and business processes through artificial intelligence, 1–18
19. Rael JM (2018) Influence of green reward and compensation practice on environmental sustainability in selected service-based state corporations in Kenya. Euro J Bus Strat Manage 3(6):1–12
20. Ardiza F, Nawangsari LC, Sutawidjaya AH (2021) The influence of green performance appraisal and green compensation to improve employee performance through OCBE. Int Rev Manage Market 11(4):13–22. https://doi.org/10.32479/irmm.11632
21. Mamoria CB (1991) Personnel management. Himalaya Publishing House, Bombay
22. Sinha D, Shukla SK (2013) A study of employee retention in the pharmaceuticals sector in Dehradun City. Int J Educ Psychol Res 2(1):30–39
23. Barber S (2022) Unilever: using AI to freeze out competition & discover "breakfast for dessert." Digital Innov Transform. https://d3.harvard.edu/platform-digit/submission/unilever-sugary-sweet-moves-to-freeze-out-competitors-game-their-way-to-success/
24. Find the right app, Microsoft AppSource (n.d.) Appsource.microsoft.com. Retrieved 20 Aug 2023, from https://appsource.microsoft.com/en-us/product/office/wa200002365?tab=overview
25. IBM Online Courses (n.d.) Coursera. Retrieved 20 Aug 2023, from https://www.coursera.org/ibm
26. Sonja Schurig (2016) How to connect the world's best learning experiences. Degreed Blog. https://blog.degreed.com/how-to-connect-the-worlds-best-learning-experiences/
27. LinkedIn learning and glint integration to help managers better connect with employees (n.d.) News.linkedin.com. https://news.linkedin.com/2020/may/linkedin-learning-and-glint-integration-to-help-managers-better
28. Goodwin B (2018) Capgemini turns to tech startup to identify "most engaged" teams. Computer Weekly. https://www.computerweekly.com/news/252437584/Capgemini-turns-to-tech-startup-to-identify-most-engaged-teams
29. Thangjam T (2021) Top15 compensation management tools in 2022. Nurture an Engaged and Satisfied Workforce, Vantage Circle HR Blog. https://blog.vantagecircle.com/top-compensation-management-tools/

30. Johansson J, Herranen S (2019) The application of artificial intelligence (AI) in human resource management: Current state of AI and its impact on the traditional recruitment process
31. Upadhyay AK, Khandelwal K, Upadhyay AK, Khandelwal K (2018) Applying artificial intelligence: implications for recruitment. Strat HR Rev 17(5):255–258
32. Baldegger R, Caon M, Sadiku K (2020) Correlation between entrepreneurial orientation and implementation of AI in human resources management (HRM). Technol Innov Manage Rev
33. Mathapati CM (2013) Green HRM: a strategic facet. Tactful Manage Res J 2(2):1–6

Archival of *Rangabati* Song Through Technology: An Attempt to Conservation of Culture

Jayasmita Kuanr and Deepanjali Mishra

1 Introduction

The *Rangabati* folk song is a priceless piece of Indian culture that comes from the state of Odisha. Its significant aesthetic and historical importance reflect the customs, values, and legacy of the neighbourhood. *Rangabati*, like many other traditional folk tunes, has difficulties being preserved and passed along to future generations. The goal of this study is to examine how technology might be used to preserve and archive the *Rangabati* folk song while also fostering greater accessibility to and appreciation for it among a variety of audiences. Similar to many other folk tunes, it might be difficult to preserve and transmit *Rangabati* to new generations [1]. The purpose of this research is to look at how technology may be used to conserve and archive the *Rangabati* folk song while also increasing its accessibility and enjoyment across a wide range of listeners. The song praises Odisha's customs and cultural past, showing the region's creative capability. *Rangabati*, despite its historical significance, presents preservation challenges owing to oral transmission, rendering it prone to alterations and deterioration through time.

Social and cultural developments, urbanisation, and the impact of modern music all pose further challenges to the preservation and continuous transmission of these tunes. This study investigates cutting-edge methods that use technology to preserve and spread old folk tunes like *Rangabati* as a solution to these problems. It aspires to secure the protection of *Rangabati's* cultural history while promoting better accessibility and appreciation among broad audiences by using technical improvements like as digital archiving, internet platforms, and interactive tools. The outcomes of this study will add to a greater knowledge of traditional folk music preservation in the digital era, emphasising the relevance of cultural preservation, technical innovation, and inclusive distribution techniques. The preservation and promotion of *Rangabati*

J. Kuanr · D. Mishra (✉)
KIIT (Deemed to be University), Bhubaneswar, India
e-mail: Deepanjalimishra2008@gmail.com

M. S. Kaiser et al. (eds.), *ICT: Smart Systems and Technologies*, Lecture Notes in Networks and Systems 878, https://doi.org/10.1007/978-981-99-9489-2_23

and other folk songs would assist to conserve Odisha's cultural past and allow future generations to understand and engage with this priceless piece of Indian culture. The research on the archival of the *Rangabati* song through technology has the potential to significantly impact of cultural preservation by making cultural heritage more accessible, implementing research and education, and promoting and preserving oral traditions.

2 Background and Significance of *Rangabati* Folk Song

Midway through the 1970s, the song was initially produced for All India Radio, and Odisha and surrounding states soon became fans of it. The lyrics of *Rangabati* underline traditional melodies, rhythmic beats, and Sambalpuri language and creates a unique and mesmerising musical experience. The song honours the traditional customs of the area while showcasing Odisha's rich cultural legacy and creative prowess.

2.1 *Historical Context of* Rangabati

An important part of India's Odisha state's cultural history is the folk song *Rangabati*. All India Radio made the first recording of it in the middle of the 1970s. When it started to be played often during wedding processions and during the immersion of Murtis (idols), the song rose to prominence in the 1970s and 1980s. It immediately gained popularity in the area due to its memorable melody and stirring lyrics [2].

2.2 *Cultural Importance and Symbolism of* Rangabati

In Odisha as well as numerous regions of West Bengal, Bihar, Jharkhand, Andhra Pradesh, and Chhattisgarh, *Rangabati* is of utmost cultural significance. It depicts the passionate sentiments of two lovers in a wooded village and is an example of traditional Sambalpuri/Odia folk song. The song serves as an ambassador for Odisha's culture by representing the region's history, language, and customs.

2.3 Challenges in Preserving Traditional Folk Songs

Rangabati confronts preservation issues, much like many other traditional folk tunes. Folk songs are handed down orally through the generations, rendering them susceptible to deterioration and change over time. The continuing transmission and preservation of these songs face difficulties due to elements including social and cultural change, urbanisation, and the impact of current music. Investigating novel approaches to preserve and promote traditional tunes like *Rangabati* is so essential.

3 Utilising Technology for Preservation

Utilising technology for the preservation of folk songs has revolutionised the way we safeguard and promote cultural heritage. With the advent of advanced recording equipment, digitisation techniques, and online platforms, traditional folk songs can now be captured, preserved, and shared with a global audience.

3.1 Audio and Video Recording Techniques

Folk tunes like *Rangabati* are greatly preserved because of technology [3]. Techniques for audio and video recording make it possible to capture the subtleties of melody, rhythm, and lyrics in high-quality recordings of performances. Future generations can use these recordings as important archives to research, deconstruct, and appreciate the song's original performance.

3.2 Digitisation and Online Archives

Analogue recordings may be transformed into digital forms by digitisation, assuring its accessibility and lifespan. Digital recordings are stored in online archives, where a worldwide audience may easily access them. *Rangabati* and like folk songs may be preserved and made accessible to scholars, enthusiasts, and artists by having a dedicated online repository for them.

3.3 Virtual Collaborations and Online Platforms

Technology makes it possible for folk performers, artisans, and lovers from many regions to collaborate virtually and share information and techniques [4]. Musicians may collaborate, interpret, and produce new versions of *Rangabati* through online sites. Through such partnerships, the song's basic elements are preserved while remaining relevant and adaptable to current musical trends.

3.4 Preservation of Lyrics and Musical Notations

Technology can help in the preservation of *Rangabati's* lyrics and musical notations in addition to audio and video recordings. The original lyrics and notations' correctness and accessibility for future use are ensured by digitising and storing them. This explanation makes it easier to comprehend and analyse the song's lyrical complexity, melody, and structure.

4 Enhancing Accessibility and Dissemination

Enhancing accessibility and dissemination of folk songs is an essential aspect of preserving and celebrating cultural heritage. Technology plays a pivotal role in expanding access to folk songs, breaking down geographical barriers, and reaching a wider audience. Digital platforms and streaming services have made it possible to share folk songs with people around the world, transcending physical limitations. Online archives and databases provide a centralised repository for storing and organising folk songs, ensuring their easy retrieval and exploration.

4.1 Global Reach Through Online Platforms

Through internet distribution channels like social media, music streaming services, and video sharing websites, technology permits the global propagation of *Rangabati*. It improves appreciation and knowledge of this cultural treasure among many cultures throughout the world by making the song available to a larger audience outside the confines of Odisha [5]. *Rangabati's* accessibility on online platforms has made it possible for those who may not have previously been exposed to Odia folk music to find it. By exhibiting the distinctive musical traditions of Odisha to a worldwide audience, this exposure contributes to fostering the diversity and depth of India's cultural legacy. Additionally, it opens doors for cross-cultural contact and cooperation between enthusiasts and artists.

4.2 *Translation and Transcription for Broader Understanding*

It is essential to translate and transcribe *Rangabati's* lyrics into many languages in order to improve accessibility and foster a deeper understanding of the song. People who don't speak the native Odia language can comprehend the meaning and core of the songs by providing translations. This makes it possible for a larger audience to relate to the feelings and cultural importance the song has. For those interested in analysing the song's structure, lyrical devices, and linguistic intricacies, transcriptions of the lyrics in addition to translations are an invaluable resource. The ability to analyse lyrics in-depth thanks to transcriptions makes it easier to do academic study and explore different cultures [6]. Care should be taken during the transcription and translation processes to guarantee accuracy and maintain the original lyrics' intended meaning.

To capture the subtleties, analogies, and cultural allusions included in the song, intensive collaboration with language specialists, linguists, and cultural consultants is necessary. *Rangabati's* translation and transcription make it possible for collaborations and modifications in various linguistic and cultural situations. It is possible for musicians from various backgrounds to relate to the song, interpret it using their own musical idioms, and produce versions that are popular in their individual communities. *Rangabati* can help people to communicate across languages and cultures by offering translations and transcriptions. It makes it possible for people from various language backgrounds to recognise the song's beauty, relate to its themes, and acquire understanding of the rich cultural history it symbolises.

4.3 *Adaptation of* Rangabati *in Contemporary Contexts*

Technology can make it easier to adapt *Rangabati* for use in modern settings, enabling creative partnerships and interpretations. While retaining the essential elements of the original song, musicians might experiment with other musical genres, styles, and arrangements. *Rangabati* can attract new listeners and remain relevant in the competitive music market of today by adopting modern aspects, such as fusion with other musical genres or the use of modern instruments.

4.4 *Cultivating Interest Among Younger Generations*

It is essential to include younger generations in the promotion and preservation of *Rangabati*. Technology which offers ways to pique young people's attention and curiosity [7]. Younger people can be exposed to the music in an approachable and captivating way through social media promotions, online competitions, and interactive platforms. The history, cultural importance, and performance skills of *Rangabati* may be taught through educational efforts like workshops and online tutorials, ensuring its continuance and relevance for future generations. *Rangabati* may break down generational barriers and become a shared cultural experience that appeals to people of all ages by using technology to adapt and engage younger audiences.

5 Review of Literature

To investigate the preservation and promotion of this traditional folk song *Rangabati*, it is necessary to analyse current literature on the issue. This literature review will give insights on *Rangabati's* cultural significance, the difficulty of preserving traditional folk songs, and the role of technology in their preservation. *Rangabati's* cultural importance and symbolism have been thoroughly examined in many sources. It is recognised not just in Odisha but also in West Bengal, Bihar, Jharkhand, Andhra Pradesh, and Chhattisgarh as a representation of the Sambalpuri/Odia folk song heritage. *Rangabati* depicts the intense feelings of two lovers in a remote community while displaying Odisha's rich cultural heritage, language, and customs [8]. Its popularity skyrocketed in the 1970s and 1980s, and it became a favourite song for wedding processions and religious events [9]. Keeping ancient folk songs like *Rangabati* alive presents a number of obstacles. Folk songs are generally passed down orally, rendering them prone to change and degradation over time. Social and cultural changes, urbanisation, and the influence of modern music all have an impact on the transmission and preservation of these songs. These problems demand novel techniques to ensuring the survival and accessibility of traditional folk melodies.

The role of technology in the preservation and dissemination of cultural material, particularly folk tunes, is critical. Digital archiving, internet platforms, and interactive technologies have emerged as important resources for the preservation of traditional music (Coca-Cola Deutschland). The use of technology can allow for the preservation and digitisation of historical recordings, the dissemination of traditional melodies to a larger audience, and the creation of interactive experiences that engage and educate people. However, it is critical to strike a balance between technology interventions and the authenticity and integrity of folk tunes. Collaborations between technology professionals and folk musicians are encouraged in order to ensure that technological advances respect the artistic purpose and cultural peculiarities of the songs. Folk music preservation relies heavily on ethical issues. *Rangabati's* originality and integrity must be respected by faithfully preserving its lyrics, melodies, and traditional components in any digital recordings or replicas. In order to achieve a balance between innovation and cultural preservation, inclusive decision-making procedures that include folk musicians and stakeholders are required [10].

6 Methodology

In order to get insights and fully comprehend phenomena, qualitative analysis is my research method that involves gathering and interpreting non-numerical data. The following actions have followed while doing qualitative analysis in the context of conserving folk music like *Rangabati*.

6.1 Data Gathering

Rich and thorough data have gathered for qualitative analysis using techniques including focus groups, observations, document analysis, and interviews. Interviewing folk musicians, local residents, and subject-matter specialists helped me gather information about the preservation of folk tunes. Observations of rituals, performances, and cultural events offer insightful information. The folk song's historical records, recordings, and archive materials also examined in this research.

6.2 Data Coding

After collection of data, it is coded in order to be subjected to a methodical analysis to locate and label common themes, ideas, or patterns in the data. This has done by axial coding, where interactions and connections between categories are built, and open coding, where initial categories are produced.

6.3 Data Analysis

Following the coding of the data, we have done data analysis to find recurrent themes, patterns, and meanings to analyse the connections between various codes and creating interpretations and justifications. Depending on the size and complexity of the study, we have analysed the data manually.

6.4 Interpretation and Findings

By developing insightful interpretations and conclusions from the data through the analytical process by understanding the folk song's cultural relevance, figuring out what influences its preservation, and examining how technology affects its promotion are all part of this. These interpretations are backed by direct quotes or examples from the data and are based on the knowledge of the subject.

6.5 Validity and Reliability

This qualitative analysis is valid and reliable. Reliability is the consistency and repeatability of the analytical process, whereas validity is the correctness and veracity of the interpretations and results were used triangulation to increase validity by gathering information from several sources and viewpoints. Additionally, by keeping a record of the research process helped increase accuracy and dependability.

A comprehensive and complex knowledge of cultural assets and preservation activities I have gained through qualitative analysis. It enables academics to dive deeply into the nuances and meanings connected with folk songs like *Rangabati*, guiding future study areas, preservation methods, and policy decisions. The current educational landscape places significant emphasis on instilling appropriate religious, moral, spiritual, cultural, and conventional values within elementary-level youngsters. This imperative is particularly challenging due to the rapid advancements in artificial intelligence and materialistic dynamics that characterise our contemporary world. Folklore, which is an interdisciplinary phenomenon encompassing conventional knowledge, art, literature, and practise, has historically been transmitted primarily through oral communication and more recently through digitisation. It can be argued that folklore serves a culturally responsive function by functioning as a network of diverse cultural expressions and serving as a foundation for common human behaviour [11].

7 Ethical and Cultural Consideration

Ethical and cultural considerations are of paramount importance when dealing with folk songs. Folk songs are deeply rooted in the traditions, values, and identities of specific cultures and communities. It is crucial to approach the collection, documentation, and dissemination of folk songs with respect, sensitivity, and a commitment to cultural preservation. This entails obtaining informed consent from the performers and communities involved, ensuring that they are aware of the purpose and potential impact of sharing their songs. By upholding ethical and cultural considerations, we can foster cultural appreciation, preserve authenticity, and contribute to the vitality and diversity of folk song traditions.

7.1 *Authenticity and Integrity of* Rangabati

The *Rangabati* folk song's purity and originality must be preserved at all costs. Any technical alterations or changes should make an effort to preserve the song's cultural importance and original spirit. Any digital recordings or replicas must accurately capture the lyrics, melodies, and other traditional components [12]. *Rangabati's* original lyrics, melodies, and cultural components must be preserved while converting it for digital recordings or reproductions. This entails keeping the song's unique rhythmic structures, voice timbres, and traditional instruments. In order to ensure that any modifications or remixes maintain the originality and integrity of *Rangabati*, collaborations between technological professionals and folk musicians can assist strike a balance between innovation and cultural preservation. Technology may be a potent instrument in preserving and promoting this cultural asset by recognising and protecting the authenticity of *Rangabati*, allowing future generations to enjoy and interact with this folk song while embracing technological improvements.

7.2 *Collaboration Between Technology Experts and Folk Artists*

Collaboration between technologists and folk artists is essential to navigating the area where technology and cultural preservation meet. This partnership can make sure that technical advancements respect the *Rangabati* folk song's creative intent and cultural specifics. Folk musicians ought to actively participate in decision-making processes, contributing their knowledge and ideas to maintain the song's authenticity. Folk artists' invaluable ideas can direct technology interventions by including them in the decision-making processes [12]. Folk musicians have a thorough awareness of the *Rangabati* folk song's background, distinctive characteristics, and performance strategies. Their engagement guarantees that any technical advancements or efforts maintain the song's originality and core meaning. Experts in the field of technology can provide their knowledge in fields like audio recording, digitalisation, and web platforms. They may offer information on the most recent technology developments and resources that can be used to successfully promote and conserve *Rangabati*.

They can also assist in resolving any technological issues that come up along the procedure. Technology specialists and folk artists may have fruitful dialogues while exchanging information and ideas via open and respectful collaboration. This cooperative method enables a comprehensive comprehension of the *Rangabati* folk song's cultural relevance and artistic integrity, ensuring that any technological interventions are in line with the goals and values of the folk community. Technology specialists and folk artists can collaborate to create a balance between innovation and cultural preservation. They can provide digital recordings, platforms, or interactive experiences that preserve the essence and authenticity of the song while also improving accessibility and reach for *Rangabati*. By working together, the folk artists are given the opportunity to actively shape technology advancements, preserving and celebrating their cultural legacy in the digital era.

7.3 Balancing Innovation and Cultural Preservation

A crucial factor is striking the ideal balance between innovation and cultural preservation. While technological advancements provide us new options for preservation and distribution, it is crucial to keep the traditional spirit of the *Rangabati* folk song intact [13]. Innovation should be carefully considered, making sure that it strengthens rather than weakens the cultural legacy represented in the song. Engaging folk musicians and cultural specialists who have a thorough grasp of the *Rangabati* folk song is crucial to achieving this balance. Their expertise could guide technology use in a way that enhances the artistic purpose and complex cultural elements of the music that can help in the discovery of pertinent software, hardware, and technological advancements that safeguard and enhance the song's conventional components. Interactive platforms, audio and video recording techniques, digital archival materials, and online collaborations are just a few of the areas where innovation lies.

This could involve creating interactive websites or mobile applications that allow users to explore the history, lyrics, and musical elements of the song. Additionally, these platforms can facilitate community engagement by providing spaces for discussions, sharing of personal stories, and collaborative projects related to the song [14]. Throughout the process, the folk community and other significant stakeholders must also be regularly contacted and provided feedback that ensures that innovation and cultural preservation are balanced and that any adjustments or enhancements may be made to align with the goals and values of the community. By striking the right balance, innovation might be a powerful tool for maintaining, promoting, and revitalising the *Rangabati* folk song. This will ensure that it continues to have its traditional qualities and cultural value for future generations while making it more widely heard.

8 Case Studies and Success Stories

8.1 Examples of Technology-Based Interventions in Folk Song Archiving

These illustrations demonstrate how technology has helped to advance and preserve folk song archiving. Technology makes traditional music more accessible, documentable, analysable, and disseminable while also insuring its long-term preservation and cultural significance [12]. Examples include digitising recordings, building online archives, including multimedia displays, and using analytical tools.

Example 1: The Digital Library of Traditional Music (DLTM)—This is an example of how technology may revolutionise the way traditional music is preserved, made available, and developed. Modern recording methods, meticulous cataloguing, and the availability of an internet platform are used by DLTM to guarantee that folk songs from different cultures are preserved, honoured, and made accessible for future generations [12]. This online resource offers a thorough collection of traditional music, complete with audio files, supporting materials, and background information that promotes enjoyment and understanding across cultural boundaries.

Example 2: Smithsonian Folkways Recordings—In order to conserve and promote folk music from many traditions, Smithsonian Folkways Recordings, a non-profit record label connected to the Smithsonian Institution, has embraced technology. Folk music recordings, field recordings, and oral histories are all included in their sizable collection. Smithsonian Folkways has embraced digital platforms and streaming services in addition to traditional media to make its huge catalogue available to people all over the world. They have created interactive websites, mobile applications, and online educational tools in partnership with technological professionals that offer in-depth knowledge, contextual materials, and audio-visual content to improve the comprehension and appreciation of folk music.

Example 3: The Alan Lomax Archive—The renowned folklorist Alan Lomax gathered ethnographic field recordings, which are preserved and made available through the Alan Lomax Archive, a fascinating online archive. Folk songs, traditional music, and oral histories from all across the world are included in this enormous collection. The archive converts analogue recordings into digital forms using cutting-edge digitisation processes to ensure their preservation and accessibility. Users can examine the recordings, access supplementary information, and contribute to the archive through crowdsourcing programmes using the web platform, which offers an immersive experience. A worldwide community is involved in the preservation and interpretation of folk music history through this collaborative method.

These illustrations show how technological innovations have revolutionised the collection, preservation, and study of folk tunes. These projects have greatly improved access to and knowledge of classical music by leveraging digitalisation, internet platforms, interactive technologies, and crowdsourcing.

Impact on Preservation and Promotion of *Rangabati*—Analysing how technological interventions affect the promotion and preservation of the *Rangabati* folk song is essential. Digital platforms and online repositories have expanded the global audience's access to the *Rangabati* folk song [15]. Folk music aficionados and scholars now have better access to and discovery of the tune thanks to websites, audio streaming services, and internet archives. Due to its wider audience and ability to cross-generational and regional borders, this accessibility has contributed to the song's preservation. It also looks at the impact and scope of modifications and remixes, taking into account both advantages and disadvantages.

Lessons Learned and Best Practices—The effort to preserve folk music is continuing, as technology advances with time. Maintain a constant eye on technological and preservation breakthroughs in order to adjust and enhance the procedures employed. Review and update preservation strategies often to make sure they adhere to changing standards and best practises. We should make sure that the preservation efforts are with the song's original spirit and be aware of the song's history, significance, and traditional elements [16]. We should treat the preservation process with regard to the folk song's traditions and cultural sensitivity. Engage the neighbourhoods' residents, musicians, and specialists to learn more about them and get them involved in the preservation efforts. By working together, it will be possible to preserve the song's originality and subtleties. The long-term protection and enjoyment of these priceless cultural artefacts may be achieved via the use of technology tools, community involvement, and a strong focus on authenticity and cultural preservation.

9 Challenges and Future Directions

Folk songs face a range of challenges in the modern world, but there are also exciting opportunities for their future direction. One significant challenge is the risk of losing traditional knowledge and oral traditions as younger generations become increasingly disconnected from their cultural roots. The globalisation and homogenisation of popular music pose another challenge, as they can overshadow and diminish the visibility of folk songs. Additionally, the lack of resources, funding, and institutional support often hampers efforts to preserve, document, and promote folk songs. In some regions, access to the necessary technology infrastructure, such as computers, internet connectivity, and recording equipment, may be limited. To address this, the research project could explore ways to make technology more accessible, potentially through partnerships with organisations that provide technology resources to underserved communities. The *Rangabati* song is likely to be in a regional language, which may pose challenges for transcription, translation, and documentation. The research can employ local experts and involve community members who are fluent in the language to ensure accurate representation. Digital archives require ongoing maintenance and updates [17]. The research should consider the long-term sustainability of the archive, including funding, storage, and regular maintenance to ensure its accessibility for future generations. However, there is hope for the future. By addressing these challenges and leveraging the potential of technology, we can ensure that folk songs continue to thrive, evolve, and contribute to the rich cultural tapestry of our world.

9.1 Technological Barriers and Solutions

There are various technological obstacles to employing technology to preserve and archive the *Rangabati* folk song. To guarantee the efficient and correct preservation of our cultural legacy, it is essential to recognise and solve these problems. Key technological obstacles include the following:

Audio Quality: Various audio formats, such as dated recordings or live performances with subpar audio quality, exist for the folk song from *Rangabati*. This makes it difficult to preserve the song's originality and capture its actual character. It can be challenging to record fine details and delicate musical components when the audio quality is poor since it might damage the recordings' clarity and fidelity.

Solution: One can enhance the audio quality of existing recordings by using cutting-edge audio restoration methods and equipment. These methods focus on lowering noise, improving clarity, and repairing or improving damaged or deteriorating audio. By using such technologies, we may improve the listening experience and guarantee that the song's original sound is preserved.

Digitisation Challenges: Traditional music preservation frequently faces the issue of transferring analogue recordings or physical material into digital formats. To transfer the material without loss or harm from fragile or degrading formats like vintage vinyl records or cassette tapes, additional precautions and knowledge may be necessary.

Solution: These difficulties can be solved by collaborating with archival specialists and audio technologists who specialise in digitisation. They are equipped, knowledgeable, and skilled enough to handle fragile media with care, resulting in accurate and excellent digitalisation. Maintaining the integrity of the original content also benefits from adhering to industry best practises and standards for digitisation.

Specialised Tools or Software: It is necessary to use specialised equipment or software created, especially for archiving and preserving audio recordings in order to preserve the *Rangabati* folk song. The particular needs for documentation, metadata management, and long-term preservation should be addressed by these technologies, which should also be tailored to the distinctive qualities of folk music.

Solution: The creation of specialised tools or the adaptation of existing software can be facilitated by working together with technology experts and developers who have experience in cultural preservation. These programmes can help with audio file management and organisation, metadata integration, and preservation format and standard compatibility.

Technology hurdles may be surmounted and the *Rangabati* folk song can be effectively preserved by utilising these technology developments and working with specialists in the field. This cultural legacy may be made better, more accessible, and longer lasting for future generations to enjoy and value via the use of the proper tools, techniques, and knowledge.

9.2 Copyright and Intellectual Property Issues

The intricate web of copyright and intellectual property laws must be taken into account while preserving and promoting the *Rangabati* folk song. In order to protect the rights and interests of artists and the cultural community, this section examines the legal and moral implications of using technology. It also offers tips on how to navigate copyright difficulties.

9.3 Sustainability and Long-Term Preservation

It is essential to guarantee the long-term viability of technological preservation initiatives. The techniques for continuous upkeep, storage, and accessibility of digital archives are covered in this part, taking into consideration new technical developments, changing format requirements, and the requirement for frequent upgrades and backups. The continued upkeep, preservation, and accessibility of digital archives have ensured using the following tactics:

Digital migration and format migration: Changes in file formats and storage medium result from technological improvements. For digital assets to remain accessible and avoid obsolescence, regular migration to modern formats and storage methods is necessary. Updates to audio and video codecs, file conversion to open standards, and data migration to trustworthy and secure storage systems are all included in this.

Backup and redundancy: To safeguard against data loss or corruption, it is essential to implement reliable backup and redundancy solutions. Digital archives should be preserved in several copies across a variety of locations to ensure redundancy and disaster recovery capabilities. To ensure data integrity, regular backups should be made and integrity tests should be carried out.

Metadata and documentation: For the long-term preservation and discoverability of digital archives, appropriate metadata and documentation are crucial. Carefully managed metadata should describe the cultural context, language specifics, and technological requirements of the saved information. Information on digitisation procedures, preservation tactics, and any alterations or changes made to the source material should all be included in the documentation.

Collaboration with institutions and specialists: For the long-term preservation of digital archives, collaboration with archival institutions, libraries, cultural organisations, and technical specialists may be extremely helpful. In order to handle new issues and possibilities, partnerships may assist in securing financing, exchanging best practises, and engaging in continuous research and development.

Continuous monitoring and quality assurance: It is important to regularly monitor digital archives to spot and resolve any problems that can compromise their integrity or accessibility. To guarantee the authenticity and dependability of the stored material, quality assurance mechanisms such as routine audits, metadata validation, and content verification should be put into place.

Funding and Resource Allocation: Allocating sufficient funds and resources is essential for the long-term viability of technology-based preservation initiatives. The funds required for continuing maintenance, infrastructure improvements, and capacity building can be obtained via securing financial backing from governmental organisations, private foundations, or crowdsourcing campaigns.

The evaluation of interventions and representations should not be solely based on their nature as interventions and representations, even when they may result in the preservation of static, everyday cultural elements [10]. These techniques can be used by technology-based preservation projects to guarantee the long-term viability of digital archives, protecting the cultural legacy ingrained in popular songs like *Rangabati*. Future generations will have access to and be able to enjoy these priceless cultural resources thanks to ongoing monitoring, cooperation, and resource allocation. Effective technology integration in cultural preservation often requires collaboration between experts in technology, culture, history, and other fields. The research should foster interdisciplinary collaboration to address complex challenges.

10 Conclusion

The research will result in the creation of a comprehensive digital archive of the *Rangabati* song, encompassing audio recordings, lyrics, historical context, and related cultural elements. This archive will serve as a valuable resource for scholars, educators, and enthusiasts interested in the song's cultural significance. By digitising the *Rangabati* song, the research will contribute to the preservation of an important oral tradition within the cultural heritage of the region. This ensures that the song's unique melodies, lyrics, and cultural significance are safeguarded for future generations. The research is expected to introduce innovative technological solutions and methodologies for archiving and preserving cultural artefacts. These advancements may have broader applications beyond this specific project, benefiting the broader fields of cultural preservation and technology integration. In the end, this study aims to add to the larger conversation about using technology to preserve cultural heritage. These suggestions cover community involvement, ethical issues, technical improvements, and the requirement for continual study and collaboration. In conclusion, the research on the archival of the *Rangabati* song through technology is poised to make substantial contributions to the preservation of cultural heritage, technological innovation, cross-cultural understanding, and the ethical dimensions of digital archiving within the broader field of cultural preservation and technology integration.

References

1. ACM Digital Library. https://doi.org/10.1145/3383583.3398544. Accessed 18 June 2023
2. And the singer sings his song. The Hindu. 27 May 2001. Archived from the original on 4 May 2003. Retrieved 25 Mar 2016
3. Coke studio version of 'Rangabati': weird rendition of cult Oriya song sparks massive outrage. Firstpost. 6 July 2015
4. Enchanting moment with Ramesh Mahananda— ରମେଶ ମହାନନ୍ଦଙ୍କ ସହିତ ସାକ୍ଷାତକାର—Odia Music News. www.odiamusic.com. Retrieved 25 Mar 2016
5. I want to keep folk music safe?. www.telegraphindia.com. Archived from the original on 9 May 2014
6. I want to keep folk music safe. The Telegraph. Archived from the original on 30 June 2013. Retrieved 25 Mar 2016
7. Szwed J, Lomax A (2010) The man who recorded the world. Penguin, New York, p 384
8. Ojha J, Sain M, Mishra D (2019) Importance of women folklore in education: an analysis with reference from past to present. ICETT 2019, 27–29 May 2019, Seoul, Republic of Korea. © 2019 Association for Computing Machinery, 13–16 (2019). https://doi.org/10.1145/3337682.3339889
9. Smithsonian Magazine (2004) All music is folk music. Smithsonian.Com, 1 July 2004. www.smithsonianmag.com/arts-culture/all-music-is-folk-music-2580774/
10. Chatterjee P, Mishra D, Padhi LK, Ojha J, Al-Absi AA, Sain M (2019) Digital story-telling: a methodology of web based learning of teaching of folklore studies. In: International conference on advanced communications technology (ICACT), 573–578. ISBN 979-11-88428-02-1
11. Popular Odia song *Rangabati* new buzz word of Bollywood! Thanks to Nilamadhab Panda, Odisha Current News, Odisha Latest Headlines. www.orissadiary.com. Retrieved 25 Mar 2016
12. ରଙ୍ଗବତୀର ଗାୟକ ଜିତେନ୍ଦ୍ରିୟ, କ୍ରୀଷ୍ଣାଙ୍କୁ ଡକ୍ଟରେଟ ସମ୍ମାନ. www.odisha.com. Retrieved 25 Mar 2016
13. Rangabati' music composer Prabhudatta Pradhan no more. Odishatv.in
14. 'Rangabati' Promo—Ram Sampath—Coke Studio@MTV Season 4 Episode 4. Coca-Cola Deutschland. 26 June 2015. Retrieved 3 July 2015
15. Social media abuzz as Korean girls dance to *Rangabati* beats—TOI Mobile—The Times of India Mobile Site. m.timesofindia.com. Retrieved 25 Mar 2016
16. Welcome break for singer. The Hindu. 4 Apr 2007. ISSN 0971-751X. Retrieved 25 Mar 2016
17. Watch: *Rangabati* cover song by African singer Samuel Singh. 20 Feb 2018

Voice-Based Virtual Assistant for Windows Using ASR

R. Adline Freeda, V. S. Krithikaa Venket, A. Anju, Gugan, Ragul, and Rakesh

1 Introduction

Nowadays, each and every person needs an assistant to do his/her tasks. This is where AI plays a major role. There are many virtual assistants like Siri, Google assistant, Alexa, and Cortana. The first three virtual assistants are used in mobile phones and Cortana is a VA for computer systems for Microsoft OS. Siri is considered as one of the best AI assistants used in mobile phones in which it acts like a personal assistant for the user. Artificial intelligence (AI) is the capacity of a computer or a robot controlled by a computer to perform duties typically performed by humans because they call for human intelligence and judgment. Reactive memory, restricted memory, theory of mind, and self-awareness are the four subtypes of artificial intelligence. Artificial intelligence (AI) applications such as machine learning allow systems to automatically learn from their experiences and get better over time without having to be explicitly designed. The creation of computer program that can access data and use it to learn for themselves is the main goal of machine learning. AI is a bigger concept to create intelligent machines that can simulate human thinking capability and behavior, whereas machine learning is an application or subset of AI that allows machines to learn from data without being programmed explicitly. A branch of artificial intelligence called "machine learning" allows robots to pick up knowledge from past data or experiences without having to be explicitly programmed. So, our voice assistant is being built using machine learning and artificial intelligence. A virtual assistant is a program that can carry out the user's requests using voice commands in human natural language. An AI assistant or digital assistant is other names for it. Virtual assistants are often cloud-based programs that need internet-connected computers or other devices and/or software to function. Massive volumes of data are needed by the technologies that run virtual assistants, which feed platforms for

R. Adline Freeda (✉) · V. S. Krithikaa Venket · A. Anju · Gugan · Ragul · Rakesh
KCG College of Technology, Karapakkam, Chennai 600097, India
e-mail: adline@kcgcollege.com

M. S. Kaiser et al. (eds.), *ICT: Smart Systems and Technologies*, Lecture Notes in Networks and Systems 878, https://doi.org/10.1007/978-981-99-9489-2_24

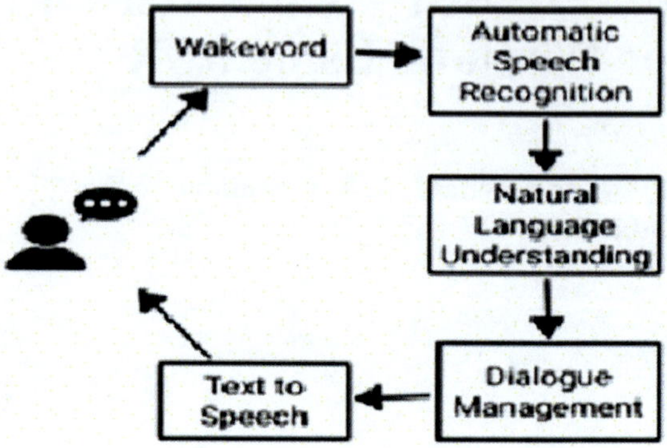

Fig. 1 Steps of a common virtual assistant

machine learning, speech recognition, and other forms of artificial intelligence (AI). For end users, the VA can handle simple tasks. In order to learn from data input and improve at anticipating the needs of the end user, AI programming employs advanced algorithms.

The working of the voice assistant as shown in Fig. 1 is given in sequential steps as shown below.

The Steps of a Virtual Assistant

1. To start the assistant, the user must say a certain word or phrase into the wake word (WW) detector, which is running on the gadget. Other methods, such as a push-to-talk button, are also available to engage the assistance.
2. Automatic Speech Recognition (ASR) turns user-provided audio into a written transcription.
3. With the use of natural language understanding (NLU), it is possible to guess the user's intention from the transcription of their words. This component is aware that users might submit the same request in numerous ways, and they should all result in the same result.
4. The Dialogue Manager (DM) decides what to say back to the user, whether to take any action, and handles any conversation.
5. Text -to speech (TTS) is the output voice of the assistant.

2 Literature Review

Yan et al. [1] say that nowadays voice assistants (VAs) have become an increasingly popular mode for human–machine interaction. They have used an attack in the name of dolphin attack in order to catch the correct words said by the user. This attack helps the assistant in hearing the inaudible commands.

Weeratunga et al. [2] say that the visually challenged persons cannot keep up with the advancement in the technologies. So they created a system named "Nethra" in order to help the visually challenged. In this work, they developed an assistant to help the visually challenged in keeping up with the advancements in the technologies.

Subhash et al. [3] developed a virtual assistant using Python programming language. In this work, Google text to speech (GTTS) is used to convert the text into English language and play sound packages from Python programming language.

Chowdury et al. [4] say that the most of the intelligent personal assistants use only one language, i.e., English, but they proposed another type where they use English and Bengali languages to process. In this paper, they used Sphinx-4 to do speech recognition and the language processing is done using a finite state automaton.

Polyakov et al. [5] say that voice control is a major factor which is changing the lifestyle of the people. They have proposed a very simple assistant which gathers human voice using a microphone, and then using GTTS, the text is converted into speech. And they have used a play sound package from the Python programming language.

Mahmood et al. [6] developed an assistant which not only can search things or schedule things, but this can process/handle multidimensional IoT as well as application of data. They have proposed singular adaptive multi-role (SAM-IPA) system to do the above operations.

Kepuska et al. [7] have proposed an assistant which does process two or more combined user input modes, such as speech, image, video, touch, manual gestures, gaze, and head and body movement in order to bring advancement in the AI technologies.

Raspberry Pi [8] has been used as a processing chip and an underlying architecture in the system. This uses ambient technologies in order to bring advancement in screen-based interaction.

The Crowdsourcing system [9, 10] essentially implants an "insight" that screens and gains from the agent structure coordinated efforts and makes novel ML models which are used to definitively anticipate replies to future subtle analytic inquiries. Crowd sourcing grants an enormous scope grants an enormous scope and versatile bring of human commitment for data parties and assessment, which presents a different universe perspective on the data mining process. It also enables the use of heterogeneous establishment data from volunteers and courses the clarification methodology to little sections of attempts from different responsibilities. The information examination process is done by means of calculated relapse.

Burbach et al. [11] have considered "privacy" as one of their key factors. So, the privacy of the user is the most important. The user never gets to know where the recorded audio might go.

Adline et al. [12] propose a survey of reliable bulk data dissemination protocols, their features, strength, and weakness which provides insights for further research.

Kumaran et al. [13] use a parser named Semantic Unification and Reference Resolution (SURR) to recognize the speech. Synthesizer is used to convert text to speech.

3 Proposed Solution

An enhanced and a better virtual assistant is being created for a Windows platform. This virtual assistant can be used by the employees of the IT field. Since they work on the computers all day, they can assign some small works to the virtual assistants. Audio tuning will be used to reduce the background noises. Understanding the correct context using grammar tuning of the Automated Speech Recognition (ASR) system.

Example: QR Code ≠ Kuh R Kot.

A. Speech to text (STT): It transcripts the speech into a text form, i.e., input.
B. Automated Speech Recognition (ASR): It allows the user to interact with the computer interface. It helps in converting the transcription audio file, i.e., the speech into a sequence of words.
C. Audio tuning: It helps in getting/catching the correct words said by the user by reducing the background noises.
D. Text to speech (TTS): It converts the transcript text into a speech, i.e., output.

The system will help in getting the knowledge of a virtual assistant which can perform user's tasks by understanding the commands given by the user. The virtual assistant developed reacts to the user's commands as required. The user can find it easier when it comes to frequently assigned tasks to the virtual assistant. The user's commands are received through an external microphone. The virtual assistant displays "done listening" after listening to the user's commands.

In our system, "Google Text-To-Speech (GTTS)" package is being installed to have a human voice from a computer. Then, it will be like a person–person conversation. The work of the GTTS [8] is to analyze the command given by the user and give a response according to it. That response is being converted into text. Now, that is done by the following:

$$\text{tts} = \text{gTTs (text} = \text{audio\_ string, lang} = \text{'en').} \tag{1}$$

The basic work of a GTTS is to convert the audio to string. This audio string is nothing but the response which needs to be given to the user. The natural language of the GTTS is English and the code is written as "en". The extension of the audio would be ".mp3". A random number is chosen between 1 and 20,000,000 to save the audio file. The code for saving the audio file in ".mp3" extension is as follows:

$$\text{tts.save(audiofile)} \tag{2}$$

1. The following are the particular tasks done by the virtual assistant.

   - The name of the assistant can be changed by any users of it. It depends on the user for the name of the assistant.
   - The assistant can search for a content in Wikipedia/Google when said to "search" or "by saying the word for which the information is needed".
   - The assistant can set reminders when the user says "set a reminder".

- The assistant can send mail to a particular person or a group. This is done when the user says send mail.
- The assistant can take a screenshot of your screen. The command for this is "take a screenshot" or "capture my screen" or just "capture".
- Daily news around the world can be shown by the assistant when it is intimated to do so.
- Name of the user can be remembered by the virtual assistant till the same user uses it.
- Playing a song from YouTube or downloading can also be done by the assistant.
- The user can find out his/her exact location by mentioning "show my location" to the assistant.
- The assistant can tell about the climatic conditions of the day when the user says "How's the day?" or "today's weather".
- The assistant can only understand English. But it can translate whatever that is said by the user in English to whatever language needed or mentioned by the user.
- The assistant can shut down or restart the computer when prompted to do so.

These are some of the tasks done by the assistant. But we are working on bringing more and more advancement to our assistant.

2. Working:

The following is the methodology of the assistant:

- The requested task is first broken into parts so that the assistant can understand the commands given by the user.
- The commands are already present in the database and the given request is searched through the list.
- Once the request given is found in the database. The assistant finally receives the request. It works on it then.
- If the assistant could not understand the user's request, it will ask again.
- If the assistant can understand the user's request, it will perform the task.

Working of ASR

ASR stands for Automatic Speech Recognition. It is one of the main principles behind virtual assistants which recognizes the speech said by the user. As shown in Fig. 2 first it records the user's request as a wave file and the wave file is being broken into parts. This breaking takes place in a sequence so that the assistant can understand the request. Now, the background sound is being flattered or being reduced so that the assistant gets the user's requests correctly. And then, the voice, i.e., sound of the wave file is being raised so that the assistant gets the user's requests correctly.

1. *Acoustic Analysis*

Acoustic analysis is a method for measuring sound waves as shown in Fig. 3. Subash et al. [11] have used this analysis in the following three modules:

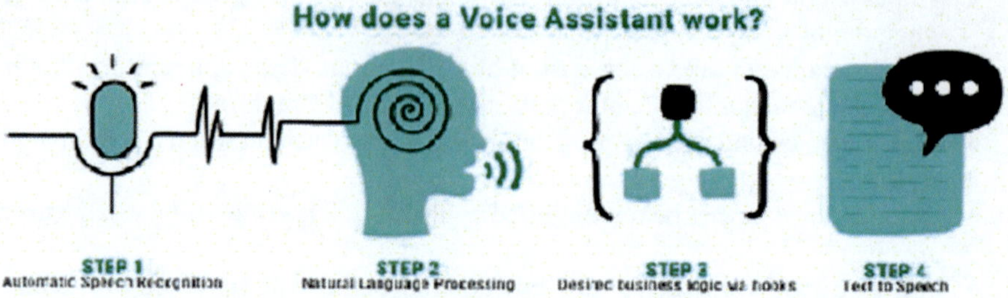

Fig. 2 Working of a voice assistant

Fig. 3 Acoustic analysis

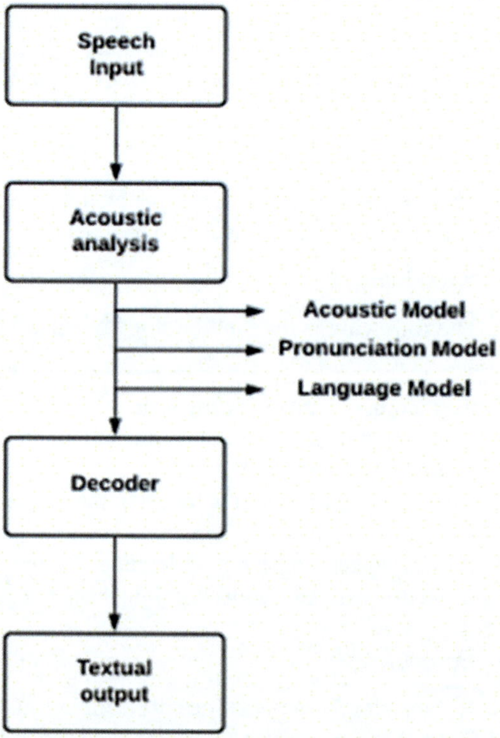

- Acoustic modeling.
- Pronunciation modeling.
- Language modeling.

The first, i.e., acoustic modeling is something where the spoken words need completion or it checks whether the words were actually pronounced or not. Pronunciation modeling checks where these words are spoken, and then, it checks for any accent or peculiarities. Language modeling is something which finds contextual probabilities counting on which the elements were captured.

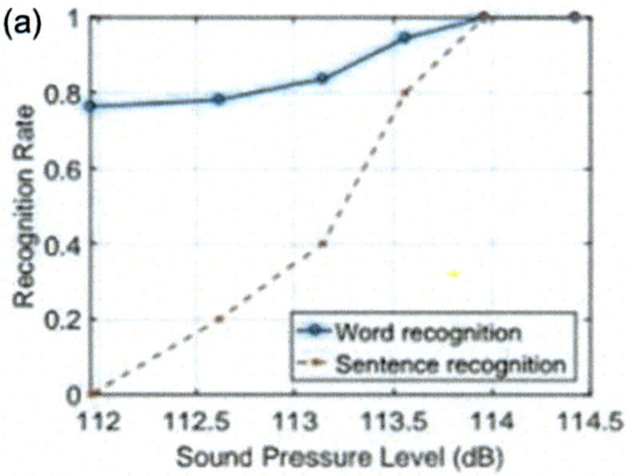

(b)

f_c (kHz)	20	21	22	23	24
Word recognition rate	80%	100%	15%	100%	0%
Sentence recognition rate	8/10	10/10	0/10	10/10	0/10

Fig. 4 **a** Virtual assistant recognition rate. **b** Efficiency of the virtual assistant

4 Experimental Analysis

The following is the graph Figs. 4a, b which portrays the recognition rate of an Android mobile. Samsung Galaxy S6 Edge.

From the above graphs, the recognition rate of the virtual assistant which is being developed is more efficient than the other assistant used in an Android mobile. The following table shows the efficiency of the virtual assistant in word recognition and sentence recognition.

5 Conclusion

The goal of the paper is to develop a voice-based virtual assistant for Windows using Automated Speech Recognition (ASR). This can help in building a relationship between the user and the assistant. The test results show that the assistant cannot work in an area without internet facilities. We are working on this AI technology more and more to bring much more features to this current world.

References

1. Yan C, Zhang G, Ji G, Zhang T, Zhang T, Xu W (2021) The feasibility of injecting inaudible voice commands to voice assistants. IEEE Trans Dependable Secure Comput 18(3):1108–1124
2. Weeratunga AM, Jayawardana SAU, Hasindu PMAK, Prashan WPM, Thelijjagoda S (2015) An intelligent assistant for the visually disabled to interact with internet services. In: IEEE 10th international conference on industrial and information systems (ICIIS), pp 55–59
3. Subhash S, Srivatsa PN, Siddesh S, Ullas A, Santhosh B (2020) Artificial Intelligence based voice assistant. In: Fourth world conference on smart trends in systems, security and sustainability, pp 593–596
4. Chowdury SS, Talukdar A, Mahmud A, Rahman T (2018) Domain specific Intelligent personal assistant with bilingual voice command processing. IEEE
5. Polyakov EV, Mazhanov MS, Voskov AY, Kachalova MVLS, Polyakov SV (2018) Investigation and development of the intelligent voice assistant for the IOT using machine learning. In: Moscow workshop on electronic technologies
6. Mahmood K, Rana T, Raza AR (2018) Singular adaptive multi role intelligent personal assistant (SAM-IPA) for human computer interaction. In: International conference on open source systems and technologies
7. Kepuska V, Bohota G (2018) Next generation of virtual assistants (Microsoft Cortana, Apple Siri, Amazon Alexa and Google Home). In: IEEE conference
8. Vashishta P, Singh JP, Jain P, Kumar J (2019) Raspberry PI based voice-operated personal assistant. In: International conference on electronics and communication and aerospace technology. ICECA
9. Krishnan S, Sathish G, Narayanan V, Freeda A (2021) Web application for RTO verification system, 690–693. https://doi.org/10.1109/ISPCC53510.2021.9609460
10. Balakrishnan V, Bhaskar N (2020) Optimized database using crowdsourcing. In: Ranganathan G, Chen J, Rocha Á (eds) Inventive communication and computational technologies. Lecture notes in networks and systems, vol 89. Springer, pp 1217–1221
11. Burbach L, Halbach P, Plettenberg N, Nakyama J, Ziefle M, Valdez AC (2019) Ok google, Hey Siri, Alexa. Acceptance relevant to virtual voice assistants. In: International communication conference. IEEE
12. Adline Freeda R, Sharmila RN (2016) A review of bulk data dissemination protocols for reprogramming in WSN. In: International conference on information communication and embedded systems (ICICES), pp 1–4
13. Kumaran N, Rangaraj V, Siva Sharan S, Dhanalakshmi R (2020) Intelligent personal assistant—implementing voice commands enabling speech recognition. In: International conference. IEEE

Music Recommendation Systems: Techniques, Use Cases, and Challenges

Shaktikumar V. Patel, H. B. Jethva, and Vishal P. Patel

1 Introduction

MRSs or music recommender systems are becoming more and more popular. Many people are using this to find music that they enjoy on the internet and other services these days. Nowadays, majority of music streaming offers their services to listen online music. Platforms like Gaana, Spotify, Apple Music, etc., are contributing a lot in this field. All these offer access to millions of songs to listen to. Users of these platforms may apply different filters as per their preferences to narrow down their choice of songs. The backend music recommender system algorithms deal best with the user preferences to generate best suitable list to the users. However, MRS algorithms are yet to be perfect to achieve higher user satisfactory levels. The facts behind this are due to multiple factors like different tastes for different users at different locations and time. The traditional MRS approaches do not consider all these factors while generating recommendations.

Music recommendation is a specialized area of recommendation systems that involves recommending music tracks or playlists to users based on their preferences, listening history, and behavior [1]. Here are some particularities of music recommendation [2].

Personalization: Music recommendation needs to be personalized to the individual user's tastes, preferences, and listening habits. This requires the system to collect and

S. V. Patel (✉)
Gujarat Technological University, Ahmedabad, Gujarat, India
e-mail: shakti.infredible@gmail.com

H. B. Jethva
Department of Computer Engineering, GEC Patan, Katpur, Gujarat, India
e-mail: hbjethva@gmail.com

V. P. Patel
Department of Computer Engineering, SPCE Visnagar, Visnagar, Gujarat, India
e-mail: Vppatel.ce@spcevng.ac.in

M. S. Kaiser et al. (eds.), *ICT: Smart Systems and Technologies*, Lecture Notes in Networks and Systems 878, https://doi.org/10.1007/978-981-99-9489-2_25

analyze data on the user's listening behavior, such as their favorite genres, artists, songs, and playlists.

Context: Music recommendation also needs to take into account the context in which the music will be listened to, such as the time of day, mood, and activity. For example, a user may want upbeat music for a workout or relaxing music for studying.

Diversity: Music recommendation systems need to provide a diverse range of music to keep the user engaged and interested. This includes recommending music from different genres, artists, and time periods.

Exploration: Music recommendation should also encourage users to explore new music and expand their musical horizons. This can be achieved by recommending music that is similar to what the user already likes but also slightly different, introducing them to new artists and genres, and suggesting popular or trending songs.

Accuracy: Finally, music recommendation systems need to be accurate in their recommendations. This means using sophisticated algorithms and machine learning techniques to analyze user data and make accurate predictions about what music the user will enjoy.

Music recommendation systems can be used in a variety of ways, including:

Streaming services: Music recommendation systems are commonly used by streaming services such as Spotify, Apple Music, and Pandora to suggest new songs and playlists to users based on their listening history and preferences.

Radio stations: Music recommendation systems can be used by radio stations to suggest songs to play based on their audience's preferences and current trends.

Music discovery apps: Music discovery apps such as Shazam, Soundhound, and Musixmatch use music recommendation systems to suggest similar songs or artists based on the user's identified song or lyrics.

Social media platforms: Social media platforms like TikTok and Instagram use backend algorithms of music recommendation systems to suggest songs and soundtracks to use in user-generated content.

Music e-commerce: Music recommendation systems can be used by e-commerce websites that sell music or merchandise related to musicians to suggest products to users based on their musical preferences and interests.

Overall, music recommendation systems can enhance the user experience by providing personalized and diverse music recommendations, encouraging music exploration, and improving engagement and retention on music platforms.

2 Methods

There are several methods/techniques that can be used in music recommendation systems, some of which are as follows:

2.1 Collaborative Filtering

Collaborative filtering [3] is a popular technique used in music recommendation systems that leverages user behavior and preferences to make music recommendations.

In collaborative filtering, the system identifies users with similar musical tastes and preferences and suggests music that these users have enjoyed in the past. This is done by analyzing data on user listening history, playlists, and ratings to identify patterns and similarities between users.

One way to implement collaborative filtering in music recommendation is through user-based collaborative filtering. In this approach, the system first identifies a target user and then finds other users who have similar music preferences to the target user. The system then recommends music that these similar users have enjoyed in the past, but the target user has not yet listened to.

Another approach is item-based collaborative filtering. In this approach, the system identifies a target music track or playlist and then recommends similar music tracks or playlists to the user based on the listening history of other users who have enjoyed the same music.

Collaborative filtering has several advantages in music recommendation systems. First, it can provide personalized recommendations that reflect the user's musical preferences and tastes. Second, it can help users discover new music that they might not have found otherwise. Finally, it can help to increase user engagement and retention on music platforms by providing a continuous stream of personalized recommendations.

2.2 Content-Based Filtering

Content-based filtering [4] is another common technique used in music recommendation systems that focuses on the characteristics of the music itself rather than user behavior and preferences.

In content-based filtering, the system analyzes the musical features of tracks such as tempo, genre, rhythm, instrumentation, and lyrics to generate music recommendations that are similar to the music tracks that the user has already enjoyed.

To implement content-based filtering in music recommendation systems, the system first extracts relevant features from music tracks using audio analysis techniques such as signal processing, machine learning, and natural language processing. Then, the system computes the similarity between the features of the music tracks to generate a list of recommended tracks that have similar features to the music that the user has enjoyed.

Content-based filtering has several advantages in music recommendation systems. First, it can provide recommendations for niche or lesser-known music that may not have a large user base or listening history. Second, it can provide recommendations that are not biased toward popular music or trends. Finally, it can help users discover new music that is similar to what they already enjoy but with different musical features, thereby expanding their musical horizons.

However, content-based filtering may have some limitations, such as the inability to capture the subjective preferences of users and the difficulty of accurately characterizing the complex and subjective nature of music. Therefore, a combination of content-based filtering and collaborative filtering can often lead to more accurate and effective music recommendations.

2.3 Hybrid Filtering

Hybrid filtering [5] technique is a blended mode of collaborative filtering and content-based filtering techniques in music recommendation systems. This approach is often used to overcome the limitations of individual filtering techniques and to provide more accurate and diverse music recommendations.

In hybrid filtering, the system first generates recommendations using both collaborative filtering and content-based filtering techniques independently. Then, the system combines these recommendations using a variety of weighting and ranking algorithms to provide a final set of personalized and diverse music recommendations.

There are several advantages of using hybrid filtering in music recommendation systems. First, it can provide more accurate and diverse recommendations by combining the strengths of collaborative filtering and content-based filtering techniques. Second, it can provide more personalized recommendations by taking into account both user behavior and musical features. Finally, it can help to overcome the limitations of individual filtering techniques, such as the cold start problem or sparsity of data.

Overall, hybrid filtering is a powerful approach to music recommendation that can provide users with more accurate, personalized, and diverse recommendations, thereby enhancing their music listening experience and engagement on music platforms.

2.4 Knowledge-Based Recommendation

Knowledge-based recommendation [6] is another approach used in music recommendation systems that leverages domain knowledge and user preferences to provide music recommendations.

In knowledge-based recommendation, the system uses explicit user preferences and constraints, such as genre preferences or tempo requirements, to generate music recommendations that match the user's preferences. The system may also use additional domain knowledge, such as music theory, to provide more specific recommendations that match the user's preferences.

To implement knowledge-based recommendation in music recommendation systems, the system first collects explicit user preferences through surveys, questionnaires, or interviews. The system then uses this information to generate a list of recommended music tracks that match the user's preferences and constraints.

Knowledge-based recommendation has several advantages in music recommendation systems. First, it can provide recommendations that are tailored to the user's specific preferences and requirements. Second, it can provide recommendations that are not influenced by other users or trends. Finally, it can help users discover new music that matches their musical tastes and preferences but that they may not have found otherwise.

However, knowledge-based recommendation may have some limitations, such as the difficulty of collecting accurate and comprehensive user preferences and the limited scope of recommendations based on explicit user preferences. Therefore, a combination of knowledge-based recommendation and other filtering techniques, such as collaborative and content-based filtering, can often lead to effective and accurate music recommendations.

2.5 *Context-Aware Recommendation*

Context-aware recommendation [7] is an approach used in music recommendation systems that takes into account contextual information, such as time of day, location, weather, and user activity, to provide more relevant and personalized music recommendations.

In context-aware recommendation, the system collects information about the user's context and then uses this information to generate recommendations that match the user's current situation. For example, if the user is jogging, the system may recommend high-energy music with a fast tempo. If the user is working in a coffee shop, the system may recommend calming music with a low tempo. Same way the preference of music would be different for fitness centers and reading halls or libraries [8].

To implement context-aware recommendation in music recommendation systems, the system first collects contextual information through sensors, user input, or external APIs. The system then uses this information to generate recommendations that match the user's current situation and preferences.

Context-aware recommendation has several advantages in music recommendation systems. First, it can provide more relevant and personalized recommendations that match the user's current situation and preferences. Second, it can help users discover new music that is tailored to their current activity or environment. Finally, it can enhance the user's music listening experience by providing music that matches their mood or situation.

However, context-aware recommendation may have some limitations, such as the difficulty of accurately collecting contextual information and the limited scope of recommendations based on context. Therefore, a combination of context-aware recommendation and other filtering techniques, such as collaborative and content-based filtering, can often lead to more accurate and effective music track recommendations.

3 Challenges

There are several challenges that music recommendation systems [9] face, some of which are as follows.

3.1 Cold Start Problem

The cold start problem [10] is a common issue in music recommendation systems that occurs when the system has little or no information about a new user or a new item. This can make it difficult for the system to provide accurate and relevant recommendations to the user.

In music recommendation systems, the cold start problem can occur in several scenarios. For example, when a new user signs up for a music platform, the system may not have any information about the user's musical preferences or listening history. Similarly, when a new music track is released, the system may not have any information about the track's musical features or how it relates to other tracks.

To overcome the cold start problem in music recommendation systems, several techniques can be used. For example, one approach is to use content-based filtering to provide recommendations based on the features of the new item. This can be especially useful for recommending new music tracks or artists that may not have a large user base or listening history.

Another approach is to use knowledge-based recommendation to provide recommendations based on explicit user preferences and constraints. This can be useful for recommending music to new users who may not have a large listening history or for recommending music for specific contexts or situations.

Finally, one more approach is to use hybrid filtering techniques that combine multiple filtering approaches like collaborative filtering and content-based filtering, to provide more accurate and diverse recommendations. By combining these techniques, the system can provide accurate recommendations even when there are limited data available about the user or the item.

Overall, the cold start problem is a significant challenge in music recommendation systems, but by using a combination of techniques, it is possible to provide accurate and relevant recommendations to users even when there are limited data available.

3.2 Data Sparsity

The data sparsity [11] problem is a common issue in music recommendation systems that occurs when there are a large number of items and users but only a small amount of data available about their preferences or interactions. This can make it difficult for the system to provide accurate and diverse recommendations to users.

In music recommendation systems, the data sparsity problem can occur when users listen to only a small subset of the available music tracks, resulting in a sparse user-item matrix. This can make it difficult for collaborative filtering techniques to identify similarities between users or items and provide accurate recommendations.

To overcome the data sparsity problem in music recommendation systems, several techniques can be used. One approach is to use matrix factorization techniques. Among them, the popular techniques like singular value decomposition (SVD) or matrix factorization with Bayesian personalized ranking (BPR) are utilized to reduce the dimensionality of the user-item matrix and identify latent factors that can be used to generate recommendations.

Overall, the data sparsity problem is a significant challenge in music recommendation systems, but by using a combination of techniques, it is possible to provide accurate and diverse recommendations to users even when there is limited data available.

3.3 Subjectivity

The subjectivity issue [12] is a common challenge in music recommendation systems that occurs due to the subjective nature of music preferences. Different users may have different tastes in music, and what one user likes may not be preferred by another user. This can make it difficult for music recommendation systems to provide accurate and relevant recommendations that are tailored to individual users.

To overcome the subjectivity issue in music recommendation systems, several techniques can be used. One approach is to use collaborative filtering techniques that identify similarities between users based on their listening history or preferences. By identifying users with similar preferences, the system can provide recommendations that are more likely to be relevant to the user.

Another approach is to use content-based filtering techniques that recommend items based on their features, such as genre, tempo, or mood. This can be especially useful when the user has specific preferences or is looking for music that matches a particular mood or context.

Additionally, incorporating user feedback and explicit ratings can help to improve the accuracy of recommendations and address the subjectivity issue. This can be done by allowing users to rate songs or provide feedback on recommended items, which can then be used to improve the accuracy of future recommendations.

Finally, it is important to consider the diversity of recommendations and provide a variety of options to cater to different user preferences. This can be done by incorporating serendipity and novelty into the recommendation process, providing users with recommendations that are unexpected or outside of their usual preferences.

Overall, the subjectivity issue is a significant challenge in music recommendation systems, but by using a combination of techniques, it is possible to provide accurate, relevant, and diverse recommendations to users with different preferences.

3.4 Diversity

The diversity issue [13] is a common challenge in music recommendation systems that occurs when the system provides recommendations that are too similar or repetitive. This can result in users becoming bored or dissatisfied with the recommendations and can lead to a decline in user engagement.

To overcome the diversity issue in music recommendation systems, several techniques can be used. One approach is to use hybrid filtering techniques that combine multiple filtering approaches, such as collaborative filtering and content-based filtering, to optimize the outcome list and provide more diverse recommendations. By incorporating different filtering approaches, the system may suggest recommendations that are more varied and cater to different user preferences.

Another approach is to incorporate serendipity [14] into the recommendation process. This can be done by providing recommendations that are unexpected or outside of the user's usual preferences. For example, recommending a new artist or genre that the user may not have heard before can help to increase the diversity of recommendations and provide a more engaging experience for the user.

Additionally, it is important to consider the context in which the recommendations are being made. For example, recommending music for a party or workout may require different recommendations than recommending music for relaxation or focus. By taking into account the context, the system can provide recommendations that are more appropriate and diverse for the user's current situation.

Finally, incorporating diversity metrics into the recommendation process can help to ensure that the system provides diverse and engaging recommendations. For example, measuring the diversity of recommended items based on their genre, artist, or other features can help to identify areas where the recommendations may be too repetitive and adjust the recommendations accordingly.

Overall, the diversity issue is a significant challenge in music recommendation systems, but by using a combination of techniques, it is possible to provide diverse and engaging recommendations to users that cater to their individual preferences and context.

3.5 *Scalability*

The scalability [15] issue is a common challenge in music recommendation systems that occurs when the system is unable to handle large volumes of data and users. As the number of users and items in the system grows, the computational and storage requirements of the system also increase, making it difficult to provide timely and accurate recommendations.

To overcome the scalability issue in music recommendation systems, several techniques can be used. One approach is to use distributed systems and parallel computing techniques that allow the system to process large volumes of data more efficiently. For example, using a distributed computing framework like Apache Spark can help to distribute the workload across multiple nodes and improve the scalability of the system.

Another approach is to use machine learning techniques that can scale to large volumes of data. For example, using deep learning models like convolutional neural networks (CNNs) or recurrent neural networks (RNNs) can help the system process large volumes of audio data and extract meaningful features that can be used for recommendation.

Additionally, it is important to consider the storage and retrieval of data in the system. Using a scalable and distributed database system like Apache Cassandra or Amazon DynamoDB can help to improve the scalability of the system by allowing for efficient storage and retrieval of large volumes of data.

Finally, the trade-off between scalability and personalization should also be accounted in the system. Providing highly personalized recommendations may require more computational and storage resources, which can limit the scalability of the system. By balancing the level of personalization with the scalability of the system, it is possible to provide accurate and timely recommendations to a large number of users.

Overall, the scalability issue is a significant challenge in music recommendation systems, but by using a combination of techniques, it is possible to overcome this challenge and provide timely and accurate recommendations to a large number of users and items in the system.

4 Future Scope

There are several future scopes for music recommendation systems, some of which are as follows.

4.1 Personalization

The future of music recommendation systems is in providing highly personalized recommendations that take into account the user's context, preferences, and listening habits.

4.2 Multimodal Recommendation

The future of music recommendation systems is in incorporating multiple modalities, such as audio, video, lyrics, and user-generated content, to provide more diverse and engaging recommendations.

4.3 Explainability

The future of music recommendation systems is in providing explanations for why certain music items are recommended, to increase user trust and transparency.

4.4 Interactivity

The future of music recommendation systems is in providing interactive features, such as music discovery tools, social sharing, and playlist customization, to enhance the user experience.

4.5 Integration

The future of music recommendation systems is in integrating with other music-related services, such as streaming platforms, concert tickets, and merchandise, to provide a more comprehensive music experience.

4.6 Real-Time Recommendation

The future of music recommendation systems is in providing real-time recommendations [9] that adapt to the user's changing context and preferences.

Addressing these future scopes requires continued research and development in advanced algorithms and techniques, as well as collaboration between music industry stakeholders, user communities, and regulatory bodies.

5 Conclusion

In conclusion, music recommendation systems have become an integral part of the music industry, providing users with personalized music experiences and helping music service providers increase engagement and customer loyalty. These systems use a variety of algorithms and techniques to analyze user–music interaction data and generate recommendations. However, there are several challenges that must be addressed, such as the cold start problem, data sparsity, subjectivity, diversity, scalability, shilling attacks, and privacy concerns. Moreover, there is a need for more user-centric and explainable music recommendation systems that take into account user context and preferences. The future of music recommendation systems lies in providing highly personalized, multimodal, and interactive recommendations while also addressing ethical considerations. To achieve this, continued research and development in advanced algorithms and techniques, collaboration between music industry stakeholders, user communities, and regulatory bodies are necessary. It is hoped that this paper has provided a comprehensive overview of the techniques, use cases, and challenges of music recommendation systems and will stimulate further research in this exciting field.

References

1. Baltrunas L, Ludwig B, Ricci F (2011) Matrix factorization techniques for context aware recommendation. In: Proceedings of the fifth ACM conference on recommender systems, pp 301–304
2. Schedl M, Knees P, Gouyon F (2017) New paths in music recommender systems research. In: Proceedings of the 11th ACM conference on recommender systems (RecSys 2017), Como, Italy
3. Schafer JB, Frankowski D, Herlocker J, Sen S (2007) Collaborative filtering recommender systems. The adaptive web: methods and strategies of web personalization, 291–324
4. Aggarwal CC (2016) Content-based recommender systems. In: Recommender systems. Springer, pp 139–166
5. Aggarwal CC (2016) Ensemble-based and hybrid recommender systems. In: Recommender systems. Springer, pp 199–224
6. Burke R (2000) Knowledge-based recommender systems. Encyclopedia Libr Inf Syst 69(Supplement 32):175–186
7. Verbert K, Manouselis N, Ochoa X, Wolpers M, Drachsler H, Bosnic I, Duval E (2012) Context-aware recommender systems for learning: a survey and future challenges. IEEE Trans Learn Technol 5(4):318–335

8. Cheng Z, Shen J (2014) Just-for-me: an adaptive personalization system for location-aware social music recommendation. In: Proceedings of the 4th ACM international conference on multimedia retrieval (ICMR), Glasgow, UK
9. Schedl M, Zamani H, Chen CW, Deldjoo Y, Elahi M (2018) Current challenges and visions in music recommender systems research. Int J Multimedia Inf Retrieval 7:95–116
10. Lika B, Kolomvatsos K, Hadjiefthymiades S (2014) Facing the cold start problem in recommender systems. Expert Syst Appl 41(4):2065–2073
11. Schedl M, Knees P, McFee B, Bogdanov D, Kaminskas M (2015) Music recommender systems. Recommender systems handbook, 453–492
12. Song Y, Dixon S, Pearce M (2012) A survey of music recommendation systems and future perspectives. In: 9th international symposium on computer music modeling and retrieval, vol 4, pp 395–410
13. Kunaver M, Požrl T (2017) Diversity in recommender systems—a survey. Knowl-Based Syst 123:154–162
14. Zhang YC, Séaghdha DÓ, Quercia D, Jambor T (2012) Auralist: introducing serendipity into music recommendation. In: Proceedings of the fifth ACM international conference on web search and data mining, pp 13–22
15. Zahra S, Ghazanfar MA, Khalid A, Azam MA, Naeem U, Prugel-Bennett A (2015) Novel centroid selection approaches for KMeans-clustering based recommender systems. Inf Sci 320:156–189

Obstacle Detection Using Arduino Board and Bluetooth Control

N. Pavitha, Rohit Dardige, Vaibhav Patil, Ameya Pawar, and Bhavesh Shah

1 Introduction

In the contemporary world, a myriad of diverse occupations is being undertaken by automated machines, thereby reducing the effort exerted by human beings. The technologies such as artificial intelligence (AI) has further facilitated the simplification of tasks, thereby perpetuating this trend. Two prominent examples of the multifarious navigation strategies currently employed are edge detection and line following. It should be noted that the identification of obstacles relies heavily on the determination of a suitable path, necessitating intricate calculations to precisely quantify optimal routes devoid of hindrances [1].

Following this course of action, the robot must continue to explore without interruption until an obstacle is sense. Consequently, a guiding algorithm is implemented to ensure that when an impediment is detected, the robot does not halt but instead seeks an alternative pathway. When the sensor detects the obstacle, the sensor transmits a signal to the microcontroller, thereby enabling the robot to navigate around it [2]. These sensors find utility not only in robot navigation but also in obstacle detection systems such as radars. In the realm of robot navigation, image detection is commonly utilized for the purpose of obstacle detection. This involves capturing images and subsequently processing them to determine the appropriate course of action. In general, infrared (IR) sensors are employed for this task. However, it should be noted that IR sensors may yield inaccurate results when confronted with simple materials and are susceptible to the influence of atmospheric conditions, thereby decreasing precision due to decreased humidity and moisture. Moreover, IR sensors are capable of detecting various radiations, thus leading to errors. IR sensors offer advantages such as low power consumption.

N. Pavitha (✉) · R. Dardige · V. Patil · A. Pawar · B. Shah
Department of Computer Engineering, Faculty Science and Technology, Vishwakarma University, Pune, India
e-mail: pavitha.nooji@vupune.ac.in

M. S. Kaiser et al. (eds.), *ICT: Smart Systems and Technologies*, Lecture Notes in Networks and Systems 878, https://doi.org/10.1007/978-981-99-9489-2_26

The model also incorporates a Bluetooth module (HC-05), which facilitates control of the robot vehicle via voice commands and mobile devices. With configurations for both master and slave, this module employs wireless communication utilizing a serial port protocol [3]. The master module establishes connections and controls devices once they are connected. Autonomous and robotic vehicles are poised to assume a significant role in the future. The attainment of accuracy in object recognition and subsequent actions is of utmost importance. Enhancing precision entails refining both sensors and motors. It is projected that these robotic vehicles will assist humans in a multitude of tasks. In terms of object detection, the Arduino Uno of the robot vehicle serves as the central processing unit, giving commands to the motors based on information obtained from ultrasonic sensors [4].

When there is detection of an obstacle, the vehicle must navigate through the path. By employing components such as ultrasonic sensors, Bluetooth modules, an L293D driver, and DC motors, the ultimate objective is to develop an all-rounder robot vehicle that is both cost-effective and reliable. The core aim of this strategy is to guide the robot back to its original navigation path upon encountering hindrances. By proactively redirecting the robot, this method circumvents collisions and divergences prompted by encountered obstacles, preserving the intended navigation trajectory [1]. A novel framework formulated by researchers, targeting the decision-making process for autonomous vehicles. This framework encompasses a nonholonomic motion planner, localization techniques, obstacle avoidance mechanisms, and effectively addresses uncertainties. It is designed to handle both safety considerations and task-oriented assignments [5].

Distinct unmanned vehicles—a ground vehicle (UGV) and a micro-aerial vehicle (MAV). These vehicles, outfitted with cameras, collaborate to navigate around obstacles. Importantly, this navigation occurs within an indoor setting where Global Positioning System (GPS) signals are absent [6]. Vision-based obstacle avoidance can be very effective way to detect obstacle and avoid them. That helps to navigate through the path without any kind of disturbance [7]. We can control the robot by using simple gestures. For that we have to use hand gesture device then the robot can be control by using human gesture [8]. The sensor assumes a pivotal role in facilitating control and interaction with the module.

It's important to highlight that this innovative module not only complements Arduino functionalities but also capitalizes on the precision and responsiveness offered by the ultrasonic sensor [2]. Image processing can be used for obstacle avoidance. A camera-based technology is used to detect the obstacle give alert to the system and avoid accidents [9]. Deep learning techniques to enhance the obstacle detection and avoidance capabilities of driverless cars. With the rapid advancement of autonomous vehicle technology, the ability to accurately perceive and react to obstacles on the road is of paramount importance [10]. This system can spot obstacles to prevent crashes and can identify objects using a special sensor that works in three directions: up, down, and in front of it. This sensor uses sound waves to do this [11]. Detecting obstacles is a crucial aspect of environmental awareness technology. When we talk about obstacles above the ground, it's relatively easier to identify them. However, obstacles concealed beneath the ground pose a greater challenge for detection and can inflict more harm on unmanned vehicles. Therefore, conducting research into different methods of detecting these hidden obstacles holds significant importance and value [3].

Addressing the challenge posed by the limited computing capabilities of current computing equipment to accommodate emerging algorithms, an enhanced obstacle detection method called YOLOX is introduced. This algorithm incorporates an attention mechanism into the YOLOX detection method, building upon the original approach. This addition serves to reduce the computational demands on hardware devices while achieving both high-speed and precise obstacle recognition [12]. Losing the ability to perceive objects in one's surroundings is a profoundly challenging situation for individuals. It not only hampers their capacity to navigate independently within familiar places and environments but also poses significant obstacles to their mobility. Throughout the past decade, numerous devices have been developed to assist visually impaired individuals in navigating various types of surroundings [13].

Visually challenged individuals encounter numerous difficulties when navigating through unfamiliar public spaces. To offer them a sophisticated and practical solution, the development of a smart walking cane emerges as a promising option. Legally visually impaired individuals often grapple with challenges related to perceiving elements that others might take for granted, such as street signs, traffic signals, and objects directly in front of them. This diminished awareness increases their vulnerability to stumbling and encountering accidents due to the limitations in their ability to discern their immediate surroundings [4]. A smart automobile control system was created with a dual focus on obstacle avoidance and engine temperature management.

The primary aim was to significantly diminish car accidents, which often occur due to unforeseen road obstacles and sudden engine overheating breakdowns. To achieve the goal of detecting and evading road obstacles effectively, a cost-effective smart car control system was devised, built upon the Arduino Uno platform, specifically using the ATmega328P microcontroller [14]. Sonar technology plays a crucial role in underwater search and rescue missions, particularly in cases involving aircraft, shipwrecks, and individuals in distress. However, in scenarios where these missions endure for an extended duration, fatigue can set in among sonar operators, potentially causing them to overlook important objects. In such situations, the utilization of object identification or characterization techniques can provide valuable assistance in maintaining a high level of vigilance and ensuring that no potential objects are ignored [15].

The alert system is composed of a smartphone's rear-mounted camera affixed to a wheelchair. To detect objects effectively, the YOLOv3 model was employed for object recognition. In addition, the researcher has devised an algorithm that employs edge detection techniques to enhance the system's ability to identify obstacles like pillars, doors, or the edges of walls. This improvement significantly bolsters the efficiency of obstacle detection within the system [16]. Object detection represents a computer vision method that empowers the identification and pinpointing of labeled objects within an image or video. This specialized technique involves recognizing objects within a given scene and accurately determining their respective positions [17].

By amalgamating the data from two grid map images, we categorize obstacles that consistently appear as fixed obstacles, while those obstacles that shift as the grid map updates are labeled as dynamic obstacles. This innovative obstacle detection system was integrated into the Sea Sword USV, a creation of LIGNex1, and underwent testing in a marine setting [18]. Stereo vision-based obstacle detection can also be the efficient way in autonomous vehicle [19] obstacle detection on runway can be done by using the vision-based model, in which we can use actuators to detect obstacle on vision based such as IR sensors [20].

2 Methodology

The entire system's connections are in the block diagram above, with Bluetooth module connected to the Arduino board, which is connected to the motor shield, and the battery providing power to the motor so that it can start working. Presently we will examine the working. Ultrasonic sensor constantly checking for deterrent by sending radio waves in climate, when it sense any article in explored way it conveys that message to Arduino that there is obstruction to keep away from breakdown. At the point when Arduino board gets that signal, it will begin the ringer and sends data to L293D board that sensor sense deterrent. The DC motors that are connected to the wheels will stop rotating when this signal is sent to the L293D motor driver. Also, the vehicle stops. All the equipment is associated with one another [2] (Fig. 1).

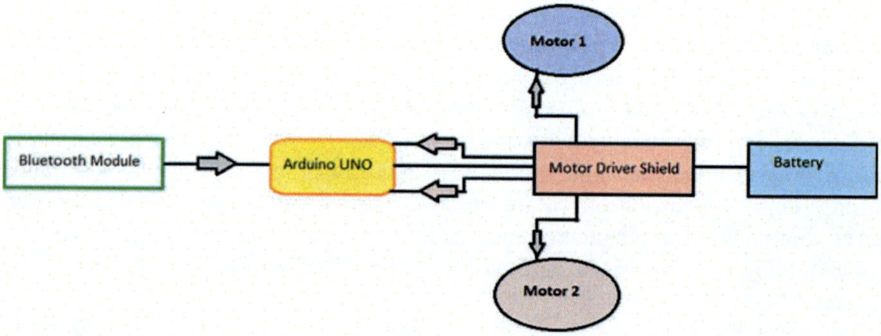

Fig. 1 Block diagram

Fig. 2 Ultrasonic sensor

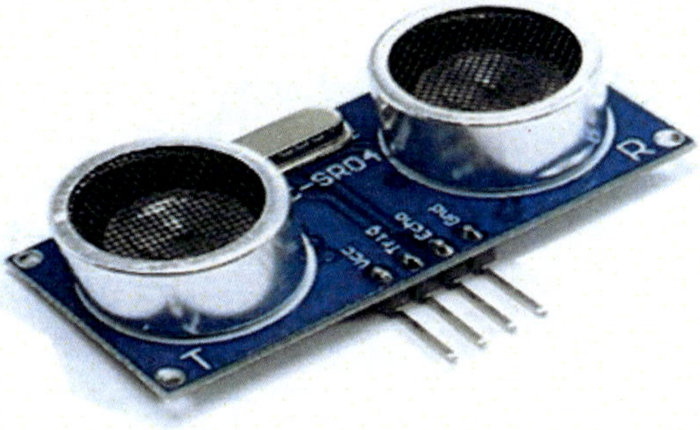

2.1 *Sensors for Object Detection*

Numerous sensors are available for object detection, with some widely recognized ones like infrared sensors, ultrasonic sensors, camera sensors, and image detection sensors. In robotics, ultrasonic sensors are commonly employed. They engage in a continuous scan of the surroundings, searching for objects. Upon detection, they transmit signals back to the microcontroller board. A radar system utilizes sound waves, which fall outside the range of human hearing. Radar emits these waves into the air, and upon detecting an obstacle or requiring evasion, the signals are relayed to the microcontroller. To counteract potential interference with the ultrasonic signals from the sensors, certain common obstructions like dust, snow, ice, and dirt need to be addressed. This sensor is majorly used to detect object or to gain information about environment [1] (Fig. 2).

- Features of Sensor:

  Accuracy: The sensor is quite precise, with the ability to measure details as small as 0.3 cm.

  Energy Consumption: It uses a small amount of current, around 15 mA, to operate.

  Power Supply: The ultrasonic sensor runs on + 5 V DC power.

  Detection Range: It can sense objects at varying distances, from as close as 2 cm to as far as 400 cm (which is about 1 in. to 13 feet).

2.2 *Module for Voice Control*

The Bluetooth module HC-05 is a versatile component utilized for wireless communication. It facilitates the establishment of connections between devices through Bluetooth technology. This module is particularly valuable for enabling communication and control between electronic devices, often used in applications such as robotics, automation, and remote-control systems. The HC-05 module functions by

Fig. 3 Bluetooth module

employing a serial port protocol for data transmission and reception. It offers two primary configurations known as master and slave. And in our research, we have use this for voice control and connectivity. We have used this Bluetooth module in this research to connect to voice control and app control application to control vehicle [1, 2] (Fig. 3).

2.3 For Message Processing

Arduino UNO board, which holds a total of 14 pins. Within these pins, there are six designated as input pins. Additionally, there is a USB connector, a power slot, an ICSP header, and a reset button. ICSP pins allows users to program using firmware of Arduino. Reset button is to add reset button to connections. GND pins have zero voltage. It is used to know the input voltage. When the ultrasonic sensor identifies an object along the intended path, it forwards signals to the Arduino board. Subsequently, these signals undergo internal processing within the board itself. Once processed, these signals are then transmitted to the motors to execute the necessary avoidance actions. Arduino board sends the specific message to the system on the basis of code that we have install in it through computer [2] (Fig. 4).

- Features of the Arduino UNO

  Easy to Use: Arduino Uno is designed for beginners and hobbyists, making it simple to start working with electronics and programming.

Fig. 4 Arduino board

Microcontroller: It has a small computer called a microcontroller that can be programmed to perform various tasks.
Digital Pins: Arduino Uno has pins that you can use to connect to other devices, like buttons, LEDs, and sensors.
Analog Pins: It also has pins that can sense and measure things like light and temperature.

2.4 L293D Board

The L293D board is equipped with multiple H-bridge circuits, which are electronic circuits that enable bidirectional control of motors. Each H-bridge consists of four transistors that control the motor's direction (forward or reverse) and speed. By appropriately toggling these transistors, the motor driver board can control the voltage polarity applied to the motor, determining its movement direction. This motor driver works on basic principle of H-Bridge. It helps to rotate DC motors in both directions as it allows the voltage to flow in any direction. This single IC consists two H-Bridge inside that can rotate two DC motors. L293D board can control small as well as big motors. This driver is smaller in size that why we have used in vehicle to save space on chassis. This driver mainly used in robots [16] (Fig. 5)

Fig. 5 L293D board

3 Results

We created this module to identify obstacles and avoid them. Where an ultrasonic sensor continually reads the surroundings and stops the automobile if it detects an obstruction within 30 m of it. The Arduino board, the system's brain, processes all of the data coming from the ultrasonic sensor. Through an application and a gyro sensor, the Bluetooth module allows us to operate robots. Additionally, we have an L293D board that can open motor gates or control motors. To put it simply, an Arduino board can deliver a certain amount of current, but if a motor requires a higher current to be controlled, an L293D board can supply that current. DC motor rotation is made easier by this (Fig. 6).

3.1 *For Calculating Distance*

$$\text{Distance} = \frac{\text{speed of sound} \times \text{time taken}}{2} \tag{1}$$

For calculating the distance between vehicle and obstacle this is the equation

$$\text{Speed of sound} = \text{Sound that Captured by Ultrasonic sensor}$$

The statistics in Table 1 show how a vehicle scans for obstacles, finds them, and acts correctly. The sensor works with greater accuracy the closer it is to the vehicle to avoid collisions with obstacles. The distance in centimeters between the object and the vehicle is measured. And when the car detects an obstruction, it comes to a stop.

Fig. 6 Vehicle completed framework

Table 1 Observation of robot heading to obstacles

Speed of the robot (cm/s)	Distance from obstacle (cm)	Outcomes
80.50	75	Robot is in motion
80.50	60	Robot is in motion
80.50	55	Robot is in motion
80.50	45	Robot is in motion
80.50	40	Robot is in motion
80.50	35	Robot is in motion
0	30	Buzzer start vehicle at rest

4 Conclusion

The focal point of this study, titled "Obstacle Detection and Bluetooth-Controlled Autonomous Vehicle," lies in substantiating the practical implementation of obstacle detection and avoidance within a robot's trajectory. This objective is achieved through the utilization of an ultrasonic sensor, which diligently senses obstructions and subsequently relays this information to the Arduino board. This component interlinked with DC motors responsible for wheel rotation. Crucially, the entire prototype operates within a natural environment, necessitating exceptional accuracy, real-time responsiveness, and informed decision-making. The embodiment of accuracy, real-time responsiveness, and decision-making process coalesce to define our "Obstacle Detection Using Arduino Board and Bluetooth control." This research initiative, characterized by its simplicity and cost-effectiveness, holds the potential to undertake diverse tasks with remarkable precision and efficiency [11].

References

1. Al-Qaraawi S, Jamal H (2019) Obstacle avoidance based on ultrasonic sensors and optical encoders (IJARCET) 1–8
2. Widyotriatmo A, Hong K-S (2008) Decision making framework for autonomous vehicle navigation SICE annual conference, pp 1–6
3. Tsai W-H, Chen K-H (2000) Vision-based obstacle detection and avoidance for autonomous. Elsevier, pp 1–25
4. Lwowski J, Sun L, Pack D (2016) Heterogeneous bi-directional cooperative unmanned, USA, pp 5–9
5. Tiwari A, Ghanwar RK, Madhava Red S (2019) Gesture controlled robot with obstacle avoidance using Arduino. Int Res J Eng Technol (IRJET) 6:1–3
6. Chinmayi R. Jayam YK, Tunuguntla V, Venkat Dammuru J (2008) Obstacle detection and avoidance robot. IEEE, pp 1–6
7. Athira S (2019) Image processing based real time obstacle. In: IEEE conference, pp 3–6
8. Sanil N, Venkat PAN, Rakesh V, Mallapur R, Ahmed MR (2019) Deep learning techniques for obstacle detection. IEEE, pp 1–4
9. Lolo YS, Ohammah KL, Alfa AN, Moham SA (2022) Design and implementation of obstacle detection and warning system for visually impaired people. In: IEEE Nigeria 4th international conference on disruptive technologies for sustainable development (NIGERCON), pp 1–5
10. Han J, Zhang Z, Gao X, Li K, Kang X (2022) Research on negative obstacle detection method based on image enhancement and improved anchor box YOLO. In: IEEE international conference on mechatronics and automation (ICMA), pp 1216–1221
11. Khalid Z, Mohamed EA, Abdenbi M (2013) Stereo vision-based road obstacles detection. IEEE, pp 1–6
12. Zhou Y, Dong Z (2017) A vision-based autonomous detection scheme for obstacles on the runway. IEEE, pp 1–7
13. Suraj DM, Prasad VA, Lokesh S, Hebbale SG, Salis VE (2019) Obstacle detection for the visually impaired using Arduino. IEEE, pp 260–265
14. Tejaswini S, Sahana MS, Bhargavi K, Jayamma SS, Bhanu HS, Praveena KS (2021) Obstacle sensing assistance for visually impaired person using Arduino. IEEE, pp 767–771
15. Shabani H, Razif MRM, Muhsin MF (2021) Smart car control system based Arduino Uno for obstacles avoidance and engine temperature control. IEEE, pp 1–6

16. Kiruba Devi T, Ritu, Polepaka S, Rastogi R, Wategaonkar SR, Achuthan A (2023) An advanced analyzation of ultrasonic sound detection and ranging method using Arduino. IEEE, pp 489–494
17. Patthanajitsilp P, Chongstitvatana P (2022) Obstacles detection for electric wheelchair with computer vision. In: 14th International conference on knowledge and smart technology (KST), pp 97–101
18. Jabade V, Nahata U, Jain N, Pandey A, Paratkar T (2022) Obstacle detection and walkable path detection. In: 2022 IEEE Delhi section conference (DELCON), pp 1–5
19. Gao J-S, Zhang P-N (2022) Research on obstacle detection algorithm based on YOLOX. In: 2022 7th International conference on intelligent computing and signal processing (ICSP), pp 378–384
20. Ha J-S, Im S-R, Lee W-K, Kim D-H (2021) Radar based obstacle detection system for autonomous unmanned surface vehicles. In: 21st International conference on control, automation and systems. IEEE, pp 863–867

Green ICT: Exploring the Role of IoT-Enabled Technologies in Small-Scale Businesses

Subhashree Rout and Swati Samantaray

1 Introduction

In the present era, Information and Communication Technology (ICT) has facilitated the integration of e-services via internet connectivity, network infrastructure, wireless communications, and mobile devices into everyday life, which has led to a rise in greenhouse gas (GHG) emissions. The observable consequences of GHG emissions result in climate change, which includes a rise in temperature, elevated sea levels, and more frequent occurrences of floods and storms that have significant implications on the equilibrium of ecosystems as well as the availability of water and food resources, public health, industrial operations, agricultural practices, and infrastructure. Addressing climate change requires the pursuit of strategic goals, including enhancing energy efficiency, augmenting the proportion of renewable energy in a nation's energy consumption while ensuring energy supply reliability, guaranteeing the provision of dependable energy products and services, and promoting green goods and sustainable manufacturing.

The scope of Green Communication and Network Systems encompasses several key areas for decreasing energy consumption, the promotion of environmental consciousness, the facilitation of effective communication regarding environmental awareness, and the development of environmental tracking and monitoring systems aimed at safeguarding and revitalising natural ecosystems. ICTs have the potential to significantly contribute to environmental protection and sustainability, which depend

S. Rout (✉) · S. Samantaray
School of Humanities, KIIT Deemed to be University, Bhubaneswar, India
e-mail: subhashreerout103@gmail.com

S. Samantaray
e-mail: ssamantrayfhu@kiit.ac.in

S. Rout
School of Management, Centurion University of Technology and Management, Bhubaneswar, India

M. S. Kaiser et al. (eds.), *ICT: Smart Systems and Technologies*, Lecture Notes in Networks and Systems 878, https://doi.org/10.1007/978-981-99-9489-2_27

on the adoption of green behaviour to achieve these goals. The concept of green networks has emerged as a means to address the pressing need for sustainability and reduce the carbon footprint associated with wireless communication. These networks aim to provide reliable and efficient connectivity while minimising the environmental impact. Green Informatics refers to the application of ICT resources, products, tools, and technologies, along with green behaviours like virtualisation, recycling, and telecommuting, thereby enabling the optimisation of resource consumption, waste reduction, and the promotion of sustainable practices.

The persistent escalation of pollution levels and the exhaustion of natural resources have compelled both businesses and governments to accord utmost importance to environmental preservation as an unparalleled and urgent concern. The effective enhancement of customer value and strengthening of brand image can be achieved through the strategic implementation of sustainable practices. Technologies encompassing the realms of artificial intelligence (AI), big data analytics, the Internet of Things (IoT), and the Information and Communication Technology (ICT) are being increasingly utilised to enhance environmental sustainability. Companies are currently implementing novel products and platforms that leverage ICT to cultivate ecologically conscious solutions with the objective of attaining reduced energy costs. Green ICT and Green IoT are two interrelated domains that share a mutual objective of mitigating the ecological consequences of technology while concurrently improving its efficacy and sustainability. Green ICT encompasses the practise of employing information and communication technology in a manner that is both sustainable and environmentally beneficial. Building upon this notion, Green IoT extends the principles of Green ICT to the realm of the Internet of Things, which entails the interlinking of various devices and systems for the purpose of gathering and exchanging data. IoT can be defined as "a course of action of interrelated enrolling contraptions, mechanical and progressed machines, things, creatures or people that are outfitted with exceptional identifiers (UIDs) and the ability to move data over an association without anticipating that human-to-human or human-to-computer cooperation." [1]. The advent of the IoT has presented unprecedented opportunities for small-scale enterprises to transform their approaches to energy management. By exploring IoT-enabled sensors and actuators, businesses can now gain access to real-time energy data, automate control systems, and utilise predictive analytics to enhance resource consumption and improve operational efficiency. This proactive approach helps to prevent equipment failures, reduce downtime, enhance overall operational effectiveness, and attain elevated levels of productivity and cost-effectiveness in their operations. Another significant advantage of ICT-based solutions is their potential to integrate renewable energy sources, like solar panels or wind turbines, with the existing energy infrastructure. "The Green IoT aims to bring significant improvements to the environment and human well-being to make the world smarter" through the implementation of environmentally conscious sustainable technologies that would foster the "reduction of carbon dioxide emissions', which directly affects climate change and reduces global pollution." [2]. By optimising resource consumption, organisations can reduce their operational costs and improve their overall financial performance. Additionally, this optimisation can contribute to the preservation of the environment by minimising waste generation and reducing greenhouse gas emissions.

The emergence of Industry 5.0 has the potential to serve as a facilitator rather than an impediment to an organisation's capacity to evolve into a sustainable business. Businesses have a crucial responsibility to encourage the adoption of Green IoT by making investments in IoT devices, sensors, and platforms that prioritise sustainability. In order to create and execute personalised IoT solutions that support a company's green objectives, organisations need to work with IoT-enabled suppliers and startups that have a sustainability-focused business model. The promotion of awareness and education among employees and stakeholders regarding the benefits and significance of Green IoT technologies can effectively foster behaviour shifts and enthusiastic involvement in sustainability initiatives. Green IoT solutions can be accomplished by employing diverse strategies, which can be broadly categorised into classes as depicted in Fig. 1.

The study aims to investigate the potential of an IoT-based management system for small-scale businesses and its function in enhancing resource consumption and efficacy. Through a comprehensive analysis of the challenges encountered by small-scale businesses in the realm of resource management as well as the advantages offered by Green IoT, the present study endeavours to furnish valuable discernments and suggestions for owners, managers, and interested parties of small-scale businesses to optimise their resource utilisation, minimise expenses, bolster sustainability, and attain a competitive advantage in the marketplace. The objective of the manuscript is to elucidate forthcoming trajectories and plausible developments in energy management facilitated by IoT solutions, encompassing integration with emerging technologies and the establishment of intelligent energy grids.

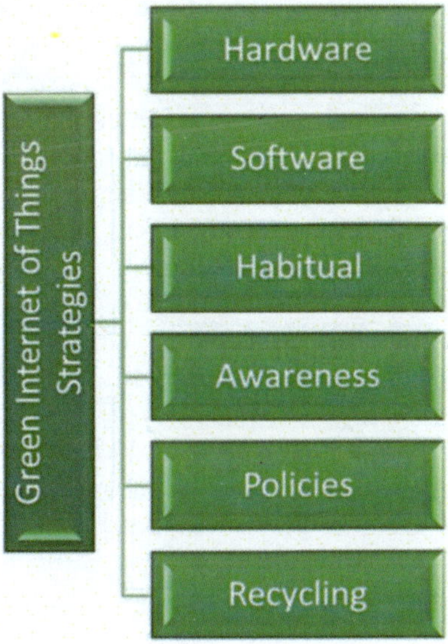

Fig. 1 Implementation methods for Green IoT [3]

2 The Significance of the Internet of Things in the Business Sphere

The role of IoT in small-scale businesses goes beyond enhancing operational efficiency and improving customer experiences. Additionally, it has the potential to enhance eco-friendly practices by monitoring and alleviating businesses' environmental footprints. The following are several ways in which the Internet of Things can prove advantageous in the pursuit of sustainability objectives within small business settings.

2.1 Energy Management: Optimising Efficiency and Sustainability

The use of IoT devices enables the monitoring of energy consumption patterns in real-time, helping high-energy enterprise users to execute energy-saving strategies. The implementation of intelligent thermostats, light controls, and power monitors enables the automatic adjustment of energy consumption in response to occupancy levels, thereby optimising energy efficiency. This results in a reduction in energy waste and a subsequent decrease in carbon emissions. IoT offers significant advantages for smart grid operators and utility companies due to its versatile technology, which facilitates the integration of diverse sensors, actuators, and other storage and secure components. The utilisation of electric vehicles as storage and the integration of renewable energies as distributed generators significantly enhances the efficiency, reliability, and cost-effectiveness of smart microgrids.

The application of this technology extends to multiple industries, including healthcare facilities and smart towns, to effectively manage and accommodate higher levels of demand thereby preventing power outages and disruptions. The proposed model of a smart power system incorporates both renewable and non-renewable energy generating systems, an Energy Storage System (ESS), and electric vehicles (EVs), which are formulated using linear programming techniques. Multiple case studies have been conducted to examine its effectiveness, and the findings [4], as depicted in Fig. 2, demonstrate a notably cost-efficient solution for both utility providers and end-users.

In his investigation, Mishra has successfully devised the Electronic Metre Automation Device (EMAD), which "include the raspberry pi, Arduino, Colour Sensor TCS 3200 and a 4channel relay" specifically designed to be seamlessly installed onto existing metering systems, which enables them to function as smart metres [5]. These metres can be conveniently accessed through customised web pages or a smart application, and notifications can be sent via SMS. Such metres have the capability to be activated and deactivated automatically when a user reaches their prepaid threshold value or by using the smart app.

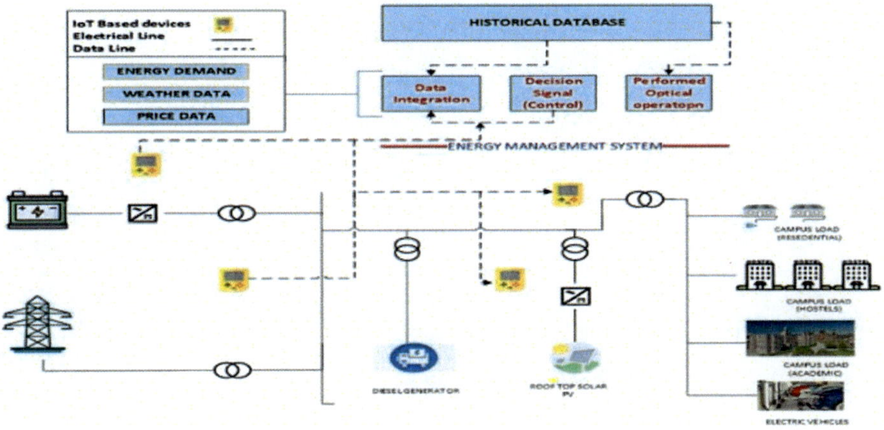

Fig. 2 A proposed model for an IoT-based energy management system [4]

Strategic Energy Management Systems (SEMS) can prove to be a highly beneficial tool for small-scale businesses aiming to minimise energy expenses, enhance energy efficiency, achieve sustainability objectives, and bolster operational resilience. As SEMS offers real-time data on energy consumption, a retail store owner can now identify peak periods of energy usage and utilise this data to optimise its lighting system and minimise energy consumption during periods of high demand. A motion sensor system can be implemented in a small retail store to activate lighting exclusively when customers are detected within the vicinity. Similarly, a small bakery can utilise IoT energy monitors for the purpose of monitoring oven usage and optimising baking schedules to enhance energy efficiency. SEMS are essential in advancing environmental sustainability as they effectively minimise energy waste, decrease greenhouse gas emissions, optimise equipment performance, and actively contribute to sustainability objectives by mitigating the necessity for additional power generation from fossil fuel sources.

2.2 *Revolutionising Waste Management: Embracing Change*

The implementation of IoT sensors and intelligent waste bins has the potential to effectively manage waste by monitoring waste levels, optimising waste collection pathways, and minimising unnecessary pickups. The concept of a smart bin has the capability to function as a Wi-Fi hotspot and can be seamlessly integrated into various container sizes, ranging from small receptacles to larger storage bins. A similar intelligent waste management solution has been developed by Pardini called "My Waste Bin". "My waste bin includes a container with a lid, and its enclosure is equipped with sensors such as the HC-SR04 module, an ultrasonic sensor responsible

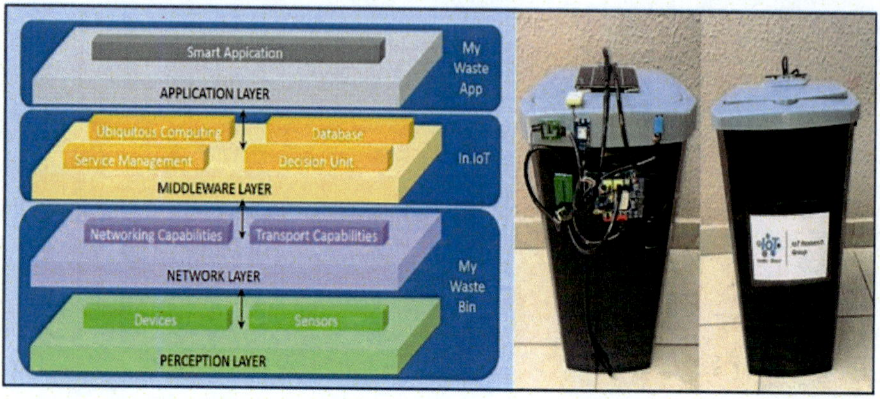

Fig. 3 A demonstrative model of a smart waste bin [6]

for measuring the level of waste filling present inside the compartment." [6]. A model of "My Waste Bin" is shown in Fig. 3.

The system incorporates "a load cell module (load sensor) that measures the weight of the residues" by using HX711 to amplify "the signal emitted by the load-cell" and to facilitate "smart bin tracking, a GPS module (model Neo-6M) was used to print geographic coordinates that represent the exact location of each bin" to indicate the distances between the bins and the nearest compartment in their vicinity [6]. The middleware and My Waste App have a significant role in the collection and archiving of all data transmitted by the containers, which enables the user to access the data efficiently through the app. In his research, Asyikin gathered data from a sample of 50 respondents, out of whom "95% of respondents feel that this smart dustbin is very useful and its performance must be further improved." [7], which can serve as a valuable asset for small-scale industries.

For instance, a small-scale restaurant business can use IoT-enabled bins, which can effectively alert waste management teams when the bins are approaching their maximum capacity. This functionality serves to encourage and enforce proper waste management practices, enabling them to effectively reduce their truck fleet, minimise fuel usage, and optimise pick-up schedules. In addition, the sensors installed in recycling bins have the capability to detect and determine whether the appropriate items are being recycled. More importantly, Smart dustbins have the potential to significantly contribute to the advancement of eco-friendly practices by reducing litter, enhancing recycling rates, optimising waste collection routes, and offering real-time data on waste levels. Thus, it has the potential to facilitate the development of cleaner, healthier, and more sustainable communities.

2.3 The Green Supply Chain: A Sustainable Approach to Business Operations

A smart supply chain management system includes the sourcing of raw materials from vendors, manufacturing goods efficiently, and the safe and intelligent transportation of goods within a closely monitored framework to a retail establishment for the purpose of being distributed to final consumers. IoT has the potential to offer a wide range of advantages to small-scale businesses within the supply chain domain. Here are several ways in which the Internet of Things can provide assistance.

Smart Transportation: The hybrid techniques of the IoT encompass wireless communication, geographical positioning, and integrated programmes, which collectively enable the precise detection of a vehicle's location within a short time frame after traversing specific geographical coordinates. In his paper, Khaizer introduces a real-time transport tracking system using Google Earth that incorporates "a transmitting embedded module to interface in-vehicle GPS and GSM devices in order to determine and send automobile location and status information via SMS" [8]. This capability holds potential for cost reduction, prevention of cargo loss, and decreased fuel consumption.

Smart Warehousing: Due to the constant evolution of technology and business practices, the field of warehousing has become increasingly intricate and vital. It is essential to prioritise space optimisation, closely monitor the warehouse environment, and implement enhancements in the product management process. Khan suggested that "the integration within the smart house is mainly possible with a short-range communication system, and ZigBee" based on IoT technology, which enhances overall control over the logistic system [9]. Warehousing can be effectively managed through the use of an IoT-based system that enables efficient inventory management, streamlined warehousing operations, and enhanced product management capabilities. For example, a furniture retailer uses IoT sensors to effectively monitor and track the precise location of its products within its warehouse facility. This system facilitates efficient identification and retrieval of products required for customer orders, enabling retailers to mitigate the risk of stockouts.

Smart Distributors: Efficient and cost-effective operations necessitate smart approaches to the distribution of items, including loading and unloading, storage, and stacking processes. The use of smart technology, such as a Radio Frequency Identification (RFID), in conjunction with IoT programmes can prove advantageous in accurately determining the precise location, facilitating the subsequent delivery of goods within a warehouse, and working "as a bridge for warehouse keepers" [9]. For instance, a food manufacturer utilises IoT sensors to effectively monitor and regulate the temperature of its products throughout the transportation process. This process aids the manufacturer in guaranteeing the integrity of its products during transportation and adhering to food safety regulations.

Smart Packaging: In order to optimise the utilisation of smart packing designs, a wireless IoT-enabled device is integrated into the product. An advanced robotic system facilitates the seamless transfer of information in real-time, ensuring high levels of versatility and velocity. At present, "NFC and RFID are mainly utilised in IoT" for "real-time monitoring," which will enhance packaging quality significantly [9]. By implementing intelligent packaging solutions equipped with embedded sensors, enterprises have the capability to effectively monitor various critical factors such as temperature, humidity, and light exposure throughout the entire process of transportation and storage, which will further aid in the protection of products, thereby minimising the risk of spoilage or damage.

Smart Logistics: Through the use of tracking and monitoring systems, businesses have the ability to optimise logistics routes, thereby minimising transportation emissions and fuel consumption. In the logistics sector, logistics enterprises use it to acquire real-time information regarding the location of vehicles and goods. "GPS is used to locate the vehicle's location; RFID information collection devices installed in the vehicle enable the vehicle to have the capacity of acquiring its real-time physical status, i.e., load, volume, and list of loading tasks." [10]. This initiative enhances workplace safety and promotes transparency in value chains and business models for commercial partners, thereby bolstering the reputation of industries and fostering stronger cooperation within the logistics industry. Additionally, green logistics practices, along with an increased emphasis on reducing, reusing, repairing, recycling, remanufacturing, and redesigning, have the potential to enhance organisational profitability. Simultaneously, these practices can also lead to increased market shares by improving customer satisfaction and sustainable performance.

IoT-enabled supply chain solutions have the potential to increase operational efficiency, exposure, security, and consumer experience and additionally contribute to the promotion of green practices within small-scale businesses. By incorporating such devices and sensors into their products, businesses can now monitor their usage patterns, gather data on product performance, and identify opportunities for enhancement, thereby allowing them to prolong the lifespan of their products, minimise waste, and adopt strategies such as product refurbishment or recycling. IoT has the potential to provide valuable support to small-scale businesses in effectively observing the entire lifecycle of their products, accumulating and using the data to enhance product development, and optimising waste reduction through product lifecycle management. This can potentially result in enhanced profitability, customer delight, and a distinct competitive advantage within the market.

2.4 *Smart Irrigation: Efficient Water Management for Sustainable Agriculture*

"Sustainable agriculture farming is a method of preserving nature without compromising the future generation's basic needs, whilst also improving the effectiveness

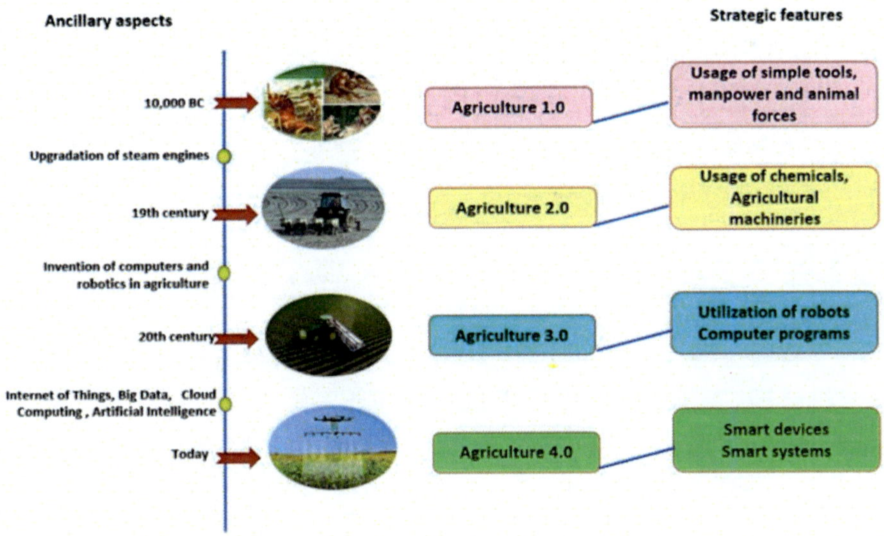

Fig. 4 Evolution of the era of agriculture [11]

of farming" [11]. Sustainable farming is of great importance in the preservation of natural resources in an economically feasible manner, the prevention of biodiversity loss, the reduction of greenhouse gas emissions, and the creation of a natural and healthy environment. Figure 4, as presented by Dhanaraju et al. [11], illustrates the evolution of the agricultural era, resulting in the advancement and prosperity of farmers and agricultural lands.

Advancements in machinery have significantly enhanced the scale, speed, and productivity of farm equipment, thereby enabling more efficient cultivation practices. "Smart farming promotes precision agriculture with modern, sophisticated technology and enables farmers to remotely monitor the plants." [11]. Innovations in seed technology, irrigation systems, and fertilisers have significantly enhanced agricultural practices, thereby facilitating farmers in augmenting their crop yields.

IoT-enabled irrigation systems have the capability to effectively monitor water usage, detect potential leaks, optimise overall water consumption, and help businesses effectively preserve water resources, minimise waste, and actively contribute towards the attainment of water sustainability. Such technologies can monitor soil moisture levels and weather conditions to optimise watering schedules and can also identify leaks within water supply lines and plumbing systems, which allows for further prompt detection, thereby reducing water waste.

In the context of mango tree farms, the implementation of IoT technology can be utilised to properly monitor plant development and general health. This can be achieved by collecting and analysing data pertaining to crucial factors, including weather conditions, water availability in the leaves, and plant well-being. Having access to this information can be helpful in mitigating the adverse effects of diseases and pests on crop productivity. In the same vein, IoT sensors can be utilised within

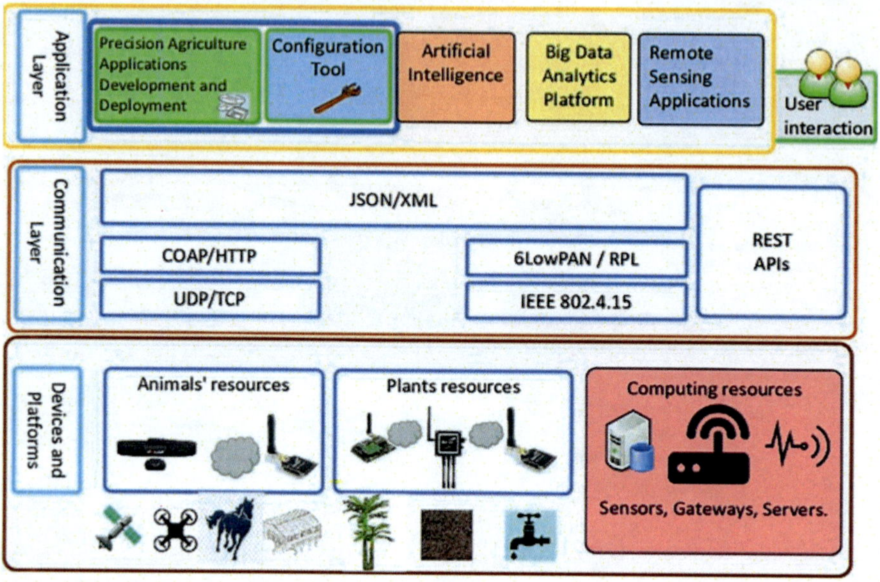

Fig. 5 A suggested framework for precision agriculture [12]

a chicken farm setting to gather data pertaining to multiple facets of individual chickens. This data encompasses factors such as temperature, health, activity levels, and nutrition.

The implementation of technologies such as drones allows farmers to efficiently monitor and separate diseased animals from the group, effectively preventing contamination and reducing labour costs. In their paper, [12] present a conceptual framework for an IoT-enabled model designed specifically for precision agriculture, illustrating interconnections among sensors and agricultural supplies in order to facilitate real-time monitoring and management, as shown in Fig. 5.

With the implementation of IoT and AI, an agricultural sector can be developed that possesses significant potential for identification and exploitation owing to its operational simplicity and cost reduction benefits. It has the capacity to enhance crop productivity, facilitate informed crop management decision-making, mitigate the environmental impact of agricultural practices by minimising chemical usage, and reduce costs associated with electricity, water, and fuel consumption.

2.5 *Monitoring Carbon Emissions: Promoting Sustainable Practices*

IoT solutions have the capability to monitor and quantify carbon emissions across diverse operational procedures within enterprises, thereby furnishing businesses with

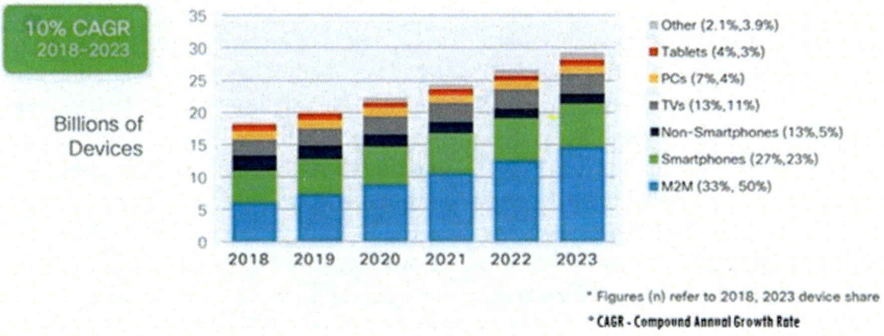

Fig. 6 Annual increase in global networking trends [13]

valuable information regarding their ecological footprints. This information has the potential to assist businesses in implementing strategies to reduce emissions, making valuable contributions to sustainability initiatives, and aligning their operational activities with green practices to yield advantages for both the business and the environment.

The exponential increase in internet traffic, escalating computational demands, expanding data storage needs, and proliferation of networks for communication have significantly contributed to environmental emissions. It is worth noting that data centres emerge as the primary source of emissions with the highest rate of contribution. Consequently, this has resulted in the generation of substantial quantities of diverse technological waste, which in turn has had a detrimental impact on both human well-being and the world's ecology. Based on the findings of the Cisco 2020 report on global networking trends, the increasing interconnectedness of the world presents a significant challenge for organisations in terms of environmental sustainability, as presented in Fig. 6.

Businesses are expected to adopt new benchmarks, including the reduction of greenhouse gas emissions, the preservation of natural resources and biodiversity, the minimisation of manufacturing waste, and the utilisation of recycled electronic waste. The adoption of the concept of the Green Internet of Things aims to mitigate carbon emissions and minimise power consumption in order to facilitate the advancement and sustainability of a more intelligent global environment.

3 Conclusion

The scope of Green Communication and Network Systems is subject to constant evolution due to the emergence of new technologies, standards, and practices. In light of the growing global interconnectivity, the significance of these systems in promoting environmental sustainability is elevated, necessitating continuous study and innovation to effectively tackle the environmental issues prevalent in our contemporary digital era. Green ICT and Green IoT refer to approaches to computing and networking that minimise their negative effects on the natural world. The Internet of Things has successfully facilitated the convergence of the physical and digital realms, enabling intelligent decision-making processes that operate autonomously without the need for human intervention. The collection and processing of a substantial volume of real-time data are significantly enhancing efficiency, productivity, accuracy, and quality of service across various business sectors. During the planning phase of IoT application-based systems, it is crucial to meticulously choose and develop the hardware electronics, incorporating a wide range of sensors, communication and data-processing procedures, data storage infrastructures, and computational needs. This selection and design process must be optimised to effectively meet the demanding power budget goals. In order to achieve the goals of the Green IoT, the study focuses on certain methods at different levels of IoT-enabled businesses that contribute towards a comprehensive mitigation of energy consumption and help in the establishment of a sustainable environment for future generations. The increasing presence of new digital technologies in the market has compelled organisations to undergo digital transformation for their businesses. Technological transformation is a multifaceted process that goes beyond simply enhancing specific functions within organisations. Instead, it entails fundamental alterations that have the potential to generate new possibilities for improvement. Additionally, the process of transformation is not centred around any particular organisation; rather, it is a phenomenon that initiates societal and economic shifts.

References

1. Rout S, Samantaray S (2022) Interplay of artificial intelligence and ecofeminism: a reassessment of automated agroecology and biased gender in the tea plantations. In: 2022 International conference on advancements in smart, secure and intelligent computing (ASSIC), 19 Nov 2022. https://doi.org/10.1109/assic55218.2022.10088342
2. Memić B, Hasković Džubur A, Avdagić-Golub E (2022) Green IoT: sustainability environment and technologies. Sci Eng Technol 2(1):24–29. https://doi.org/10.54327/set2022/v2.i1.25
3. Albreem MA, Sheikh AM, Alsharif MH, Jusoh M, Mohd Yasin MN (2021) Green internet of things (GIoT): applications, practices, awareness, and challenges. IEEE Access 9:38833–38858. https://doi.org/10.1109/access.2021.3061697
4. Fei L, Shahzad M, Abbas F, Muqeet HA, Hussain MM, Bin L (2022) Optimal energy management system of IoT-enabled large building considering electric vehicle scheduling, distributed resources, and demand response schemes. Sensors 22(19):7448. https://doi.org/10.3390/s22197448
5. Mishra JK, Goyal S, Tikkiwal VA, Kumar A (2018) An IoT based smart energy management system. In: 2018 4th International conference on computing communication and automation (ICCCA), Dec 2018. https://doi.org/10.1109/ccaa.2018.8777547
6. Pardini K, Rodrigues JJ, Diallo O, Das AK, de Albuquerque VHC, Kozlov SA (2020) A smart waste management solution geared towards citizens. Sensors 20(8):2380. https://doi.org/10.3390/s20082380
7. Asyikin AN, Syahidi AA, Subandi (2020) Design and implementation of different types of smart dustbins system in smart campus environments. In: Proceedings of the international

joint conference on science and engineering (IJCSE 2020). https://doi.org/10.2991/aer.k.201124.001

8. Khaizer M, Santha K (2015) Design and implementation of vehicle tracking system using GPS/GSM/GPRS technology and smartphone application. Int J Sci Eng Technol Res 4(35)
9. Khan Y, Su'ud MBM, Alam MM, Ahmad SF, Ahmad (Ayassrah) AYAB, Khan N (2022) Application of internet of things (IoT) in sustainable supply chain management. Sustainability 15(1):694. https://doi.org/10.3390/su15010694
10. Liu S, Zhang G, Wang L (2018) IoT-enabled dynamic optimisation for sustainable reverse logistics. Procedia CIRP 69:662–667. https://doi.org/10.1016/j.procir.2017.11.088
11. Dhanaraju M, Chenniappan P, Ramalingam K, Pazhanivelan S, Kaliaperumal R (2022) Smart farming: internet of things (IoT)-based sustainable agriculture. Agriculture 12(10):1745. https://doi.org/10.3390/agriculture12101745
12. Atalla S, Tarapiah S, Gawanmeh A, Daradkeh M, Mukhtar H, Himeur Y, Mansoor W, Hashim KFB, Daadoo M (2023) IoT-enabled precision agriculture: developing an ecosystem for optimized crop management. Information 14(4):205. https://doi.org/10.3390/info14040205
13. Cisco Annual Internet Report (2020) Cisco annual internet report (2018–2023) white paper. Cisco, 10 Mar 2020. Retrieved from https://www.cisco.com/c/en/us/solutions/collateral/executive-perspectives/annual-internet-report/white-paper-c11-741490.html. Accessed on 16 July 2023

Unravelling Obfuscated Malware Through Memory Feature Engineering and Ensemble Learning

K. M. Yogesh, S. Arpitha, Thompson Stephan, M. Praksha, and V. Raghu

1 Introduction

Malware, which is short for harmful software, was initially developed for the personal computer in the year 1986 and propagated via floppy discs [1]. Since then, a lot of progress has been achieved both in the creation of malicious software and in the technologies that are used to protect against it. Traditional methods frequently rely on signature-based detection [2] which involves comparing unclassified software to previously discovered malicious software. On the other hand, methods have developed that can conceal and manipulate malware in order to thwart the classification process. The signature-based recognition method fails as a consequence of this, despite the fact that the underlying malware may have been identified in the past. Obfuscation of malware is the name given to this approach. There is a steady stream of previously unidentified "Zero day" malware being developed, hence malware protection systems need to be able to recognize malware that hasn't been seen before. This demands that these systems include malware detection capabilities. Due to this, a number of studies have proposed machine learning classification models that are based on the extraction of characteristics from malware that is thought to be dangerous while it is functioning in a secure environment [2, 3, 5].

The aim of this study is to examine and analyse the classification performance of two different ML techniques for obfuscated malware. The basis for this comparison is the CICMalMem 2022 data set which is accessible to the general public. This data collection contains elements that were extracted from a system's memory while it

K. M. Yogesh (✉) · V. Raghu
Ramaiah University of Applied Sciences, Bangalore, Karnataka, India
e-mail: yogeshkm.mrj@gmail.com

S. Arpitha · M. Praksha
Mangalore University, Mangalore, Karnataka, India

T. Stephan
Graphic Era Deemed to be University, Dehradun, Uttarakhand, India

M. S. Kaiser et al. (eds.), *ICT: Smart Systems and Technologies*, Lecture Notes in Networks and Systems 878, https://doi.org/10.1007/978-981-99-9489-2_28

was running both good and bad software. The drawbacks of both static and dynamic analysis approaches may be addressed by memory analysis [4]. The limitations placed on malware signatures created as a direct result of static analysis can be overcome with the use of memory analysis. Some of the limitations of dynamic analysis, such as the concealed behaviour of malware while it is being analysed, can be overcome with the use of memory-based features. Memory analysis and static analysis are generally equivalent, albeit it is known that more recent malware does not exhibit certain behaviours during static analysis. On the other hand, memory analysis enables the discovery of such concealed actions and, as a result, delivers far better virus detection than more traditional static analysis. There are usually some signs of malware in the memory [6]. By using data like DDL records, registries, active network connections, operating system registries, and terminated processes, memory analysis enables one to learn certain things about the behavioural characteristics of malware and operating services. This information can be retrieved using information like [7].

2 Related Work

Attacks brought on by malware have clearly increased in frequency during the past few years. The first half of 2022 alone saw almost 2.8 billion instances of malware assaults, according to data given by Statista (2023a). These assaults are happening more frequently, are more severe, and are getting more complex. With a daily detection rate of 560,000 new malware cases and a 62% increase in malware variants year over year, malware has emerged as one of the biggest cybersecurity dangers [9]. Malicious actors frequently use malware as a very effective attack vector. It is an excellent option for a range of various attacks since it can deliver a wide range of payloads and achieve a range of objectives. The findings of Check Point's Cyber Security Report 2021 (Check Point Software 2021) show that the most frequent attacks on corporate networks in 2021 were perpetrated by the following classes of malicious software: The detection of malware presents a number of obstacles; nevertheless, machine learning is a potentially useful solution [9]. Dealing with obfuscated malware, on the other hand, makes an already difficult process even more difficult. Virus is classified as obfuscated if it modifies its source code in such a way that it is unintelligible to humans. This makes it feasible for the virus to evade detection through normal channels. A database of known malware signatures is necessary for signature-based detection since it can make identification easier and contain unfinished scripts, hashes, or other indications that were gleaned from the infection. Obfuscation, on the other hand, has emerged as a popular method for bad actors to conceal their actions and avoid detection, which makes the process of detection more challenging. Malware employs a wide range of obfuscation methods, from straightforward ones like dead code insertion to more intricate ones like packing [10]. The level of sophistication of these techniques can range from very simple to very complex. The following is a list of some of these methods of obfuscation that can be found in (Table 1).

Table 1 Malicious software

Botnets	Descriptio
Banking malware	Malware intended specifically to obtain unauthorized access to financial data, which is subsequently transmitted back to the bad actor. Even banking transactions can be made by this malware's ore sophisticated variants without the target's knowledge or agreement
Spyware (Infostealers)	Malware designed to gather data on the affected device and its user(s). This data may consist of login passwords, financial information, keystrokes, and other private data
Mobile malware	Virus developed to collect information about the impacted device and its user(s). Keystrokes, financial data, login passwords, and other sensitive data may be included in this data
Ransomware	Malware that locks or encrypts a user's data, then demands payment in exchange for the release of the data. In recent years, there has been a sharp increase in these attacks. With 304 million documented assaults in Ransomware (2022), ransomware attacks have rapidly increased

There has been a recent uptick in the incidence of hostile actors employing disguised malware. Obfuscation is used in more over 25% of JavaScript-based malware, according to [8]. This concerning trend in the concealment of malware demonstrates the critical need for detection methods that are both very efficient and highly accurate. Given the difficulty that signature-based malware detection systems have in detecting obfuscated malware, machine learning has emerged as a trustworthy technique for its identification. In most cases, machine learning-based solutions perform significantly better than signature-based methods when it comes to detecting malware that has "never been seen before" [5–10]. The research on cybersecurity threat trends that was produced by Cisco (Cisco Umbrella 2022) states that certain families of malware have the ability to rapidly spread if they are not discovered and stopped in a timely manner. This highlights the need for detection systems that are both speedy and lightweight and have the ability to swiftly identify malware with a limited amount of information (Table 2).

3 Proposed Approach

The majority of currently published works suffer from high levels of both complexity and time consumption, rendering them unfit for use in the real world. By utilizing the most useful characteristics gleaned from memory analysis, this paper presents a solution for obfuscated malware detection that is quick, effective, and simple to construct. This approach is intended to address the problem presented above.

Table 2 Malware families

Malware families	Description
Trojan Horse	Malware Users may not be aware that their computer is being infected with malicious software even though the program appears to be safe. Animal, a question-and-answer game that was initially published in 1974, is generally regarded as the very first Trojan horse. Users participated in a game in which they chose an animal and then posed questions in an attempt to determine the animal's name. Without the user's knowledge or consent, it was replicating itself in the background to all of the directories to which the user had write access. In today's world, the capabilities of Trojan horses are practically limitless. Every day, new families of Trojan horses are being discovered. In this study, there are five different types of Trojan horses described
Zeus	Zbot is another name for this thing. It made its debut in the year 2007. It is a form of banking Trojan that can steal banking credentials via keylogging and logging what is typed on the keyboard. Another essential purpose is to establish communication with the command and control server in order to set up a botnet. In the years that followed its initial appearance, open-source code was distributed, and other versions of the software, such as Citadel, GameoverZeus, Ice IX, and KINS, were developed
Emonet	It was first discovered in 2014 and is known as a Trojan horse. It is malicious software that targets financial institutions and is designed to steal sensitive data by monitoring network traffic. In the years that have passed since its introduction, it has morphed into a platform that enables the installation of further malicious software. It is able to do things like create and organize botnets, among other things. Additionally, it possesses some worm-like qualities that allow it to spread
Refroso	It is a Trojan horse that functions as a backdoor that was originally discovered in the year 2009. It does this by deleting items from the registry, which allows it to change the settings of the firewall. It has the ability to initiate and hide memory processes. It has the ability to carry out a variety of tasks, including masking undesired activities in the browser and routing users to dangerous websites. Through the provision of a configuration that enables access from the outside, it can facilitate access attacks
Scar	On the device that it infects, it acts as a Trojan horse that enables various forms of malicious software to be installed. It will download a list of URLs that connect to files with the extension exe in order to enable malware to download further files. It is also capable of doing activities such as modifying the settings of the operating system and gathering confidential information that is stored on the device
Reconyc	On the device that it infects, it is a Trojan horse that is responsible for downloading various types of malware. As is the case with the vast majority of malicious software, it is spread either through websites that cannot be trusted or as an attachment to another file. It is also possible for it to restrict access to other essential tools that are included in the operating system, such as the command prompt, task manager, and registry editor

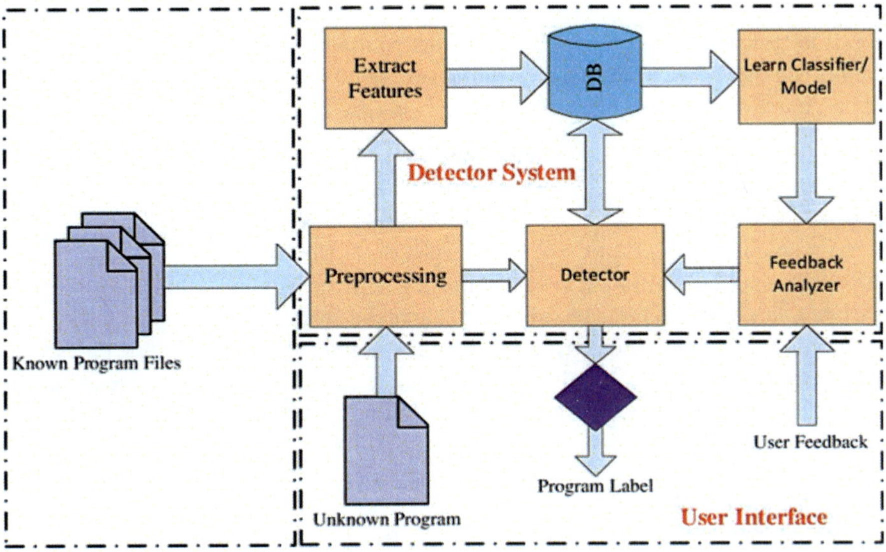

Fig. 1 Malware memory analysis process

3.1 General Overview

Figure 1 presents an abstract of this obfuscated malware detection technology for your viewing pleasure. The proposed framework comprises several components, including memory dump files. These memory dumps can be acquired through programs like MAGNET RAM, Forensic Tool Kit (FTK), ManTech Memory DD, or virtual machine managers equipped with memory-capturing capabilities. Alternatively, a memory-capturing tool can be employed. (MAG, 2021) (Man, 2021) (For, 2021). These files represent a snapshot of the activities that occurred in the system's memory. Volatility is a completely open collection of tools to extract digital artifacts from volatile memory (RAM) samples. It was written in Python and released under the GNU General Public License. (Vol. 2016). VolMemLyzer-V2 is the memory feature extractor for learning-based solutions. It includes the 26 new features that have been incorporated as a part of the proposed model to target obfuscated and concealed malware [12]e VolMemLyzer will extract the features by making use of volatility plugins and will then output a CSV file.

Feature File in CSV Format: This file serves as the output generated by the VolMemLyzer feature extractor. It takes the form of a concise comma-separated values (CSV) file encompassing all the extracted features. Ensemble learning, as described by Ens (2021), is a machine learning approach that places a strong emphasis on employing multiple classifiers to offset each other's limitations. This strategy minimizes the influence of these limitations on the overall results. Some classifiers are particularly susceptible to outliers, while others exhibit significant bias. To construct the framework, consisting of two tiers of classifiers, we harnessed the stacking ensemble technique. The binary classification output, categorized as malicious or benign, is assigned to each memory dump file, denoting the presence of malicious or benign activities.

3.2 Proposed Features

In this model, the initial step involves acquiring a memory dump suitable for volatility's latest version, which can be subjected to analysis by the VolMemLyzer [11]. Subsequently, these memory dumps are processed by the VolMemLyzer feature extractor, an extension powered by Volatility responsible for extracting 58 distinctive features. Among these, 26 newly added features are specifically designed to target obfuscated and concealed malware, a topic explored in the following section.

These 58 features fall into five distinct categories as organized by volatility. The first category, known as Malfind, focuses on identifying potentially hazardous executables, often DLLs associated with Trojan software. The subsequent section, Ldrmodule, provides insights into any potentially injected code within the system, a common method of spyware installation. Handle, the third category, scrutinizes the diverse types of memory-stored information and their organization. Process view features, in the fourth category, present the process list and related data crucial for identifying hazardous processes. Finally, the fifth category, APIhook features, reveals the total count of API hooks linked to key types. Table 1 details the features extracted for the memory analysis framework.

3.3 Detection Model

Once these features have been successfully extracted, their preparation for utilization within the proposed ensemble learning method is the subsequent step. This method encompasses two crucial stages: the training and validation phases, integral to the learning classifiers. In the training phase, the ensemble learner engages the base learners, and their prediction outputs serve as inputs for the secondary layer classifier. After training, the ensemble learner undergoes validation, replicating the training steps to confirm dataset findings and ensure the effectiveness of training.

Several approaches are available for ensemble learning, including stacking, voting, boosting, and bagging, among others. Voting combines classifier results using user-assigned weights, while stacking achieves a similar goal with the assistance of a meta-classifier. For this proposed work, stacking was chosen as the ensemble learning strategy due to its speed and variance performance. As previously explained, the first layer operates independently, and potential collaboration between subsequent layers can expedite the categorization process. The second layer addresses minor differences in inputs, enabling rapid execution once the first layer completes its task. The diverse classifiers complement each other's weaknesses, ensuring both high accuracy and swift classification.

Given the wide array of machine learning classifiers available, it is crucial to identify the most suitable classifier compatible with the provided model. Typically, classifier selection depends on the dataset earmarked for classification. This work employs decision tree classifier, LinearSVC, logistic regression, random forest tree, and AdaBoost classifiers as the foundational learners. Various combinations of these classifiers have been systematically tested and evaluated.

The process of generating masks and extracting features was executed using efficient libraries. To assess their effectiveness, no specific parameters were considered. During the feature selection process, performance evaluation was carried out in two phases. The first phase involved selecting a base model after conducting hyperparameter tuning. The second phase encompassed performance assessment subsequent to feature selection and the application of machine learning models. Performance analysis is done by using specific parameters as mentioned below Precession, recall, $F1$ score, accuracy.

4 Experimental Results

The performance of all classifier models was enhanced through hyperparameter tuning. The evaluation of their performance relied on utilizing the score function from the sklearn library. The subsequent evaluation and the resulting outcomes are elaborated upon below. The model selection based on the better accuracy based on the hyperparameters with using boosting algorithms like decision tree classifiers, logistic regression, AdaBoost classifiers and random forest classifiers boost as described Table 3 boosting algorithms make the model to predict accurate results classification accuracy and $F1$ score measures based on model performance for the displayed Fig. 2

Table 3 displays the results of various machine learning models for the proposed classification task, using three key evaluation metrics: Accuracy, Balanced Accuracy, and $F1$ score. The top three models, nearest centroid, linear discriminant analysis, and decision tree classifier, demonstrate strong numerical performance across multiple key metrics. Nearest centroid stands out as the top performer, achieving an accuracy of 91.29%, balanced accuracy of 91.29%, and an impressive $F1$ score of 95.45%. Linear discriminant analysis closely follows with an accuracy of 89.79%, balanced accuracy of 89.79%, and an $F1$ score of 94.62%, indicating robust performance. Decision tree classifier also performs admirably, achieving an accuracy of 89.27%, balanced accuracy of 89.27%, and an $F1$ score of 94.33%. These models exhibit exceptional accuracy, balanced class handling, and a strong balance between precision and recall, making them well-suited for our classification task.

LinearSVC, BernoulliNB, and logistic regression show moderate performance, with accuracy and balanced accuracy percentages ranging from 50 to 58%, and $F1$ score percentages ranging from 66.74 to 73.90%. AdaBoost classifier, label propagation, and label spreading have fair performance, with accuracy and balanced accuracy percentages around 35%, and $F1$ score percentages around 52%. Extra

Table 3 Experimental results with parameters

Model	Accuracy (%)	Balanced accuracy (%)	$F1$ score (%)
Nearest centroid	91.29	91.29	95.45
Linear discriminant analysis	89.79	89.79	94.62
Decision tree classifier	89.27	89.27	94.33
LinearSVC	58.60	58.60	73.90
BernoulliNB	56.31	56.31	72.05
Logistic regression	50.08	50.08	66.74
AdaBoost classifier	46.26	46.26	63.26
Label propagation	35.76	35.76	52.68
Label spreading	35.76	35.76	52.68
Extra trees classifier	22.65	22.65	36.94
Extra tree classifier	20.69	20.69	34.28
Random forest classifier	7.00	7.00	13.09
SGD classifier	3.35	3.35	6.47
Passive aggressive classifier	1.15	1.15	2.27
Perceptron	0.55	0.55	1.09
Ridge classifier	0.34	0.34	0.68
Ridge classifier CV	0.22	0.22	0.44

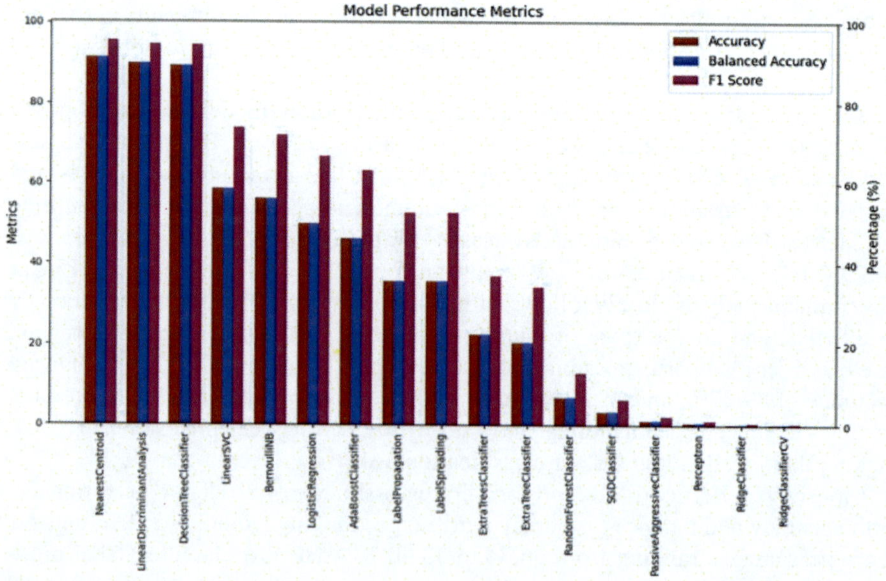

Fig. 2 Malware memory analysis process

trees classifier, extra tree classifier, random forest classifier, SGDClassifier, passive aggressive classifier, perceptron, ridge classifier, and ridge classifier CV have lower performance across all metrics, with accuracy and balanced accuracy percentages below 23%, and $F1$ score percentages below 37%.

5 Conclusion

Malware developers quickly modified their malicious code in response to the development of networking and the internet. The majority of these malicious codes are employed to exploit vulnerabilities in Microsoft Windows. Memory analysis can be used to discover obfuscated or concealed malware, and while there are various approaches available for doing so, the amount of time required and the intricacy of the work involved are both considerable. The obfuscated malware detection model that was offered as a solution to this problem collects features from memory dumps by making use of VolMemLyzer, which is a feature extractor for learning systems.

Leveraging the freshly developed dataset, an exhaustive exploration of diverse base-learner and meta-learner combinations was executed, ultimately culminating in the selection of the final model. The decision tree classifier, LinearSVC, logistic regression, random forest tree, and AdaBoost classifiers emerged as top performers among the base learners. Simultaneously, the logistic regression classifier excelled as the meta-learner, achieving an impressive accuracy rate of 91.25% for each component. This harmonious combination yielded the most favourable outcomes. Furthermore, a comparative analysis was conducted between this model and analogous works with a focus on obfuscated malware detection within memory. The outcomes of this comparison underscored the proposed model's competitive edge, showcasing shorter classification times and an overall superior performance when measured against existing models.

References

1. Milošević N (2013) History of malware. https://doi.org/10.48550/arXiv.1302.5392. https://arxiv.org/abs/1302.5392
2. Sihwail R, Omar K, Ariffin KZ (2018) A survey on malware analysis techniques: static, dynamic, hybrid and memory analysis. Int J Adv Sci Eng Inf Technol 8(4–2):1662–1671
3. Nachreiner C et al (2021) Internet Security Report-Q3 2021 | WatchGuard Technologies. https://www.watchguard.com/wgrd-resource-center/security-report-q3-2021 (visited on 05/17/2022)
4. Bazrafshan Z et al (2013) A survey on heuristic malware detection techniques. In: The 5th conference on information and knowledge technology, 2013, pp 113–120. https://doi.org/10.1109/IKT.2013.6620049.
5. Mohanta A (2020) Malware analysis and detection engineering : a comprehensive approach to detect and analyze modern malware, 1st edn. Apress, New York. ISBN: 1-4842-6193-3

6. Malware Memory Analysis | Datasets | Canadian Institute for Cybersecurity | UNB. https://www.unb.ca/cic/datasets/malmem-2022.html (visited on 03/29/2022)
7. Carrier T et al (2022) Detecting obfuscated malware using memory feature engineering. In: Proceedings of the 8th international conference on information systems security and privacy. Institute for systems, technologies of information, control, and communication (INSTICC)
8. Sewak M, Sahay SK, Rathore H (2018) Comparison of deep learning and the classical machine learning algorithm for the malware detection. In: 2018 19th IEEE/ACIS international conference on software engineering, artificial intelligence, networking and parallel/distributed computing (SNPD). IEEE, pp 293–296
9. Dataset-Malicia Project. http://www.malicia-projectcom/dataset.html (visited on 04/06/2022)
10. El Merabet H, Hajraoui A (2019) A survey of malware detection techniques based on machine learning. Int J Adv Comput Sci Appl 10(1):366–373
11. Abri F et al (2019) Can machine/deep learning classifiers detect zero-day malware with high accuracy?' In: 2019 IEEE international conference on big data (big data), pp 3252–3259. https://doi.org/10.1109/BigData47090.2019.9006514
12. Baig D et al (2021) Malware detection and classification along with tradeoff analysis for number of features, feature types, and speed. In: 2021 international conference on frontiers of information technology (FIT). 2021, pp 246–251. https://doi.org/10.1109/FIT53504.2021.00053

Quad-Port MIMO Antenna System for n79-5G and WLAN Wireless Communication Applications

Trushit Upadhyaya, Killol Pandya, Upesh Patel, Jinesh Varma, Rajat Pandey, and Poonam Thanki

1 Introduction

Wireless communication is an essential tool to survive the recent developments in the electronics era starting from home automation to satellite communication. There have been consistent efforts in the development of antenna technology to meet the technological demands of wireless communication standards. The user demand and user quantity have been substantially increasing worldwide because of this it becomes absolutely essential to improve the quality of the communication network and channel capacity. Multiple-input multiple-output (MIMO) communication fulfills such demanding necessities.

The 5G communication has a strong reliance on the MIMO communication technology. The third and fourth generations of communication technology could not sufficiently provide high-speed requirements of extremely high data rate communication, viz. live streaming, and online gaming to name a few. 5G communication has the capability of providing an extremely high data rate without sacrificing the user count or antenna parameters because of enormously improved network parameters such as delay and jitter [1]. It can play a crucial role in enabling cutting-edge applications such as 4K video streaming, augmented reality (AR), and virtual reality (VR). In addition, in WLAN and Wi-Fi communication networks, especially in high-density environments such as airports, stadiums, and shopping malls, such antennas' spatial diversity and beamforming capabilities can prove to be invaluable. 5G communication has abilities to support enhanced user count with self-effacing connections through effective frequency band allocation. The communication dependability can be enhanced by employing manifold antennas. By utilizing transmit and receive combinations, MIMO technology mitigates fading problems. Time diversity helps MIMO to support

T. Upadhyaya (✉) · K. Pandya · U. Patel · J. Varma · R. Pandey · P. Thanki
Chandubhai S. Patel Institute of Technology, Charotar University of Science & Technology (CHARUSAT), Changa, Gujarat 388421, India
e-mail: trushitupadhyaya.ec@charusat.ac.in

M. S. Kaiser et al. (eds.), *ICT: Smart Systems and Technologies*, Lecture Notes in Networks and Systems 878, https://doi.org/10.1007/978-981-99-9489-2_29

multiple access technology which in turn aids in better communication efficiency. Such diversity is also available in the frequency spectrum for enhancing transmission efficiency. MIMO technology also provides low-power communication abilities, which can be quite helpful in short-distance communication. The number of options for the multiplexing scheme also helps significantly. The spatial and temporal diversity aids in increasing the overall system gain [2]. The reception of manifold signals over multiple antennas severely reduces noise.

The typical issue with antenna technology is its existence in space-constrained environments. The surface-integrated antennas are quite space-efficient; however, they bring the disadvantage of antenna parameters. The gain-bandwidth product of such antennas remains quite low compared to other types of antennas due to archetypal printing on lossy substrates. For the MIMO antenna design prime design challenge is the presence of multiple elements in close proximity. The mutual coupling between the ports becomes vital for extracting optimal benefits from these antennas. The literature presents varied MIMO antennas having enhanced port isolation [3–16]. With the inexistence of such techniques, it becomes very important to optimally place the elements of an antenna for reduction in mutual interelement coupling. The latest research on MIMO antenna technology can be found in literature [17–21]

2 Antenna Design and MIMO Configuration

An engineered and compact MIMO resonator is etched on a low-cost lossy FR-4 laminate. The thickness of the substrate is kept at a standard 1.6 mm. The dielectric constant for FR-4 laminate is 4.4, and the loss tangent is in order of 0.026. The loss tangent becomes important for the frequencies above 1 GHz because of the skin depth at higher frequencies, however, for the mass production of the antenna, the FR-4 laminate provides a cost-effective solution compared to expensive low-lossy laminates. The antenna was designed in a full wave software simulator. The designed antenna is exemplified in Fig. 1. The resonator is fed through the strip line and coaxial connector. The strip-line connector exhibits the lowest design complication and effortlessness in implementation. The slots were introduced in the planar resonator for achieving dual-band resonance. The arranging of the slots was reiterated multiple times to achieve the targeted resonance and MIMO diversity parameters. The electrical dimensions of the slots can be further optimized for achieving the additional frequency bands. It is noteworthy that further excited mode shall impact other antenna parameters. A defective ground plane was introduced to improve upon bandwidth; however, it also comes with drawbacks in performance reduction of other antennas. The optimized model could produce dual mode at 4.5 GHz and 5 GHz bands for n79 5G and WLAN wireless communication applications. The optimized antenna dimensions are as below:

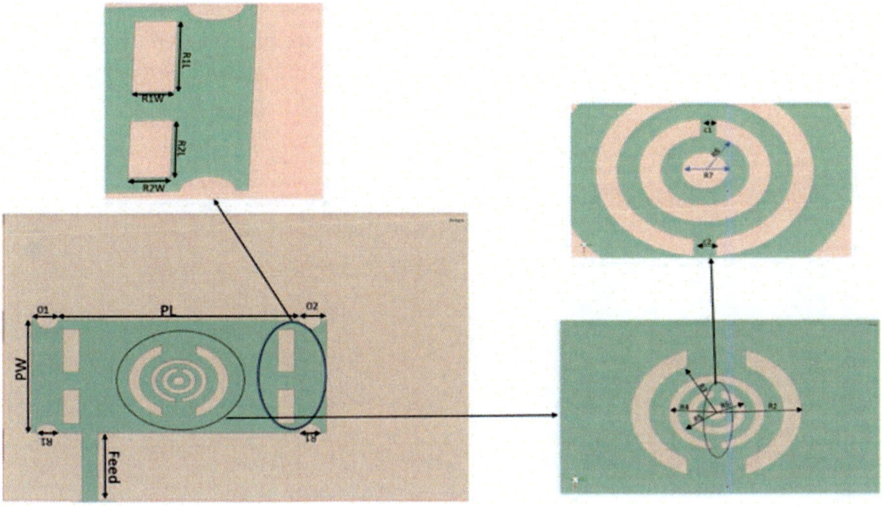

Fig. 1 Antenna view

FEED = 7.46, PL = 12.48, PW = 25.42, 01 = 02 = 2.91, R1 = 3, R2Y = 5.33, R2X = 4.4, R3Y = 4.14, R3X = 2.7, R4Y = 3, R4X = 2.4, R5Y = 2.5, R5X = 1.9, R6Y = 1.41, R6X = 1.8, R7Y = 0.412, R7X = 0.412, R8Y = 1.2, R8X = 0.92, C1 = 0.04, C2 = 0.06, R1L = 4.8, R1W = 1.8, R2L = 3.84, R2W = 1.8, SL = SL = 90, GL = GW = 90, R8 = 0.412, SH = 1.68

3 Proposed Antenna Results

The designed MIMO antenna design is illustrated in Fig. 2. The antenna has quad elements posing resonance at targeted n79 5G and WLAN wireless communication application. The electrical dimensions of the substrate were optimized to have enhanced isolation between the individual ports for improved diversity parameters. The mechanical dimensions of the antenna volume are 90 mm $\times$ 90 mm $\times$ 1.6 mm. The optimization through simulation software was carried out to effectively meet the tradeoff between electrical size and antenna parameters. The antenna was fabricated on PCB as shown in Fig. 3. The entire measurement process was carried out within a controlled environment, specifically an anechoic chamber, to minimize external interference and reflections. The antenna was securely mounted on a non-metallic pedestal, ensuring minimal interference from the mounting structure, and its positioning was arranged to maintain an unobstructed line of sight between the antenna and the measurement equipment. To eliminate potential measurement system errors, the measurement equipment, including signal generators, vector network analyzers

(VNAs), and spectrum analyzers, underwent rigorous calibration using known calibration standards. The observations of the simulation and measured S_{11} are shown in Fig. 4.

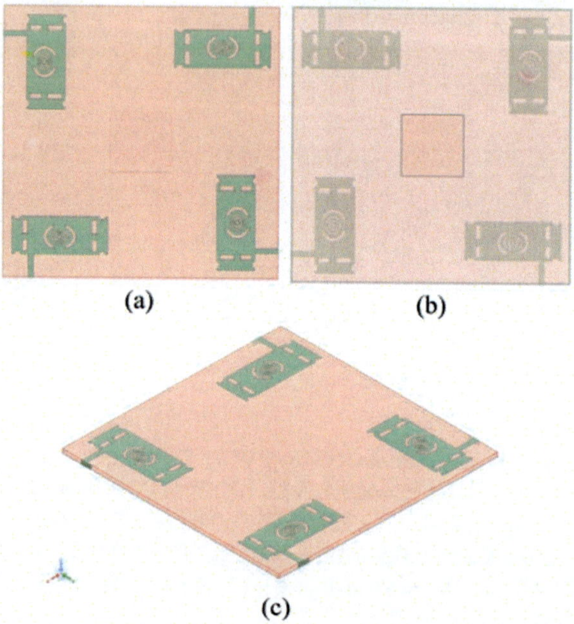

Fig. 2 Full antenna view **a** top, **b** bottom, **c** isometric

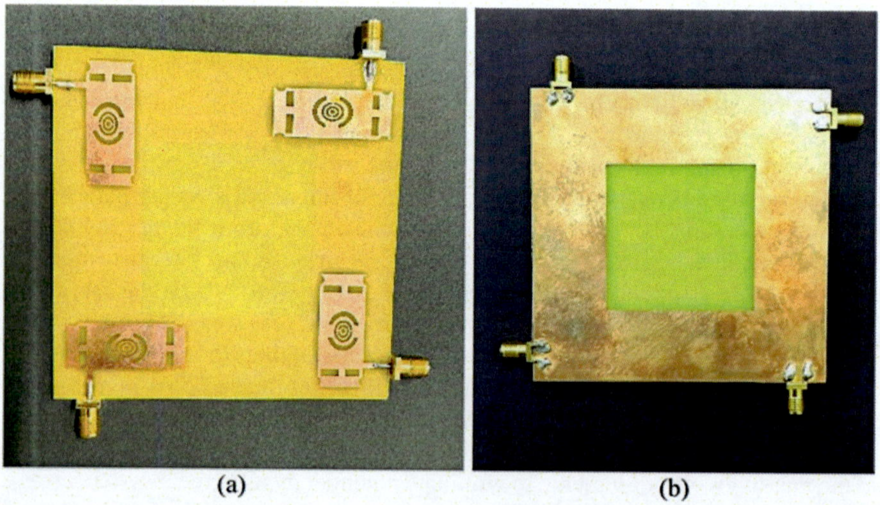

Fig. 3 Antenna **a** top, **b** bottom

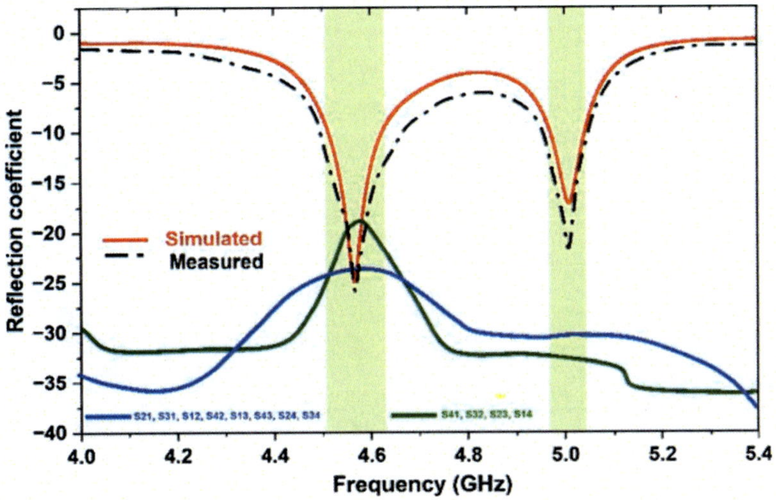

Fig. 4 Antenna reflection coefficient

It resonates at 4.51–4.63 GHz covering n79 5G communication band and 4.97–5.04 encompassing the Wireless Local Area Network (WLAN) frequency band. MIMO Antenna possesses excellent port isolation characteristics as evident from the reflection coefficient parameters. The fractional bandwidth of the first frequency band is 2.6%, and the second frequency band is 1.4%. The improved bandwidth compared to traditional microstrip antenna is the inclusion of a truncated ground plane; however, this also worsens the antenna front-to-back ratio. A strip-line-fed configuration for MIMO antennas offers several notable advantages for wireless communication systems. First, it excels at providing isolation between individual antenna elements, a crucial feature for MIMO setups to prevent signal interference and ensure reliable data transmission. Additionally, strip-line-fed antennas can be designed in a compact form factor, making them suitable for space-constrained environments, such as modern electronic devices. The antenna efficiency and gain plots are depicted in Fig. 5, and normalized radiation patterns are exhibited in Fig. 6. The antenna fares good impedance matching and exhibits gain of 2.88 dBi and 2.31 dBi at first and second resonance, respectively.

4 MIMO Diversity Characteristics

The MIMO diversity parameters were computed for 5G n79 and WLAN frequency bands as depicted in Fig. 7. Channel capacity loss (CCL), mean effective gain (MEG), and envelop correlation coefficient (ECC) were computed. CCL provides the threshold of data rate for transmission. The proposed antenna CCL at both band is lesser than 0.4 bits/s/Hz. As evident from diagram that MEG is near to the one for

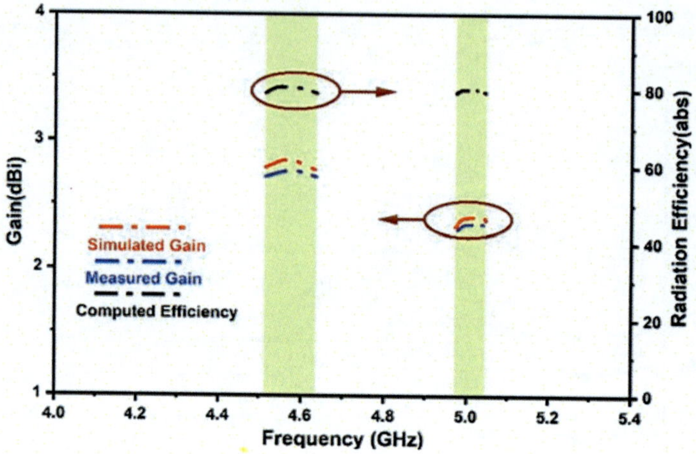

Fig. 5 Gain and efficiency

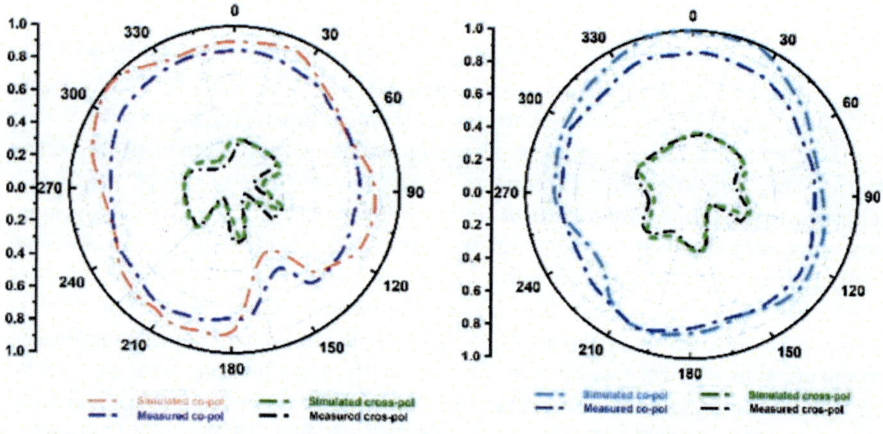

Fig. 6 Normalized radiation patterns **a** 4.6 GHz, **b** 5 GHz

both adjacent and diagonal elements. The envelope correlation coefficient provides the correlation midst the quad-element of the resonator. The ECC values were 0.03 and 0.011 showing excellent isolation. This aids in achieving optimal diversity gain.

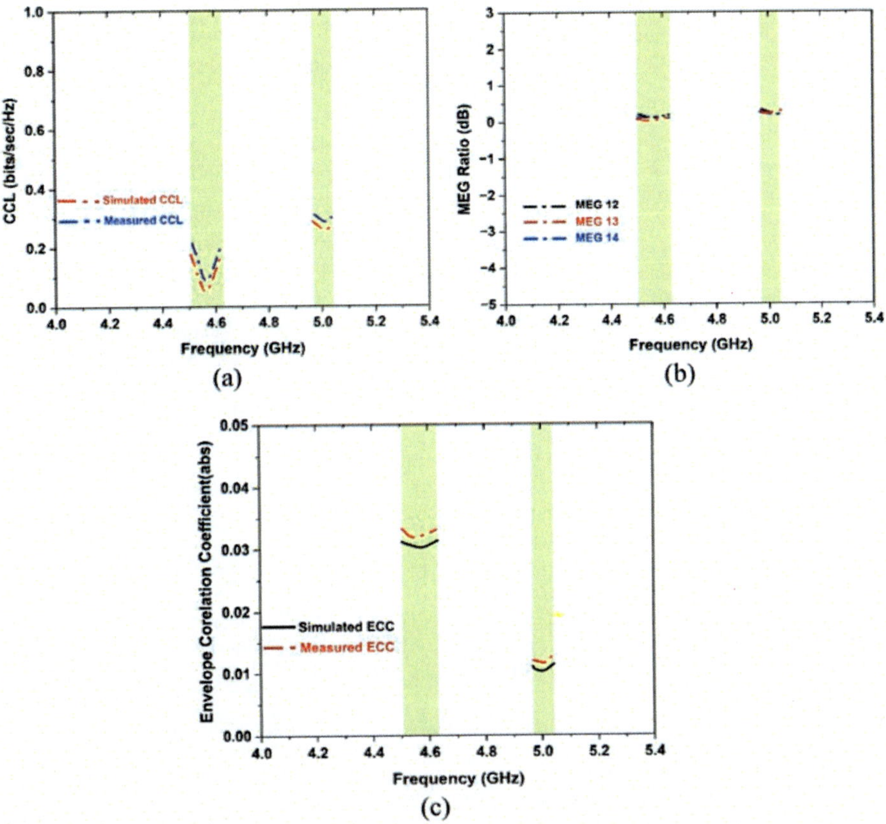

Fig. 7 MIMO parameter **a** CCL, **b** MEG, **c** ECC

5 Conclusion

A quad-port high-isolation MIMO antenna for n79 5G communication and WLAN has been proposed. The proposed antenna gain was achieved as 2.88 dBi and 2.31 dBi, respectively, with fractional bandwidth of 2.6% and 1.4%. The truncated ground plane antenna presents a fairly wide bandwidth for both targeted applications. The optimal placing of the multiple elements MIMO antenna surface presents port high isolation.

References

1. Sheikh W (2023) Jitter-sensitive data communication in emerging wireless networks. Trans Emerg Telecommun Technol e4746

2. Jaglan N, Gupta SD, Sharawi MS (2021) 18 element massive MIMO/diversity 5G smartphones antenna design for sub-6 GHz LTE bands 42/43 applications. IEEE Open J Antennas Propag 2:533–545
3. Alibakhshikenari M, Babaeian F, Virdee BS, Aïssa S, Azpilicueta L, See CH, Althuwayb AA et al (2020) A comprehensive survey on "Various decoupling mechanisms with focus on metamaterial and metasurface principles applicable to SAR and MIMO antenna systems". IEEE Access 8:192965–193004
4. Zhang XX, Ren AD, Liu Y (2020) Decoupling methods of MIMO antenna arrays for 5G applications: a review. Front Inf Technol Electron Eng 21(1):62–71
5. Girjashankar PR, Upadhyaya T, Desai A (2022) Multiband hybrid MIMO DRA for Sub-6 GHz 5G and WiFi-6 applications. Int J RF Microwave Comput Aided Eng 32(12):e23479
6. Li M, Cheung S (2020) A novel calculation-based parasitic decoupling technique for increasing isolation in multiple-element MIMO antenna arrays. IEEE Trans Veh Technol 70(1):446–458
7. Upadhyaya T, Park I, Pandey R, Patel U, Pandya K, Desai A, Pabari J, Byun G, Kosta Y (2022) Aperture-fed quad-port dual-band dielectric resonator-MIMO antenna for Sub-6 GHz 5G and WLAN application. Int J Antennas Propag
8. Patel U, Upadhyaya T (2022) Four-port dual-band multiple-input multiple-output dielectric resonator antenna for sub-6 GHz 5G communication applications. Micromachines 13(11):2022
9. Bilal M, Naqvi SI, Hussain N, Amin Y, Kim N (2022) High-isolation MIMO antenna for 5G millimeter-wave communication systems. Electronics 11(6):962
10. Li R, Mo Z, Sun H, Sun X, Du G (2020) A low-profile and high-isolated MIMO antenna for 5G mobile terminal. Micromachines 11(4):360
11. Kumar Saurabh A, Singh Rathore P, Kumar Meshram M (2020) Compact wideband four-element MIMO antenna with high isolation. Electron Lett 56(3):117–119
12. Ahmed BT, Rodríguez IF (2022) Compact high isolation UWB MIMO antennas. Wireless Netw 28(5):1977–1999
13. Sharma M (2020) Design and analysis of MIMO antenna with high isolation and dual notched band characteristics for wireless applications. Wireless Pers Commun 112(3):1587–1599
14. Aboelleil H, Ibrahim AA, Khalaf AA (2021) A compact multiple-input multiple-output antenna with high isolation for wireless applications. Analog Integr Circ Sig Process 108:17–24
15. Kapure VR, Rathod SS (2023) A two element EBG-inspired UWB MIMO antenna with triple band notched characteristics and high isolation. Sādhanā 48(1):7
16. Güler C, Bayer Keskin SE (2023) A novel high isolation 4-port compact MIMO antenna with DGS for 5G applications. Micromachines 14(7):1309
17. Morsy MM (2023) Compact eight-element MIMO antenna array for sub 6 GHz mobile applications. SN Appl Sci 5(10):261
18. Tiwari RN, Thirumalaiah R, Naidu VR, Sreenivasulu G, Singh P, Rajasekaran S (2023) Compact dual band 4-port MIMO antenna for 5G-sub 6 GHz/N38/N41/N90 and WLAN frequency bands. AEU Int J Electron Commun 154919
19. Gangwar D, Malik J, Patnaik A (2023) Enhanced isolation 4 × 4 MIMO antennas on cross-substrate for vehicular communications. IEEE Antennas Wirel Propag Lett
20. Muttair KS, Shareef OA, Mosleh MF, Zahid AZG, Shakir AM, Qasim AM (2023) A dual-element quad-port MIMO antenna modern design with ideal isolation correlation for 5G systems. AIP Conf Proc 2804(1)
21. El-Nady S, Abd El-Hameed AS, Eldesouki EM, Soliman SA (2023) Circularly polarized MIMO filtering dielectric resonator antenna for 5G sub-6 GHz applications. AEU Int J Electron Commun 154882

Socio-technical Approaches to Solid Waste Management in Rural India: A Case Study of Smart Pipe Composting in Raichur District, Karnataka

R. K. Chethan, Aniketh V. Jambha, Chirag Pathania, Mutyala Sai Sri Siddhartha, Sanjay, Arifuzzaman, K. Darshan, Souresh Cornet, and Sajithkumar K. Jayaprakash

1 Introduction

In India, rural areas are home to 65% of the population [1]. India's rural population of 800 million experiences severe sanitation and waste management problems. Inadequate access to clean and functional toilets is a major concern, resulting in open defecation, posing health concerns and environmental damage. Similarly,

R. K. Chethan · K. Darshan
Department of Commerce and Management, School of Arts Humanities and Commerce, Amrita Vishwa Vidyapeetham, Mysuru, India

A. V. Jambha
Department of Computer Science and Engineering, Amrita School of Engineering, Amrita Vishwa Vidyapeetham, Bengaluru, India

C. Pathania
Department of Electrical and Electronics Engineering, Amrita School of Engineering, Amrita Vishwa Vidyapeetham, Bengaluru, India

M. S. S. Siddhartha
Department of Artificial Intelligence, Amrita School of Engineering, Amrita Vishwa Vidyapeetham, Bengaluru, India

Sanjay
Department of Computer Science, Amrita School of Computing, Amrita Vishwa Vidyapeetham, Mysuru Campus, Mysuru, Karnataka, India

Arifuzzaman
Department of Visual Communication, School of Arts Humanities and Commerce, Amrita Vishwa Vidyapeetham, Mysuru, India

S. Cornet (✉) · S. K. Jayaprakash
Amrita School for Sustainable Futures, Amrita Vishwa Vidyapeetham, Amritapuri, India
e-mail: soureshcornet@am.amrita.edu

M. S. Kaiser et al. (eds.), *ICT: Smart Systems and Technologies*, Lecture Notes in Networks and Systems 878, https://doi.org/10.1007/978-981-99-9489-2_30

waste management is hampered by a lack of infrastructure and awareness, leading to improper disposal practices like open dumping. Non-biodegradable waste remains a problem contributing to environmental damage.

A sustainable community manages its human, natural, and financial capital to meet current requirements while preserving adequate resources for future generations [2]. Coordination of education, infrastructure development, and behavioral change initiatives is essential to sustain communities. The Government of India initiated the Swachh Bharat Mission (SBM) in 2014 to aid in developing hygienic water facilities in India. Despite increasing the number of families with access to toilets by 60% [3], it could not completely eradicate open defecation.

According to prior research, there is no one-size-fits-all answer to these complicated problems [4]. Technology has frequently been presented as a solution to tackling complicated situations, but when technology is not based on people's genuine needs, it falls short of its objectives. It is crucial to customize solutions and take a holistic approach to complete the last mile of sustainability.

This study explores this research question with two main objectives: (1) To gain insight into rural communities' challenges and (2) To outline appropriate technology solutions to improve the community's sustainability.

2 Review of Literature

Rural India produces approximately 100–200 g of garbage per person daily, a modest amount compared to the 620 g produced daily in metropolitan regions [5]. However, when factored into the sizable rural population, it results in substantial waste. Rural areas produce between 15,000 and 18,000 million liters of gray water and between 0.3 and 0.4 million metric tons of solid waste daily [6].

Due to rising populations, changing lifestyles emphasizing consumerism, and expanding economic activity, the amount of garbage produced in rural regions constantly rises and is a growing worry. Most solid waste generated in rural areas is organic and biodegradable [7]; however, this waste is not being separated in situ, which presents a serious issue. Lack of environmental cleanliness due to careless trash results in a poor quality of life.

Residential waste management in rural areas is the responsibility of the Gram Panchayat. Regrettably, only a few efficient Solid Waste Management units in rural Indian villages have been benchmarked in the past few years [8]. Successful examples in Tamil Nadu, Kerala, West Bengal, Gujarat, and Rajasthan have taught us that one model cannot work in all the 238,617 g Panchayats in India. There is a need for customized approaches that are relevant to the local culture and based on place-specific requirements [8].

Additionally, it is important to acknowledge the connection between waste and other areas. All sustainable development challenges are connected from a sustainability point of view. One issue could result from several others: waste management also impacts energy [9], and health [4]. A holistic approach to problem-solving is required to create a self-sufficient society [10].

2.1 Study Area

This study focused on the case of Korthakunda, a rural community in the Raichur district of the southern Indian state of Karnataka. It is situated approximately 32 km north of the district headquarters, Raichur. The main socio-economic features of the community are summarized in Table 1. Fields surrounds the village, which is predominantly rural. As with many villages in the region, agriculture is the primary occupation in Korthkunda. The main crops grown in the area include Paddy, Toor dal, Cotton, and Groundnuts.

The village also has a few industries, such as Rice mills. Korthkunda village has a rich cultural heritage with several festivals and cultural events celebrated yearly. They also celebrate a unique festival, called Jaatre, in their native language, which shows the village's culture.

3 Methodology

This study was part of the Live-in-Lab™ program, which promotes long-term technological solutions to difficulties faced by India's rural communities [11]. The program emphasizes cultural immersion by sending participants to remote rural communities to comprehend and identify their challenges. This immersion helped identify the key management, engineering, social, and economic restrictions to consider while designing a solution [11]. To perceive and assess the difficulties in the community, an ethnographic research approach was used. Information was collected using participant observation, field journals, and collaborative tools: Participatory Rural Appraisal (PRA) and Human Centered Design (HCD).

Table 1 Basic socio-economic profile of the community [1]

Total population	1585
Male	763
Female	822
Languages spoken	Kannada, Telugu
Communities present	OBC, SC, and ST
Occupations in general	Agriculture, daily labor, lorry drivers, electricians, plumbers, and tailors

3.1 Participatory Rural Appraisal (PRA)

In order to address the first objective of this research, which is to identify the main challenges of rural communities, PRA was used. PRA facilitates the engagement of the research team with the target population. It helps to comprehend the rural community members from their respective points of view [12]. PRA comprises several cooperative tools and activities so that everyone in the community can participate without being constrained by their age or level of education, which in turn contributes to making the challenges emerge naturally [13]. Other non-participatory research methods were not considered as they lacked flexibility, minimal involvement of the stakeholders, and couldn't be put into action in the short time of field research.

3.2 Human Centered Design (HCD)

Once the challenges have been identified, the use of HCD aims to fulfill the second objective: to identify a suitable technological solution. HCD is an approach to problem-solving that allows for incorporating the end-users' capacities as well as demands during the solution development process [14]. The HCD process begins with empathy for the people we are designing for. By considering user experiences, the HCD process is directed by the demands of current or potential users. Incorporating such a method would ensure greater user satisfaction and acceptance capacity, resulting in a more sustainable intervention [15].

HCD was mainly implemented through interviews. The research team was split in smaller teams with a native speaker, and the other team members noting the answers. Responses were centralized in a dedicated app.

4 Results

By residing in the community and using a variety of participatory tools, the research team was able to obtain first-hand information about the village's resources, socio-economic conditions, and overall daily challenges faced. It aided in developing a more thorough grasp of the pressing demands that must be met to preserve the community's quality of life. This section presents the main findings.

4.1 Participatory Rural Appraisal (PRA) Findings

Resource Map A resource map was created with the villagers to understand the location and distribution of different social and natural resources (Fig. 1). The key

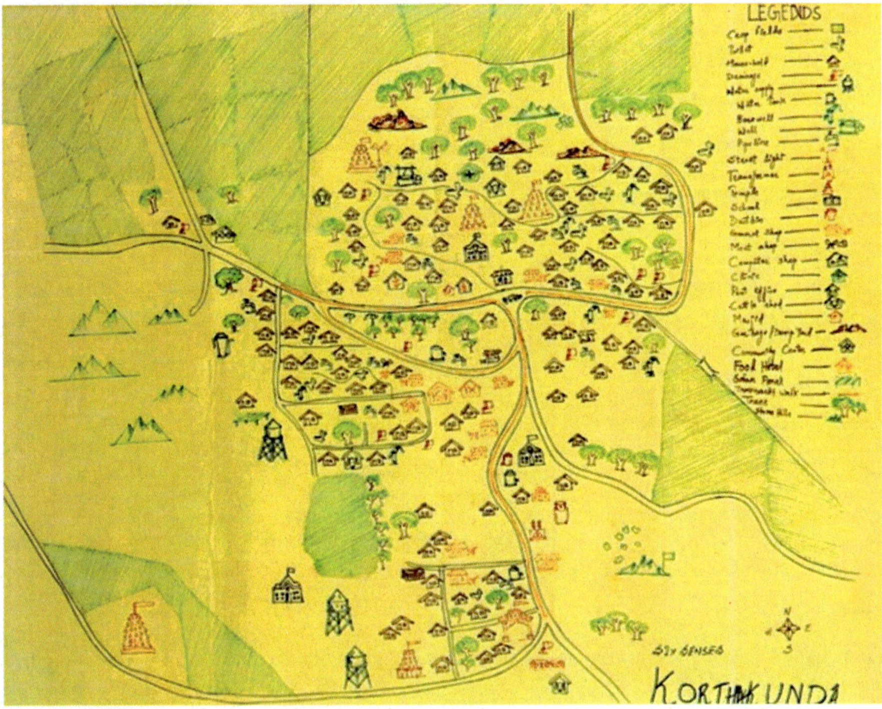

Fig. 1 Natural and social resources of Korthakunda, Karnataka

observations from this activity were the insufficiency of waste bins and collection points and the importance of garbage sites. Junk yards are prevalent in the northern part of the community. Waste is dumped anywhere in the village and accumulates in the streets due to the absence of a proper drainage system and irregular waste collection.

Income Expenditure Matrix This tool outlines the economic activities and financial flows in Korthakunda. Through interviews with key informants and informal group discussions, information on the types of goods and services produced in the village, the prices of various commodities, the sources of income, and the types of expenses were gathered and compiled in Fig. 2 (left). It shows that 60% rely on agriculture for a living.

Inflow-Outflow Diagram This participatory tool provided information on the circulation of resources in and out of the village. By determining which resources travel into the communities and which ones depart from the villages, this tool helps to understand how the community is connected to the surroundings. This activity was completed through group discussions with the villagers. The results compiled in Fig. 2 (right) show that the community imports essential items (food, milk, vegetables, gas, electricity) or work (pesticides, construction materials) and highlight their

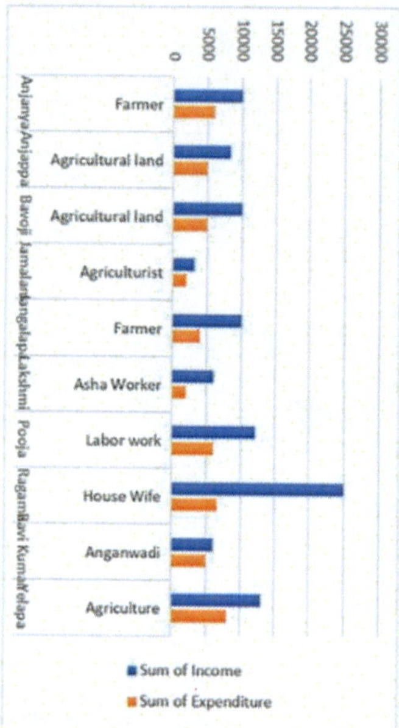

Sl. No	Inflow	Outflow
1	Labors	Rice
2	Groceries	Groundnuts
3	Ration	Dal
4	Electricity	Fruits
5	Gas	Cattles
6	Vegetables	Labors
7	Alcohol	Cotton
8	Tractor	Flowers
9	Harvest Machine	Cloths
10	Pesticides	
11	Cylinder	
12	Vehicles	
13	Fruits	
14	Cloths	
15	Construction material	
16	Clothing	
17	Electronic gadgets	

Fig. 2 Income and expenditure summary (left), and summary of the circulation of resources in the community (right)

dependency. There is no flow related to waste management, which indicates the lack of local government support to deal with that issue.

Problem Tree This tool analyzes the cause-and-effect relationship within the challenges identified [12]. Through group discussions with the community members and considering the information already collected from the previous PRA tools, the causes and repercussions of each challenge were discussed and summarized in Fig. 3. Additionally, it helps to prioritize the problems [12]. Three main problems were identified: Unemployment, Drainage problems, and Lack of waste management. Unemployment was a larger challenge that required government support, and solving the drainage issue would require significant infrastructure investments beyond this study's scope. So, waste management was prioritized, as it could be addressed immediately with technological solutions.

The main consequences of poor Solid Waste Management, observed in and around the community, were formally confirmed: water-borne diseases, air and water pollution, and emission of toxic substances that percolate into agricultural land, thereby directly affecting the community's main source of livelihood.

More importantly, the causes were discussed, providing insight on where to start addressing the challenge. Waste mismanagement is caused by the high density of the population, limited availability of collection points for garbage (confirmed with the resource map), insufficient transport system leading to the accumulation of waste in the streets, and an overall lack of awareness on waste disposal and management.

Fig. 3 Problem tree, summarizing the waste management challenge of the community, with an illustration from the village

4.2 *Human Centered Design (HCD)*

From the third day of the fieldwork, the research team conducted door-to-door HCD interviews to understand the users' pain points and their experiences with waste management. The findings confirmed some of the PRA observations and allowed to identify the requirements for a possible solution. Some notable issues were a lack of an effective drainage system and open defecation. However, the people considered not having a good waste disposal system a higher priority.

To address the causes of the waste disposal problem, the key requirements would be: (1) Having a proper way to dispose of the waste throughout the densely populated community; (2) Having the capacity to address intense waste generation and avoid the accumulation of waste in the streets.

5 Discussion

According to the findings, one of the causes of waste mismanagement is a lack of organized waste management and a transportation system to transfer garbage to dump sites. Given the community's lack of awareness and resources, implementing a centralized waste management system would be difficult. This will necessitate a significant investment in infrastructure and training. Therefore, decentralized waste management seems a more feasible solution.

Decentralized waste management is a promising solution for household waste management that addresses waste at the household rather than the municipality level [16]. To address the first key requirement, every household can be responsible for their waste, instead of dumping it near their house. With this option, aerobic pipe composting can be considered the best solution for treating household organic waste. In this aerobic composting method, two or more PVC pipes were placed over the ground where households can store their biodegradable waste, especially kitchen waste, cattle farm, solid waste residues, etc. Some microbial cultures, such as effective microorganism (EM) solution, should sprayed over the trash intermittently to expedite the composting process [17].

To address the intensity of waste production, villagers can install multiple pipes to continue composting while the first pipe is full. Once the available pipe volume is filled and concealed thoroughly after 30–35 days, solid waste will undergo aerobic composting and break down the complex organic matter into simple organic matter, which can be later used as fertilizer to enhance crop productivity [18]. This manure will benefit the community, as agriculture is their main source of livelihood.

5.1 Using IoT to Monitor the Composting Process

Aeration is a crucial aspect of composting, as it provides the necessary oxygen for the microorganisms responsible for decomposition. The composting process must be monitored closely, with parameters like temperature, moisture content, and oxygen levels controlled to ensure optimal conditions for microbial activity [19]. Sensors and control systems are often used to automate this process and make it accessible. In addition, this study proposes an Internet of Things (IoT) system for automation to ensure the efficient operation of the pipe composting process. To scale up from a laboratory setup to an industrial or community-level application, monitoring and controlling various parameters such as temperature, moisture content, and oxygen levels become crucial, especially when dealing with multiple setups. Sensors of the above parameters are to be integrated into the IoT system and continuously monitor the reactor's performance of converting waste into organic manure. If parameters exceed predefined thresholds, the IoT system will notify the operator, allowing timely intervention and preventing potential issues [20]. The IoT system will facilitate the management of these parameters remotely and help householders maintain the composting unit effectively for an extended period (Fig. 4).

6 Conclusion

This study aimed to gain better insight into the challenges faced by rural com munities in terms of Solid Waste Management and to explore how technology can help create an immediate positive change toward greater sustainability.

Using PRA, the research team could understand the community members' livelihood, their challenges, and the parameters leading to them. Though there were several issues, Solid Waste Management was their priority. Using HCD, the key requirements

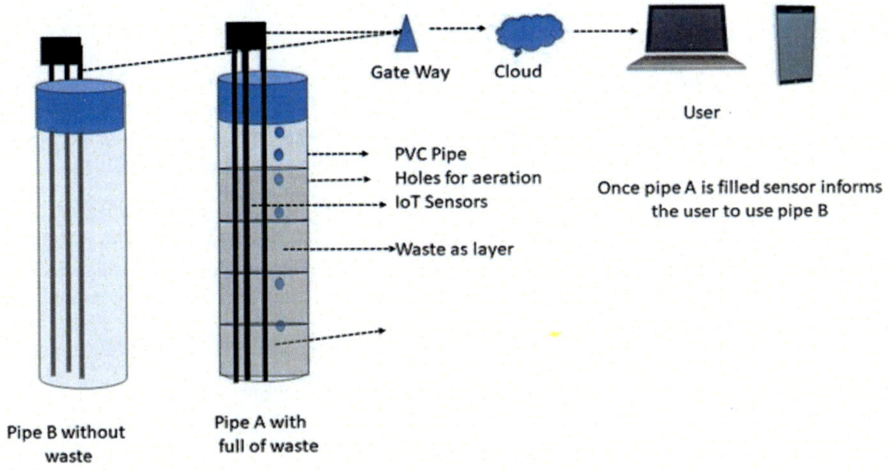

Fig. 4 Schematic diagram of IoT-based pipe composting solution

of a solution were identified. Given the lack of resources and the steep learning curve in terms of awareness, a household level decentralized solution was found feasible for an immediate application. The design of a smart pipe composting solution is proposed. This will allow each household to take care of its organic waste. Using sensors, people will be able to easily monitor the status of the composting from their phones without undergoing extensive training. This will address both the main key requirements identified through the study.

Developing an awareness campaign on waste management and good hygiene practices in the community is essential for future development. This study proposes a first-level system design that addresses only the organic waste generation. Further study is ongoing to quantify the quantity and type of waste generated weekly to refine the proposed solution's design and develop a prototype.

Acknowledgements This study was conducted as part of the Amrita School for Sustainable Futures and the UNESCO Chair on Experiential Learning for Sustainable Innovation & Development, at Amrita Vishwa Vidyapeetham. The authors sincerely thank the Amrita Live-in-Labs® Academic program for providing all the support. The authors express their immense gratitude to Sri. Mata Amritanandamayi Devi, Chancellor of Amrita Vishwa Vidyapeetham, who has inspired them in performing selfless service to society.

We acknowledge the contribution of Mr. Tejas H. in an earlier draft of this study. The authors also thank all the stakeholders of the Korthakunda community for participating in the study and guiding them throughout the process. Finally, authors thank the government officials of the Raichur district of Karnataka.

References

1. Government of India (2011) National census 2011
2. Chethan R et al (2023) The exploration of impact of pre and post covid-19 pandemic syndromes on sustainable challenges in MSMEs sector in India. J Namibian Stud Hist Politics Cult 33:827–838
3. Swachh Bharat Mission (2021) All India report. https://sbm.gov.in/sbmReport/home.aspx
4. Rengaraj V, Simmon A, Vikash S, Zel R, Adhikari H, Sherpa P, Geetha P, Cornet S (2021) Addressing sanitation and health challenges in rural India through socio-technological interventions: a case study in Odisha. In: 2021 IEEE 9th region 10 humanitarian technology conference (R10-HTC). IEEE, pp 01–06
5. Kumar S, Smith SR, Fowler G, Velis C, Kumar SJ, Arya S, Rena, Kumar R, Cheeseman C (2017) Challenges and opportunities associated with waste management in India. R Soc Open Sci 4(3):160764
6. UNICEF (2018) Financial and economic impacts of the Swachh Bharat Mission in India
7. Government of India (2023) Solid and liquid waste management in rural areas
8. NIRD (2016) Solid waste management in rural areas: a step-by-step guide for gram panchayats
9. Lokare A, Harish MS, Bhargav G, Raghul S, Thangavelu S, Cornet S (2021) Sustainable solution to address waste management and energy challenges in rural India. In: 2021 IEEE 9th region 10 humanitarian technology conference (R10-HTC). IEEE, pp 01–06
10. Showkat N (2016) Coverage of sanitation issues in India. SAGE Open 6(4):2158244016675395
11. Ramesh MV, Muir A, Nandanan K, Bhavani RR, Mohan R (2022) HCI curricula for sustainable innovation: the humanitarian focus at Amrita Vishwa Vidyapeetham. Interactions 29(1):54–57
12. Narayanasamy N (2009) Participatory rural appraisal: principles, methods and application. SAGE Publications, India
13. Das A, Aarthi R, Vijay S, Kailash M, Gogul Nithish K, Saravanan R, Cornet S (2022) Performance enhancement of photo voltaic system for rural electrification in higher altitude region: a case study in Uttarakhand, India. In: ICT analysis and applications: proceedings of ICT4SD 2022. Springer, Berlin, pp 373–381
14. Ranger BJ, Mantzavinou A (2018) Design thinking in development engineering education: a case study on creating prosthetic and assistive technologies for the developing world. Dev Eng 3:166–174
15. Varma DS, Nandanan K, Vishakh Raja PC, Soundharajan B, Pérez ML, Sidharth KA, Ramesh MV (2021) Participatory design approach to address water crisis in the village of Karkatta, Jharkhand, India. Technol Forecast Soc Change 172:121002
16. Otterpohl R, Braun U, Oldenburg M (2002) Innovative technologies for decentralised wastewater management in urban and peri-urban areas. Berichte-Wassergute Abfallwirtsch Tech Univ Munch Berichtsheft 173:109–126
17. Ismail SNS, Muttalib SAA, Praveena SM (2016) Application of effective microorganism (EM) in food waste composting: a review. Asia Pac Environ Occup Health J 2(1)
18. Mandpe A, Kumari S, Kumar S (2020) Composting: a sustainable route for processing of biodegradable waste in India. In: Organic waste composting through nexus thinking: practices, policies, and trends, pp 39–60
19. Meena AL, Karwal M, Dutta D, Mishra R (2021) Composting: phases and factors responsible for efficient and improved composting. Agric Food e-Newslett 1:85–90
20. Ramesh MV, Nibi K, Kurup A, Mohan R, Aiswarya A, Arsha A, Sarang P (2017) Water quality monitoring and waste management using IoT. In: 2017 IEEE global humanitarian technology conference (GHTC). IEEE, pp 1–7

Attendance System Using Face Detection and Face Recognition

Harsh N. Chavda, Sakshi P. Bhavsar, Jaimin N. Undavia, Kamini Solanki, and Abhilash Shukla

1 Introduction

In face recognition, system student has to register their selves once and when they click on signup, web cam starts, the system compares the data base image with current image. Automatically resolving this challenge is particularly challenging due to the various factors that can impact identification, such as changes in a person's facial expression, aging, and even lighting conditions. Web cam helps to compare the image of student. Captured image is detecting, compared with database and attendance is marked in .csv file.

2 Advantages

- Reduce paper work if attendance is taken by model.
- Save the time of teacher.
- Attendance can be stored for two times in database at the time of entrance and at the time of living the classroom.
- Check presence of student with date and time.
- Multiple student's attendance can be marked at once.

H. N. Chavda · S. P. Bhavsar · J. N. Undavia (✉) · K. Solanki · A. Shukla
MCA Department, Charotar University of Science and Technology, Changa, Anand, Gujarat, India
e-mail: Jaiminundavia.mca@charusat.ac.in

M. S. Kaiser et al. (eds.), *ICT: Smart Systems and Technologies*, Lecture Notes in Networks and Systems 878, https://doi.org/10.1007/978-981-99-9489-2_31

3 Disadvantages

- Cameras are required to take the attendance it is expensive at certain point.
- There might be technical issues after some years because of abundant data for which large storage engines are required.

4 Workflow of Image Processing

The process of face recognition is divided into two parts:

1. Prior to detection, there is a preprocessing stage that encompasses face detection and alignment, specifically pinpointing the location of the face or faces.
2. Recognition is achieved through feature extraction and matching processes, which involve making comparisons.

4.1 Face Detection

The primary objective of this component is to determine if human faces appear within a live video capture or image and to determine the location(s) of these face(s). The output of this process consists of patches or rectangular regions, each containing a detected face within the input image. To create a robust and easily recognizable face recognition system, face alignment is performed to standardize the sizes and orientations of these frames [1, 2].

4.2 Features Extraction

After the face detection, the process involves extracting human face frames from images. Subsequently, these faces are transformed into vectors with fixed coordinates [3, 4].

4.3 Recognition of Face

The final step, following the representation of faces, is their identification. To enable automatic recognition, it's essential to construct a face database. Each processed image of a person, along with their coordinates, is stored in this database. When an image or video is presented, the system performs face detection and feature extraction. The features extracted are then compared to each face class in the database.

Subsequently, the system reveals the name of the identified image and records the attendance of that identified individual [5–9].

5 Proposed System

1. We have added the feature of single face for single person using Eigen-vector that was associated with face_recognition library that scans and extracts the features of single image in 3D form giving it more accuracy and more efficiency.
2. The accuracy rate of proposed system with single image for single person is 91%.
3. In addition, we have added the GUI in this system to make in effective, i.e., first the operations were placed by manual insertion of multiple images. This process was much longer and time consuming. Therefore, we found out the way to it using GUI where we can easily add the image to database while registration. It converts the image in.jpg irrespective of any image format, which saves the task of conversion.
4. The face gets recognize with its coordinates and encodings of the features. These two things are compare with all the images present in database and the image with minimum distance number was took forward using Numpy library.
5. The identified image was taking forward and the name of that particular image displays around the face creating the green rectangle containing the name.
6. After recognition of image, the attendance gets mark for the identified person that inserts in excel sheet-containing name, time, and date.
7. We are also thinking to add the feature of RFID card detection and SMS alert while identification to improve its accuracy.

6 Flowchart of Proposed System

Figure 1 shows the flow chart of the proposed system that includes all possible inputs, output, processes, and conditional statements. The flow of the data is shown using following steps:

Step 1: For new User, he/she needs to fill all the required details.
Step 2: If he/she is already, a user go to Step-7.
Step 3: If he/she fails to enter details in any of the mentioned fields, the error message will be popped-up.
Step 4: User need to check the Checkbox of Agreeing Terms and Conditions.
Step 5: After entering all the fields and checking the Checkbox, User can click on Register Now button.
Step 6: The Entered Details will be store in the database.
Step 7: Now the User can click on Sign In button.
Step 8: New screen will pop-up that will act as video.

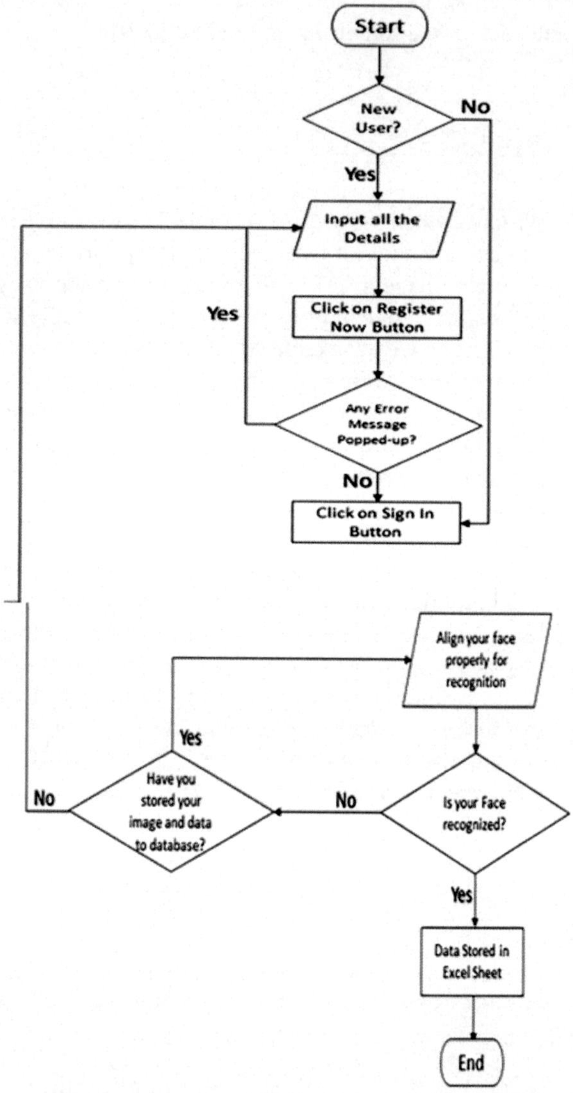

Fig. 1 Flowchart of proposed system

Step 9: It will populate the name of the person whose face is recognize on the screen under his/her face.
Step 10: The Attendance of that particular person with his/her name, date and time will be store in excel sheet.

7 Software and Libraries Used

7.1 To Implement the System in Effective Way Following Software and Libraries Are Required

- PyCharm
  - OpenCV-python
  - Numpy
  - Dlib (for smooth working of face_recognition library)
  - Face_recognition
  - Datetime
  - Os
  - Tkinter
- XAMPP
  - Pymysql

7.2 Software

PyCharm: For Designing and Taking Back-end process of face detection, face extraction, and face recognition we have used PyCharm software. It is a firm Python IDE that is providing much tools for Python operations. It nicely unified to develop better environment for creative Python and Data Science Projects. Cross-platform supports Windows, macOS, and Linux versions. It provides API so that we can write our own plugins and can extend PyCharm features.

XAMPP: For Storing the Data collected from user, we have used XAMPP control panel, which allows us to access localhost and phpmyadmin where we can create own table and can store and retrieve the necessary data. We first need to start Apache server and then we can start MySQL server where localhost can be accessed.

7.3 Libraries Used for Back-End

- OpenCV-Python: OpenCV-Python is a Python wrapper for the original OpenCV C++ implementation. It is use to read image from the given path, change the color order of the image, to on the web cam so that we can detect and recognize the face, to resize the photo so that we can interpret properly, to frame the image by displaying rectangle shape around the identified image, to write the text (the image name) around the identified image.

- Numpy: It stands for Numerical Python, and in our system it is used for recognize the minimum distance between all the images present in the database directory and the real-time image which is clicked when the web cam turns on.
- face_recognition: Recognize and manipulate faces from Python or from the command line, the world's simplest face recognition library. Built using dlib's state-of-the-art face recognition, built with deep learning. It is use to find encodings of each face, used to get location of face where it is located so that we can compare it and get result.
- datetime: It is used here to fetch current date and time.
- os: It is used here to fetch all the images from database directory and making the list of all the images. In addition, it is use to split the text so that we can use the name to display on identified image.

7.4 Libraries Used for Front-End

- Tkinter: This framework provides Python users with a simple way to create GUI elements using the widgets found in the Tk toolkit. Tk widgets can be used to construct buttons, menus, data fields, etc. in a Python application. Once created, these graphical elements can be associated with or interact with features, functionality, methods, data or even other widgets.
- Libraries Used to Store Data in Database:
- Pymysql: PyMySQL is an interface for connecting to a MySQL database server from Python.

8 Experimental Representation

- Default Screen

  Figure 2 is Default Screen that will display when program executes. It is for New Users. All the New Users have to fill the required details in this form and submit it to store the details to database.
- Wrong Credentials or Password Difference:

  Figures 3 and 4 shows that if user enters incorrect data or fails to enter data in any of the fields, error/information message will be popped-up.
- Correct Credentials:

  Figure 5 shows that if all the details are entering properly with correct input format; the data will be saved to the database as shown in Fig. 6.

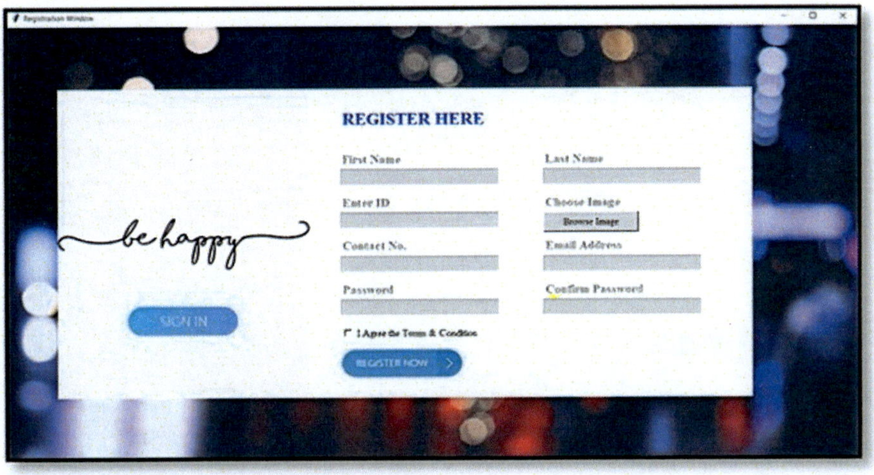

Fig. 2 Default screen for registering new user

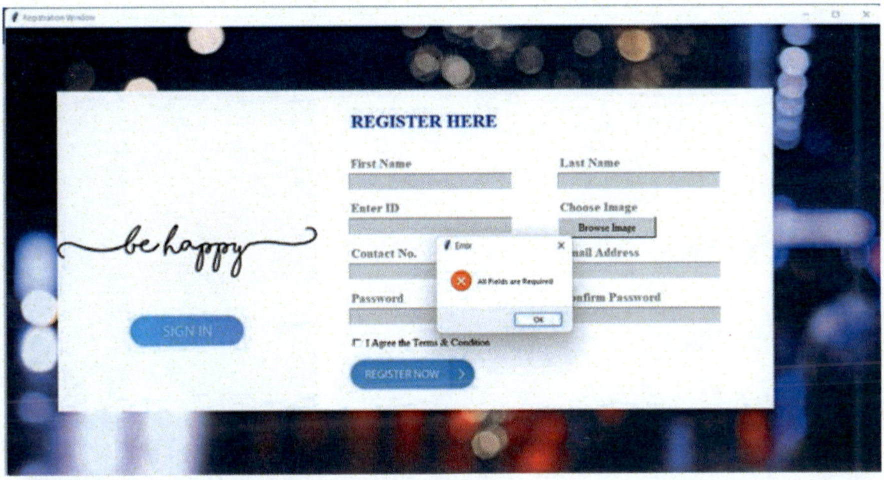

Fig. 3 When no fields are inserted

9 Registered Users

The registered user can click on sign in button and new window will be popped-up and the name/id number of that recognized employee or student will be displayed around the face as shown in Fig. 7.

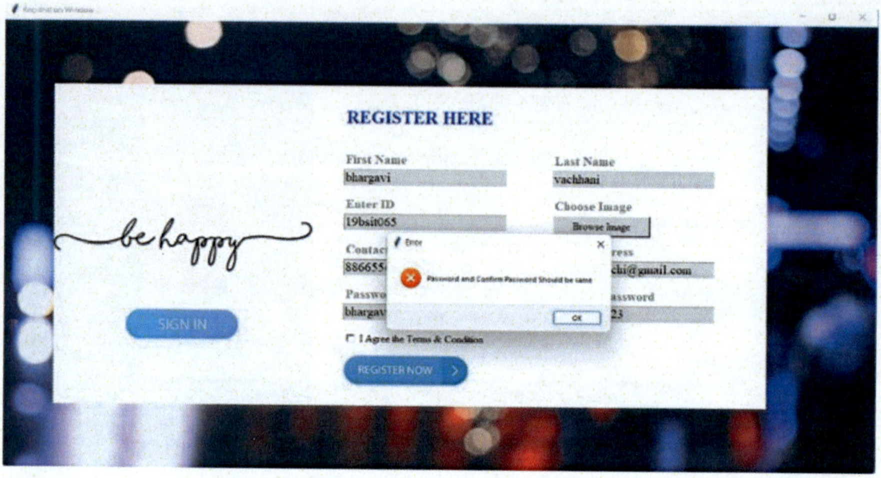

Fig. 4 When both passwords don't match

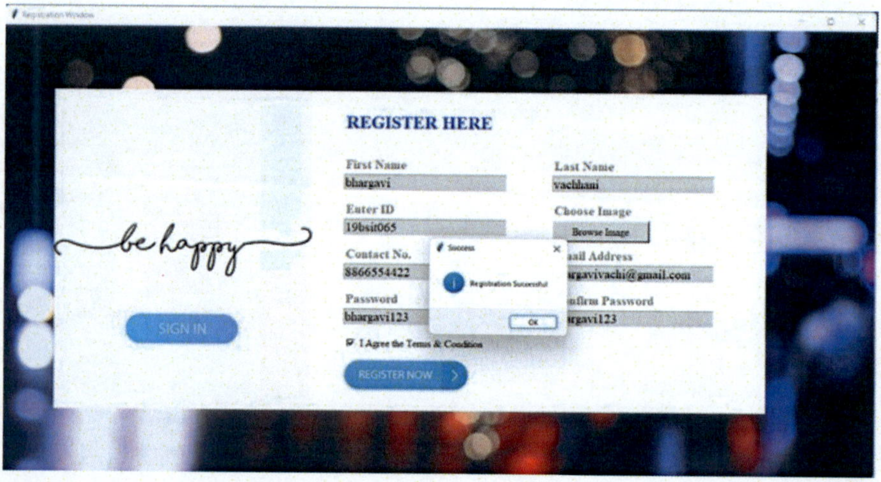

Fig. 5 Correctly entered details

10 Attendance Marking

Figure 8 shows that the attendance of a particular identified user will be marked in the excel sheet including it's ID, date, and time when it was marked.

The Alert Box pops up only when the attendance of student is marked. Figure 9 represents the output of attendance marked for Student ID: 19BSIT019.

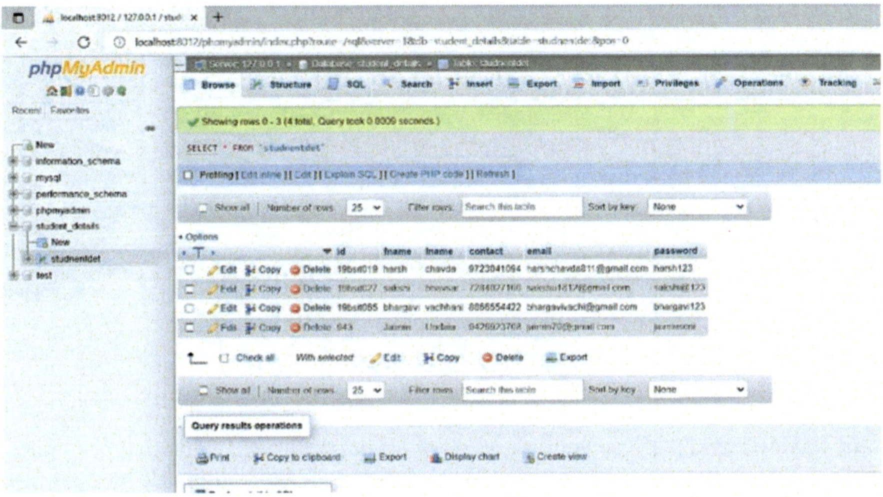

Fig. 6 The data will be stored in database

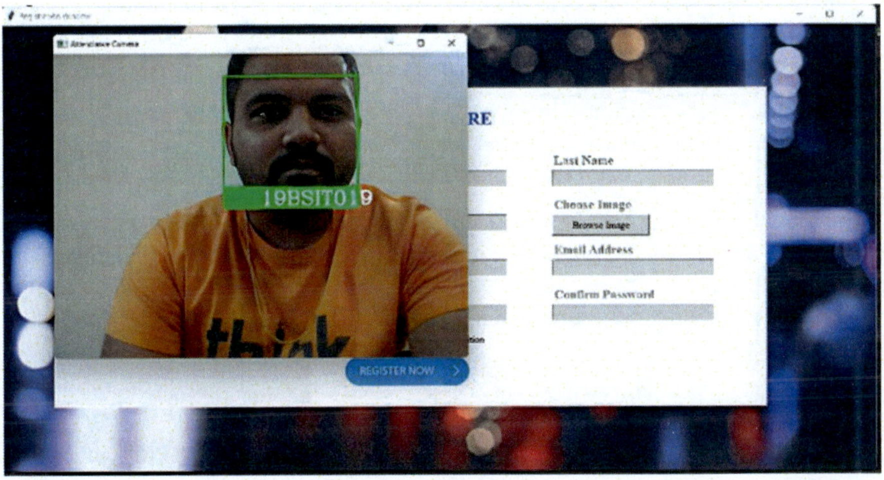

Fig. 7 Recognized image of user

Fig. 8 Attendance of recognized faces

1			
2	19BSIT086	09:34:33	10-05-2022
3	19BSIT019	09:34:34	10-05-2022
4	19BSIT065	09:49:45	10-05-2022
5	19BSIT041	09:56:22	10-05-2022
6	19BSIT061	09:57:06	10-05-2022
7			
8			

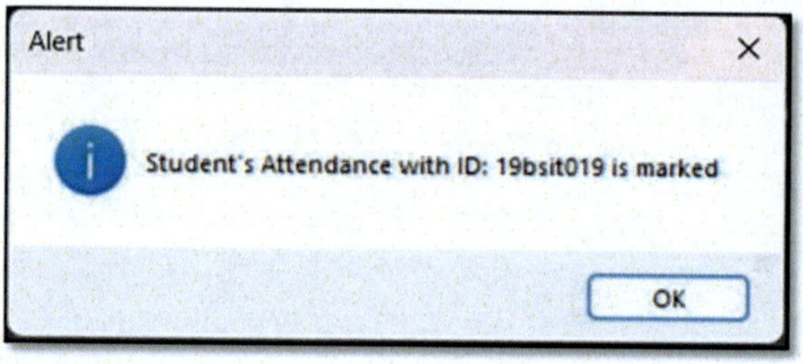

Fig. 9 Student's attendance marked

11 Challenges Faced for Facial Recognition

Light variations: Facial detection and recognition work well when we are under good lighting. Darkness or changes in lighting may make recognition a bit difficult, as co-ordinate extraction would differ. Examples like, low intensity of light or more brighter light that makes image casting a bit difficult. To resolve this challenge, we have added brightness adjustment feature in the model using 3D cameras.

Misalignment or posture change: The students/faculties may turn their heads sideways or may not be able to pose it correctly, which makes the facial recognition and identify actual entity. Variation in expressions can also lead to mismatches location of face and coordinates extraction may vary. To overcome this challenge, 3D face models are process and stored using a library.

Occlusions: Objects that covers the face partially or fully such as sunglasses, hairs covering the face, masks, or any other objects. The occlusions may not identify key features, which may make it difficult to identify and recognize them. To overcome this challenge, we can add future model of fingerprint processing.

12 Conclusion and Future Scope

The proposed system provides the attendance facility for students, faculties, and employees to mark their attendance in contactless format. So we can say that this system can reduce the time of marking attendance thereby reducing the paper work to save trees and also save time. It is helpful in every sections. In addition, the feature of encryption and decryption of Image Paths/URLs for identification and retrieval will be add to overcome security risks. Adding to it, the storage engines will have two factor authentication and admin panels to overcome server side risks.

Acknowledgements We would like to thank our principal/dean Atul Patel Sir, Jaimin Undavia Sir and the staff of Charotar University who helped us for the project, motivated and supported us to reach towards better outcome.

References

1. Manoharan JS (2019) Image detection classification and recognition for leak detection in automobiles. J Innov Image Process 1(2):61–70
2. Zhou H, Liu J, Liu Z, Liu Y, Wang X (2020) Rotate-and-render: unsupervised photorealistic face rotation from single-view images. In: Proceedings of the IEEE/CVF conference on computer vision and pattern recognition
3. Kato H, Ushiku Y, Harada T (2018) Neural 3D mesh renderer. In: Proceedings of the IEEE conference on computer vision and pattern recognition
4. Deng J, Guo J, Xue N, Zafeiriou S (2019) Arcface: additive angular margin loss for deep face recognition. In: Proceedings of the IEEE/CVF conference on computer vision and pattern recognition
5. Waingankar A, Upadhyay A, Shah R, Pooniwala N, Kasambe P. Face recognition based attendance management system using machine learning
6. Bussa S, Bharuka S, Mani A, Kaushik S (2020) Smart attendance system using OpenCV based on facial recognition. Int J Eng Res Technol (IJERT) 9(03) [Online]. Available: http://www.ijert.org, ISSN: 2278-0181, IJERTV9IS030122
7. Mohamed BK, Raghu C (2012) Fingerprint attendance system for classroom needs. In: India conference (INDICON), pp 433–438
8. Lim T, Sim S, Mansor M (2009) RFID based attendance system. In Industrial electronics & applications, IEEE symposium, vol 2, no 3, pp 778–782
9. Kadry S, Maili KS (2007) A design and implementation of a wireless iris recognition attendance management system. Inf Technol Control 36(3):323–329

Smart Health with Medi2Home: Redefining Medicine Delivery for a Safer Tomorrow

Sahal Bin Saad, Anatte Rozario, Sadi Mahmud Sagar, and Nafees Mansoor

1 Introduction

In an era of unchanging transformation, the healthcare sector remains in a state of constant evolution, with technology emerging as a climacteric motivation in enhancing the accessibility of medical services and products. Amid this dynamic landscape, a paramount predicament endured by patients revolves around the complicated process of receiving accurate medications promptly and conveniently. Conventional routes of medicine procurement often prove arduous and time-intensive, amplifying concerns about the accidental purchase of fake or faulty pharmaceuticals. As the digital realm burgeons with the ascendancy of e-commerce, the demand for a potent and efficient remedy to these challenges becomes increasingly noticeable [1].

A revolutionary smartphone app called "Medi2Home" was created to make it easier for patients to get their medications. It makes it simple to input prescription documentation, ensuring precise medicine dispensing and guarding against the sale of substandard products. Additionally, the app provides tools like reminders for medication schedules to promote greater adherence and overall wellness. It is user-friendly and focuses on scalability, dependability, and security for maximum advantage. In addition, "Medi2Home" has taken a unique approach to partnering with

S. B. Saad (✉) · A. Rozario · S. M. Sagar · N. Mansoor
Department of Computer Science and Engineering, University of Liberal Arts Bangladesh (ULAB), 688 Beribadh Road, Dhaka 1207, Bangladesh
e-mail: sahal.saad.cse@ulab.edu.bd

A. Rozario
e-mail: anatte.rozario.cse@ulab.edu.bd

N. Mansoor
e-mail: nafees@ieee.org

A. Rozario
Department of Computer Science and Engineering, Islamic University of Technology, K B Bazar Rd, Gazipur 1704, Bangladesh

M. S. Kaiser et al. (eds.), *ICT: Smart Systems and Technologies*, Lecture Notes in Networks and Systems 878, https://doi.org/10.1007/978-981-99-9489-2_32

local licensed medicine stores to improve the delivery of medicine to patients. By working closely with these stores, the app is able to offer a wider range of medicine options to patients and ensures that the medicine is delivered quickly and efficiently. This not only benefits patients but also supports local businesses and strengthens the local economy.

This study gives an in-depth analysis of "Medi2Home," stressing its design, implementation, and possible advantages. The software provides dependability, increased security, and simplicity all of which improve patient health outcomes. Collaborations with neighborhood pharmacies can enhance the impact on patients and local companies. "Medi2Home" has the ability to completely change how people obtain their medications.

The rest of this work is structured as follows: Sect. 2 addresses the existing works on medicine delivery and notification application system in brief. This highlighted the advantages and disadvantages of several current studies. Section 3 addresses the design-development and the objectives of our proposed system. Section 4 goes into detail about the system implementation and application design. Section 5 concludes the article by discussing the further implementation ideas of the application.

2 Existing Works

The study focuses on existing methods for delivering medications and materials that are critical for treating a variety of medical conditions. Current medications can be made safer and more effective by employing techniques such as dose optimization and personalized therapy [2]. The three categories into which research contributions are classified are continuous, controlled, and targeted medication delivery systems. Furthermore, there is growing interest in incorporating medication delivery devices due to a variety of developmental characteristics. These devices are gaining popularity due to their capacity to increase the accuracy and efficiency of medication administration. This broad area of medicine procurement and delivery profoundly shapes the quality of healthcare services and patient outcomes, with ongoing research and innovation making a significant contribution to its advancement [3].

The pharmaceutical procurement procedure in hospitals necessitates substantial expenditures of resources, both time and money. This process entails inherent threats particularly with regard to preservation, because a significant portion of the handled materials is sensitive, mandating regulated settings due to their delicate nature and sometimes short shelf life [4]. The development of controlled-release medication delivery technology is covered in the article [5], with an emphasis on its critical place in healthcare. It draws attention to the expansion of the market for these systems, especially oral controlled-release variants. These innovations give patients better compliance and precise control over drug release, which lowers adverse effects. Novel biocompatible polymers and different medication delivery techniques, like osmotic systems, are credited with the advancements.

The use of mobile healthcare has the potential to improve efficiency, the standard of care, patient comprehension of health conditions, medication compliance, and doctor-patient interactions. Additionally, it lowers healthcare expenditures by reducing the frequency of medical appointments. There are worries about patients receiving too much information, though, and about the necessity for specialized explanations for people with lower levels of education. Improved access to medical resources and care is one of the advantages of mobile healthcare applications, especially for patients receiving care at home. These applications can be nurse- or patient-centered, enhancing communication among medical professionals. Patient-centered applications enable direct communication between patients or caregivers and healthcare practitioners [6].

This study investigates how mHealth (mobile health) applications can help dementia patients and their caretakers. The burden and stress of care giving have a substantial impact on a person with dementia's quality of life, and carers frequently suffer unfavorable health effects; as a result the study divided mHealth apps into five major categories: screening, health and safety monitoring, recreation, sociability, and navigation. Cognitive training and daily life were included in the first category. Individuals with dementia can learn new skills and comprehend new technologies through coaching interventions. Using such apps is expected to enhance self-management abilities among dementia patients. It was discovered that these applications cover a variety of dementia care topics and offer patients and caregivers a variety of tools and techniques. Mobile health apps are thought to be a workable assistive technology intervention for people with dementia and their caregivers despite the paucity of research in this field [7].

Mobile devices are widely available, allow for real-time communication, and enable the delivery of solutions at the time and place that is most appropriate. These technologies have the potential to be used on a big scale and offer cost-effective solutions. Text messages, software programs, and multimedia are just a few of the characteristics that enable customized and interactive interventions. The paragraph also underlines the necessity of doing new systematic reviews of mobile health (mhealth) interventions because the ones that have previously been done have tended to be too narrowly focused on devices or themes that may now be outmoded. This systematic review's objective is to evaluate the impact of mobile technology-based interventions for healthcare professionals and healthcare services on various health outcomes [8].

The procedure for deploying a mobile controller to access patient data and medical imaging is described in [9]. The mobile controller is installed on the user's phone using Over-The-Air (OTA) provisioning in a dynamic deployment. Users can connect to the system through a variety of networks, including hospital Wi-Fi (WLAN or Bluetooth) and cellular networks (GSM-WAP, GPRS, and EDGE) [10]. Users of the system can explore DICOM medical images and obtain patient information, including examination data, history, and diagnosis. It enables connection between the user's mobile device and the hospital's data systems through the use of web servers, data access services, and SMS. Features for gaining access to and processing patient data and medical images are offered by the mobile controller.

Since we are electronically entering customer information into a virtual pharmacy, it will remain secure. This tool allows users to retrieve the history of any prescription, physician, reports, payments, prior details, and clients with a single click. As a result, data processing will be faster. The purpose is to make it simpler for administrators to use these systems. It simply reduces the need for human effort in purchasing, hence increasing accuracy and speed. It is designed to make system administration easier for the administrator of an online pharmacy.

3 Design and Development

The "Medi2Home" application has a client-server architecture that is cloud-based. The client, or user-facing, component is a mobile app that runs on the user's device and provides an intuitive and user-friendly interface for ordering medication and maintaining prescribed medication. The server component is a cloud-based backend that stores and analyzes data and interacts with the client component to offer real-time information and updates to users. The server also engages with licensed local pharmacists and pharmaceutical stores to complete orders and verify that consumers receive only authentic, high-quality drugs. This suggested approach takes advantage of the advantages of cloud computing to give users a safe, quick, and dependable service that assists them in managing their medicine and ensuring that they take it on time, every time.

3.1 System Overview

The system is made up of the following crucial parts: Prescription Verification for validating prescriptions, a Medication Database keeping information about medications, Delivery Coordination for purchasing and delivering, Notifications and Reminders for patient alerts, Controlled-Release Technology handling various methods of release, Implantable Devices supporting drug delivery implants, User Feedback and Analytical Data for data-driven improvements, Administration and Security controlling user roles and data security, Pharmacy Integration for purchasing medications, Regulatory Compliance assuring healthcare standards, and Back-End Services offering server hosting, APIs, and database management (Fig. 1).

3.2 Design Consideration

This is an extensive overview of the Medi2Home app's system model. For a more complete knowledge of the system flow, it may be broken into smaller phases and processes. Using their phone number, the user establishes an account. The user pho-

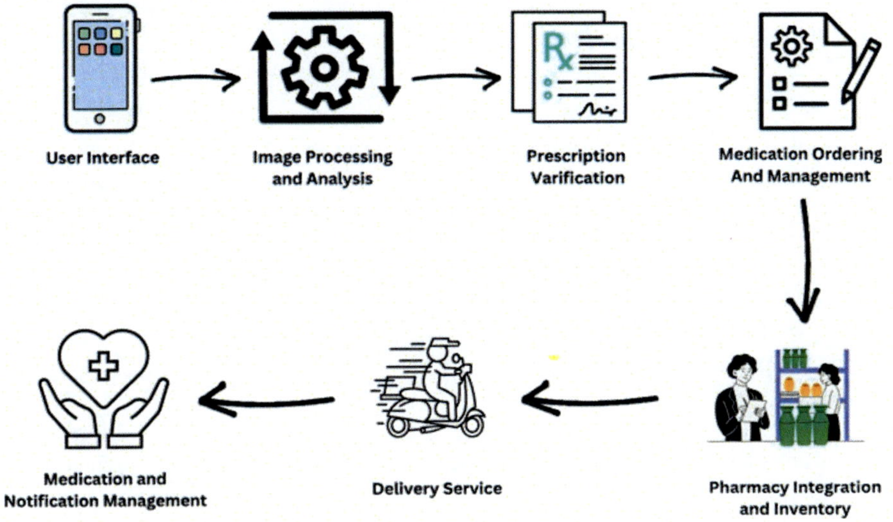

Fig. 1 System overview for Medi2Home

tographs their prescription and submits it to the app. The admin area of the software validates the prescription and updates the user's account with the relevant medication information. The user chooses a medicine from the app and places an order. The drug is delivered to the consumer by a licensed local pharmacy. The app delivers messages to the user to remind them to take their medication and also to restock. The app keeps track of the user's medication and dosages and delivers reminders to take medications and restock if necessary.

The purpose of this system is to eliminate manual complexity, and making it user-friendly is a major concern. The main features of the system are explained below:

Convenient Medicine Delivery: The application allows users to order medications with a few clicks and have them delivered to their homes.

Secure Medical Transactions: The app uses encrypted payment gateways and personal data protection mechanisms to ensure secure transactions.

Management of Prescriptions and Dosages: Users may upload and save their prescriptions in the app, which might help them keep track of their medication schedule and dosages.

Access to Quality Healthcare: The app connects users to licensed pharmacies and qualified pharmacists, guaranteeing that customers obtain high-quality healthcare services

Time-Saving: By removing the need for consumers to physically visit a pharmacy to purchase medication, the app saves them time and effort.

3.3 Requirement Analysis

The algorithm prioritizes security in the medication delivery process. It ensures that prescriptions are accurately verified, reducing the risk of patients receiving incorrect or counterfeit medications. Additionally, it employs robust encryption methods to safeguard sensitive patient data, including medical history and prescription information, guaranteeing that it remains confidential and protected from unauthorized access.

3.3.1 Secure Medicine Delivery

The algorithm prioritizes security in the medication delivery process. It ensures that prescriptions are accurately verified, reducing the risk of patients receiving incorrect or counterfeit medications. Additionally, it employs robust encryption methods to safeguard sensitive patient data, including medical history and prescription information, guaranteeing that it remains confidential and protected from unauthorized access.

3.3.2 User-Friendly Interface

The user-friendliness aspect of the algorithm is pivotal in ensuring that patients can easily interact with the application. It entails an intuitive user interface that simplifies the prescription upload process and medication scheduling [11]. Clear and concise notifications and reminders are integrated into the user experience, aiding patients in adhering to their prescribed regimens with minimal effort.

3.3.3 Medication and Dose Management

Efficient medication and dose management are at the core of this system. It allows patients to maintain a comprehensive record of their medication history, including dosages, schedules, and refills [12]. Moreover, it facilitates healthcare professionals in monitoring patient adherence and making informed adjustments to treatment plans when necessary, which can significantly enhance the quality of care provided.

3.3.4 Adaptability and Continuous Improvement

As the development of the application progresses, the algorithm remains flexible and open to modification and enhancement. This adaptability ensures that the system can evolve to meet changing patient needs, incorporate advancements in medical technology, and stay aligned with evolving healthcare regulations and standards. Regular updates and improvements are essential to keeping the application at the forefront of medication delivery solutions [13].

4 Implementation

The introduction of Medi2Home in Bangladesh is transforming access to healthcare. This smartphone app makes it easier to verify prescriptions, buy medications, and deliver them because it was developed using secure technology. With notifications and reminders, it improves patient involvement. To develop an all-encompassing healthcare ecosystem, future plans call for service expansion and cooperation with neighborhood pharmacies.

4.1 Architectural Design

The "Medi2Home" mobile application has been architecturally designed using the Flutter framework for the front end and connected to a robust database system. Flutter, renowned for its cross-platform capabilities, facilitated the development of a native-feeling mobile application compatible with both Android and iOS platforms. It ensures a seamless and responsive user interface, enhancing the overall user experience. In tandem, the database integration allows for efficient management of critical data such as patient records and medication information. This combination of Flutter and a well-structured database provides a secure, scalable, and reliable foundation for the application, ensuring data integrity and user privacy [14] (Fig. 2).

Fig. 2 Architectural diagram of Medi2Home

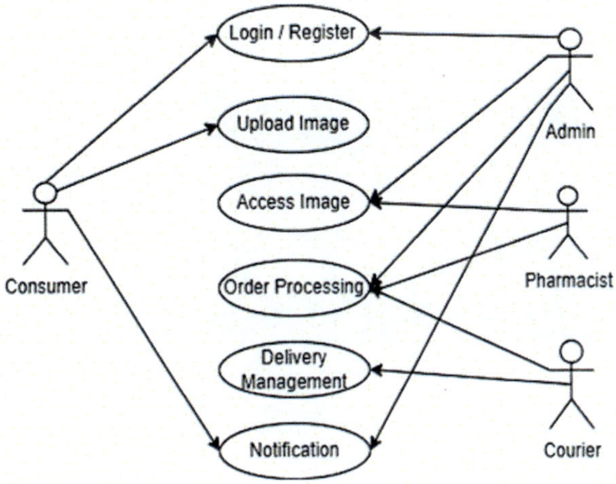

In the future, "Medi2Home" intends to increase the range of services it provides by introducing over-the-counter drugs, promotions, and subscription-based prescription support. By working together with neighborhood pharmacies, it hopes to improve the network of support for local businesses as well as patients. "Medi2Home' is a great example of how technology can transform healthcare access and delivery in the age of smart cities, moving Bangladesh closer to being a wiser and healthier country [15].

4.2 Application Design

The Medi2Home app's physical architecture requires a strong server infrastructure, mobile devices, dependable network connectivity, a secure environment, and continuous monitoring and maintenance. This enables the software to function on both the Android and iOS platforms including components for user registration, login, prescription upload, and medication scheduling (Fig. 3).

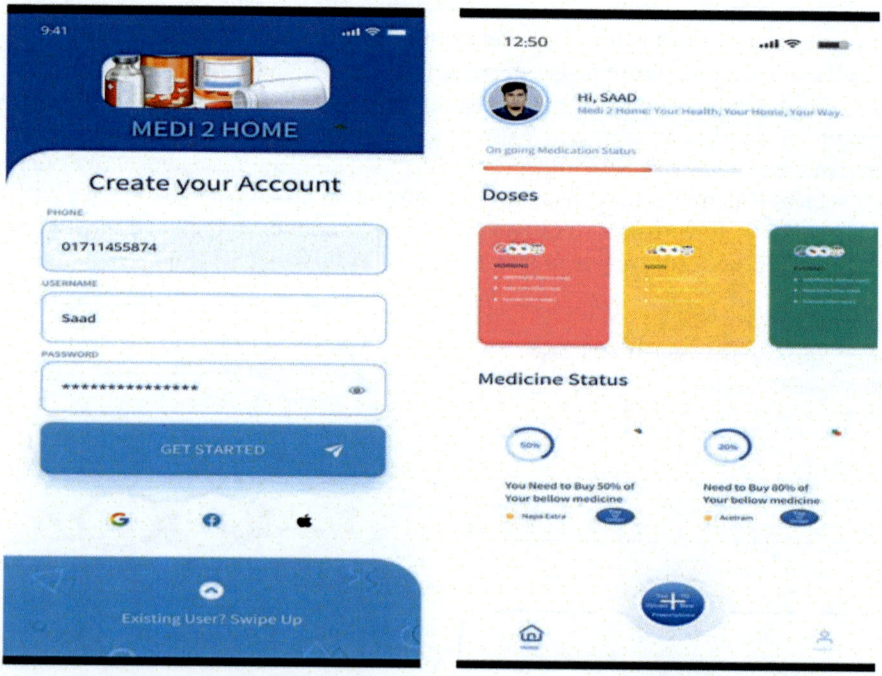

(a) Login Interface for Patient (b) Patient Account Interface

Fig. 3 User interface of Medi2Home

5 Conclusion

The development of Bangladesh's pharmaceutical and healthcare industries is greatly aided by Medi2Home. With location monitoring for faster delivery, this technology provides a safe, effective way to order and deliver pharmaceuticals to people's doorsteps. It serves as an example of how technology increases healthcare quality and accessibility. Plans for the future include an OTC drug store, notifications for patients, advertisements, subscription services, and more innovations in medical technology. For distribution at the moment, Medi2Home relies on neighborhood pharmacies in Bangladesh. There is, however, lot of space for improvement and expansion. The platform can improve the delivery system to create a more reliable and efficient distribution network while expanding its scope to encompass a wider range of pharmaceuticals and medical equipment. Potential partnerships with healthcare organizations and insurance providers can also improve the patient experience and provide all-inclusive healthcare solutions.

References

1. Barton AJ (2012) The regulation of mobile health applications. BMC Med 10:1–4
2. Mosa ASM, Yoo I, Sheets L (2012) A systematic review of healthcare applications for smartphones. BMC Med Inform Dec Making 12(1):1–31
3. Connolly TM, Boyle EA, MacArthur E, Hainey T, Boyle JM (2012) A systematic literature review of empirical evidence on computer games and serious games. Comput Educ 59(2):661–686
4. Tsai HYS, Jiang M, Alhabash S, LaRose R, Rifon NJ, Cotten SR (2016) Understanding online safety behaviors: a protection motivation theory perspective. Comput Secur 59:138–150
5. Nayak AK, Ahmad SA, Beg S, Ara TJ, Hasnain MS (2018) Drug delivery: present, past, and future of medicine. In: Applications of nanocomposite materials in drug delivery. Woodhead Publishing, pp 255–282
6. Nwabueze EE, Oju O (2019) Using mobile application to improve doctor-patient interaction in healthcare delivery system. E-Health Telecommun Syst Netw 8(03):23
7. Yousaf K, Mehmood Z, Saba T, Rehman A, Munshi AM, Alharbey R, Rashid M (2019) Mobile-health applications for the efficient delivery of health care facility to people with dementia (PwD) and support to their carers: a survey. BioMed Res Int
8. Free C, Phillips G, Watson L, Galli L, Felix L, Edwards P, Haines A (2013) The effectiveness of mobile-health technologies to improve health care service delivery processes: a systematic review and meta-analysis. PLoS Med 10(1):e1001363
9. Razzak MA, Ndiaye F, Chowdhury SA, Mansoor N (2007) Multiple description image transmission for diversity systems over unreliable communication networks. In: 2007 10th international conference on computer and information technology. IEEE, pp 1–5
10. Mansoor N, Hossain MI, Rozario A, Zareei M, Arreola AR (2023) A fresh look at routing protocols in unmanned aerial vehicular networks: a survey. IEEE Access
11. Kamal AH, Tusher MO, Ahmad SF, Farin NJ, Mansoor N (2020) Development of an expert system-oriented service support help desk management system. In: Proceedings of international joint conference on computational intelligence: IJCCI 2018. Springer, Singapore, pp 679–692
12. Farin NJ, Rimon MNAA, Momen S, Uddin MS, Mansoor N (2016) A framework for dynamic vehicle pooling and ride-sharing system. In: 2016 international workshop on computational intelligence (IWCI). IEEE, pp 204–208

13. Chen CL, Yang TT, Deng YY, Chen CH (2021) A secure internet of things medical information sharing and emergency notification system based on nonrepudiation mechanism. Trans Emerg Telecommun Technol 32(5):e3946
14. Lamooki SR (2022) Feasibility of wearable sensors for assessing the impact of repetitive task performance on occupational workers using time series analysis approaches (Doctoral dissertation, State University of New York at Buffalo)
15. Mansoor N (2020) Conceptualizing and realizing a smart city model for Bangladesh. arXiv:2012.03055

Sentiment Analysis of Product Reviews from Amazon, Flipkart, and Twitter

Padma Adane, Avanti Dhiran, Shruti Kallurwar, and Sushmita Mahapatra

1 Introduction

The aim was to perform sentiment analysis on customer tweets from Twitter and reviews from popular e-commerce platforms, such as Amazon and Flipkart, for a specific product. Here, we aimed to gain insights into the sentiments expressed by customers toward the product and provide an overall analysis of its market perception. Preprocessing procedures were used to clean up the obtained data, remove stop words, lowercase it, tokenize it, and stem/lemmatize it. A machine learning model or an NLP technique was trained on labeled data to classify the sentiment as positive or negative. This analysis was visually represented using a pie chart, showcasing the distribution of positive and negative. A word cloud image was generated to highlight the frequently occurring words and phrases associated with the product, providing further insights into customer sentiment. The outcomes provide useful information that firms may use to comprehend customer views, pinpoint areas for improvement, and make data-driven choices that will increase the product's marketability and customer satisfaction. Overall, it demonstrates the effectiveness of sentiment analysis in extracting meaningful insights from customer tweets and reviews, providing a comprehensive understanding of the sentiments associated with a specific product in the market.

Data preprocessing played a critical role in ensuring the quality and reliability of sentiment analysis. Techniques such as text cleaning, converting text to lowercase, removing stop words, and tokenization, stemming, and lemmatization [1] were applied to refine the data. Sentiment classification entails categorizing each tweet and review as positive, negative, or neutral by utilizing supervised learning models or lexicon-based methods. The chosen technique depended on specific requirements and available resources. Visualizations were used to present the findings effectively.

P. Adane (✉) · A. Dhiran · S. Kallurwar · S. Mahapatra
Shri Ramdeobaba College of Engineering and Management, Katol Road, Nagpur, India
e-mail: adanep@rknec.edu

M. S. Kaiser et al. (eds.), *ICT: Smart Systems and Technologies*, Lecture Notes in Networks and Systems 878, https://doi.org/10.1007/978-981-99-9489-2_33

A pie chart illustrated the distribution of sentiments, providing an overview of the overall sentiment landscape. This allowed stakeholders to quickly grasp the sentiment distribution and identify prevailing sentiments. Additionally, a word cloud highlighted frequently occurring words and phrases, helping to identify key themes or aspects influencing customer sentiments. The results of this project hold significant implications for businesses. The insights gained from sentiment analysis provide valuable information regarding customer perceptions, preferences, and pain points. By understanding the sentiment landscape, businesses can make data-driven decisions to improve product features, address customer concerns, and enhance their market position. In conclusion, this project demonstrated the efficacy of sentiment analysis techniques in analyzing customer sentiments expressed in tweets and reviews. The comprehensive analysis, supported by visualizations, provided a holistic understanding of customer sentiments toward the product. The findings and insights from this project offer businesses a valuable tool for decision-making, ultimately contributing to improved customer satisfaction and overall market success.

2 Literature Review

There has been several notable sentiment analysis of customer reviews that had been explored to gain the insights into the sentiment expressed by customers toward the product.

In paper [2], one of the first studies on sentiment analysis on Twitter has used a small dataset of manually categorized tweets. The authors classified tweets as good, negative, or neutral using a lexicon-based method. The researchers' accuracy was 80%. In their work, the authors propose a two-step approach for sentiment analysis. To start they construct a sentiment lexicon using distant supervision. In this approach, emoticons are used as noisy labels to represent sentiments. The second step is to train a classifier using a labeled dataset from distant supervision. They demonstrate the effectiveness of their approach through experiments and evaluations using large-scale twitter datasets. Compared to existing sentiment analysis techniques, their approach achieves competitive performance. Aside from analyzing sentiment distributions across different topics, they also observe interesting patterns in Twitter sentiment expressions. There are unique challenges and potential solutions associated with analyzing sentiment in short and noisy social media messages in this paper, highlighting the unique challenges and potential solutions for sentiment analysis on Twitter.

In paper [3], SVMs and Maximum Entropy (ME) classifiers outperformed other machine learning algorithms in the authors' comparison of performance for sentiment analysis on Twitter. The use of hashtags, retweets, and mentions as sentiment indicators is discussed, as well as various methods and procedures for gathering and preparing Twitter data. Additionally, it investigates several techniques for sentiment classification, including lexicon-based strategies and machine learning algorithms. They demonstrate the efficiency of utilizing Twitter as a corpus for sentiment analysis by presenting experimental findings and assessments of their suggested methodologies. They talk about the problems caused by noisy and casual language on Twitter and offer potential remedies to these problems.

In [4], the study offers a survey of various lexicon-based methods, machine learning methods, and hybrid models for sentiment analysis. It goes over the benefits and drawbacks of each method, as well as how they apply to sentiment analysis on Twitter. Additionally, they provide a thorough review of current sentiment lexicons and their applicability to sentiment analysis on Twitter. Based on how well they perform and how well suited they are for Twitter sentiment analysis tasks, they compare various lexicons. Additionally, they go over the uses of sentiment analysis on Twitter data, such as trend forecasting, brand monitoring, political analysis, and opinion mining. They emphasize the variety of fields in which Twitter sentiment analysis might offer insightful data.

In [5], "New ensemble methods for evolving data streams", suggested a deep learning method for sentiment analysis on Twitter utilizing a character-level Convolutional Neural Network (CNN). Using the Twitter Search API, the authors first compile a sizable dataset of tweets, concentrating on tweets about particular subjects (such as movies or books) and incorporating both positive and negative sentiment keywords in their search queries. After that, they perform preprocessing on the data by eliminating stop words, stemming words, and normalizing emoji. The authors then use a machine learning method to categorize each tweet's emotion as either positive, negative, or neutral.

3 Implementation

In the Sentiment Analysis of Product Reviews from Amazon, Flipkart, and Twitter project, sentiment analysis will be used to analyze customer tweets from Twitter and product reviews from well-known e-commerce sites like Amazon and Flipkart. This section has been divided into two parts. In the first part, the techniques that have been used for developing ML model for sentiment classification have been covered. The second part outlines user interface design and functionality. A GitHub project [6] helped in clarifying the tech stack and outlining the sequence of phases or steps required for its implementation (Fig. 1).

3.1 Data Collection

The first step is to collect the relevant data for training model [7]. Data include customer tweets related to the products and a diverse range of customer reviews and their corresponding sentiments. An already prepared dataset has been used from Kaggle online community for training the ML model. It can be found and downloaded from the cited link [8].

Fig. 1 Quick overview for training model

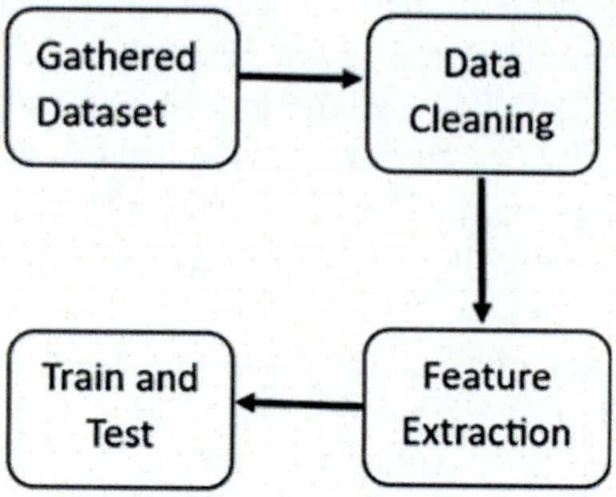

3.2 *Data Cleaning and Processing*

The dataset was altered and edited after data collection. The text must first be divided into separate words or tokens. Further analysis is made easier by tokenization because it lets you analyze each word independently. To make the writing more consistent, reduce words to their root or basic form. While stemming strips words of their suffixes, lemmatization maps words to their dictionary form (lemma). Both methods assist in combining word variations and reducing the data's dimensionality. Lemmatization, stemming, and the removal of URLs and user references are all functionalities offered by libraries like NLTK [9] or spaCy. Handling emoji and emoticons is the next step: In [10], to portray distinct feelings, users frequently use many emotions in a single tweet. Therefore, based on a positive or negative emotion, replace it with positive or negative words. Below is a list of several emotions that our method matches (Fig. 2).

In this example, the emojis " 😄", " 😆", " 😊", " 🙂", " 😍", " 😁", " 🙂", " 😮", " 😎", and " 🤗" are associated with positive sentiments, while the emojis " 😑", " 😞", " 😟", " 😭", " 😦", " 😫", " 😟", " 😕", " 🙁", and " 😒" are associated with negative sentiments [11]. When preprocessing text data that contain emojis, replace them with their corresponding sentiment labels ("positive" or "negative") before performing further analysis or feeding the data to a sentiment analysis model. This allows to consider the sentiment conveyed by emojis and incorporate it into the sentiment analysis process. After Data Cleaning, Count Vectorizer can be used for Feature Extraction [12].

3.3 *Training and Testing*

The sentiment analysis model has been trained using Support Vector Machine (SVM), a potent machine learning method frequently used for classification tasks. The SVM model's input features will be the extracted keywords from the user tweets and reviews. The target variable for training the model was the sentiment labels (positive, negative, or neutral) attached to each tweet or review. The dataset has been divided

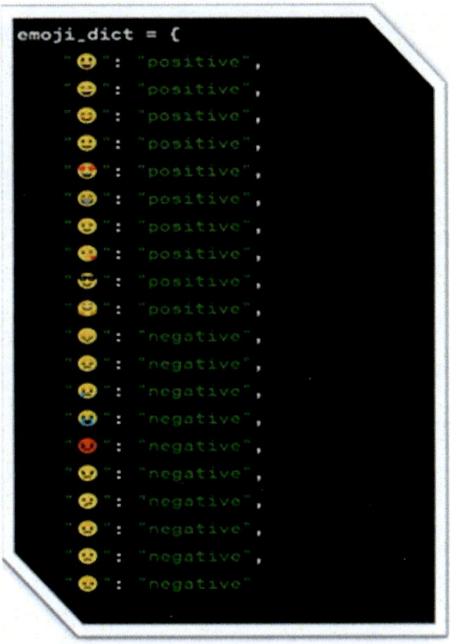

Fig. 2 List of emojis and corresponding sentiment

into training and testing sets so that the performance of the model can be assessed. The SVM model is trained using the training set, and its accuracy and generalizability are evaluated using the testing set. The accuracy of our model is calculated to be 91.37%, indicating its ability to correctly classify the sentiment of customer tweets and reviews. A detailed discussion about the results is provided in results section.

3.4 User Interface

The easy-to-use interface will provide users to extract the reviews and gain insights about the sentiment expressed by customers through visual representation. It will include features such as defining product name, viewing the reviews, downloading reviews in csv format, visual representation of sentiment analysis using pie chart, and word cloud image representing frequently occurring words and phrases associated with the product, providing further insights into customer sentiment (Figs. 3 and 4).

The following phase is sentiment classification, which categorizes tweets and reviews as positive, negative, or neutral, after extracting tweets and reviews for the chosen product and cleaning the data. The trained SVM model will then process the cleaned data as input and classify each tweet or review as either positive, negative, or neutral. After classification, a pie chart was created to show how the customer tweets and reviews were distributed among the sentiment categories (positive, negative, and neutral) [13]. The entire sentiment landscape will be clearly shown through this visual representation, which will also draw attention to any dominant sentiments.

Fig. 3 User interface to define the product name for analysis

	Product	Name	Price (INR)	Rating	Comment Heading	Comment
0	APPLE iPhone 14 (Starlight , 128 GB)	Rahul Prasad	71999.0	5	Fabulous !	Best smart phone under this price range compare to other phones in 2023 if you see overall build quality , performance and Camera with autofocus and video action mode are awesome 50 % extra RAM compared to iPhone 13 and other more features . Best time to upgrade to iPhone 14. I am so happy See Low light photos are amazing...
1	APPLE iPhone 14 (Starlight , 128 GB)	Flipkart Customer	71999.0	5	Great product	Damn such a great phone . Camera is really good , battery lasts long enough , super smooth even though its just 60 hz XDR display . Videos with action mode on are insanely stable and crisp . The night mode can take some really good shots in low light conditions. The whole apple environment itself is so seamless . I shifted from android to iPhone, and it barely took any time since I had iPad and a mac already . With a just a few clicks the phone was ready to go .
2	APPLE iPhone 14 (Starlight , 128 GB)	Flipkart Customer	71999.0	5	Best in the market !	Excellent smart phone , Good battery backup and performance is very good , Good camera quality .
3	APPLE iPhone 14 (Starlight , 128 GB)	Feecon Behera	71999.0	5	Must buy!	Just awesome !!! switch to iPhone 14 from iPhone 12. The powerful chipset gives you smooth performance . No lags , no fancy UI , just delivers the best of it . Camera is best . Battery giving a full day . Speaker sounds wide and clear . Must buy!!
4	APPLE iPhone 14	Samaresh Tikader	71999.0	5	Fabulous !	Happy to have iPhone 14. Wonderful phone specially because of Powerful camera 😍, perfect screen size and potentially battery

Fig. 4 Extracted reviews for specified product

Below is a pie chart to represent the sentiment analysis results on extracted product reviews (Fig. 5).

Fig. 5 Results of the sentiment analysis are shown in a pie chart

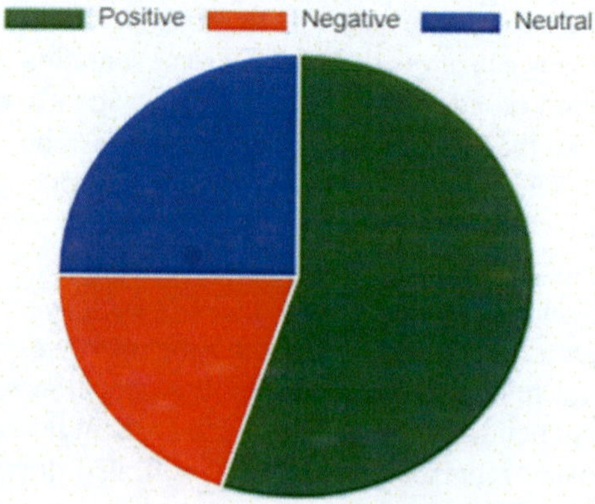

In addition to the pie chart, a word cloud graphic is also produced to show the terms and phrases that appear most frequently in the customer feedback. Project [14] helped with this visual representation. Understanding the major themes, subjects, and attitudes that customers have discussed will be made easier with the help of this visual representation (Fig. 6).

Above is a word cloud to represent the frequently occurring words on sentiment analysis results on extracted product reviews.

- Confusion Matrix

The comparison of the model's predictions to the actual sentiment labels is thoroughly broken down in the confusion matrix [15]. Helped to plot confusion matrix for proposed model (Fig. 7).

Above is the confusion matrix for test data on our trained model, here.

- The negative sentiment indicator is 0.
- The number 1 denotes neutral sentiment.
- The number 2 denotes a positive attitude.

Fig. 6 Word cloud image

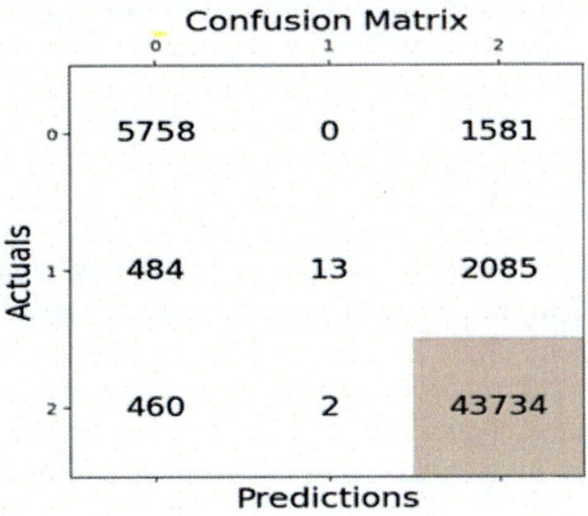

Fig. 7 Confusion matrix

Fig. 8 Accuracy scores

```
Precision Score :  0.9137913830188261
Recall Score :  0.9137913830188261
F1 Score: 0.914
Accuracy: 0.914
```

From confusion matrix, we can say that model is almost predicting correct for negative and positive sentiments, but low accuracy of model for neutral sentiment, as it is predicting neutral sentiment as positive sentiment in most of the test cases (Fig. 8).

- Accuracy
  A frequently used metric to gauge how accurate a model's predictions are overall is accuracy according to [16]. It is determined by dividing the total number of instances by the sum of true positive and true negative values. A majority of the customer tweets and reviews are correctly predicted by the algorithm, which has an accuracy rate of 91.37%.
- Precision
  Out of all instances anticipated as positive, precision is the percentage of accurately predicted positive instances. In [17] it is derived by subtracting the sum of the true positive and false positive values from the true positive value. The model is successful in accurately identifying positive sentiments when it receives a high precision score, which also suggests a low false positive rate. With a precision of 91.37%, our algorithm successfully predicts the sentiment of the bulk of user tweets and reviews.
- Recall
  A high recall score reveals a low false negative rate, which highlights the model's capacity to identify positive emotions. Using formula for recall given in [18], the result is 91.47, which indicates a strong recall score.
- F1-score
  A high F1-score denotes an effective balance between limiting false positives and false negatives while accurately detecting positive events [19]. In our situation, the result is 91.47, which indicates a strong F1-score.

4 Conclusion

In the proposed research paper, sentiment analysis on product reviews from various platforms, including Twitter, Amazon, and Flipkart, is to gain insights into the sentiments expressed by customers toward the product and provide an overall analysis of its market perception. The sentiment analysis results can be summarized in a pie chart, visually representing the distribution of positive, negative, and neutral sentiments among product reviews. The pie chart provides valuable insights into the overall sentiment landscape. Moreover, word cloud image representation helps to visualize the frequently occurring words and phrases within the customer feedback.

This visual representation provides further insights into the key themes, topics, and sentiments expressed by customers, aiding in understanding their preferences and concerns.

References

1. https://www.educative.io/answers/preprocessing-steps-in-natural-language-processing-nlp
2. Go A, Bhayani R, Huang L (2009) Twitter sentiment classification using distant supervision. CS224N Project Report, Stanford, vol 1(12)
3. Tanveer Siddiqui, U.S. Tiwari, Natural language Processing and Information Retrieval. Oxford University Press
4. Saif H, Fernandez M, Alani H (2013) Evaluation datasets for Twitter sentiment analysis: a survey and a new dataset, the STS-Gold. Lang Resour Eval 47(1):217–235
5. Bifet A, Holmes G, Pfahringer B (2010) New ensemble methods for evolving data streams. In: Proceedings of the 15th ACM SIGKDD international conference on knowledge discovery and data mining, pp 139–148
6. https://github.com/prajwalkhairnar/flipkart_product_review_sentiment_analysi
7. https://www.tableau.com/learn/articles/what-is-data-cleaning
8. https://www.kaggle.com/datasets/bittlingmayer/amazonreviews/download?datasetVersionNumber=7
9. realpython.com/python-nltk-sentiment-analysis
10. https://medium.com/geekculture/text-preprocessing-how-to-handle-emoji-emoticon-641bbfa6e9e7
11. https://unicode.org/emoji/charts/full-emoji-list.html
12. https://towardsdatascience.com/basics-of-countvectorizer-e26677900f9c
13. Twitter Sentiment analysis visualization using pie chart was taken from https://github.com/sachin2404/twitter-sentiment-analysis-using-pie-chart-
14. Word Cloud Image. https://github.com/amueller/word_cloud
15. https://www.analyticsvidhya.com/blog/2020/04/confusion-matrix-machine-learning/
16. https://www.analyticsvidhya.com/blog/2020/09/precision-recall-machine-learning/
17. https://en.wikipedia.org/wiki/Precision_and_recall
18. https://developers.google.com/machine-learning/crash-course/classification/precision-and-recall
19. https://towardsdatascience.com/the-f1-score-bec2bbc38aa6

Arithmetic Optimization Algorithm: A Review of Variants and Applications

Shivani Thapar, Amit Chhabra, and Arwinder Kaur

1 Introduction

1.1 Meta-heuristics at Large

Meta-heuristics algorithms are used to generate multifaceted solutions of real-world optimization problems [1]. Evolutionary algorithms draw inspiration from natural evolution, selecting the best solutions and discarding less favorable ones, akin to the passing on of genetic material over generations, gradually refining solutions. Prominent algorithms include the Genetic Algorithm [1], which picks the best solution over iterations taking inspiration from survival of the fittest theory of Darwin and Differential evolution (DE) [2] which mimics the process of collaboration and contest among individual solutions contained in population matrix, whereas Tree Growth algorithm [3] is based on trees seeking light through phytochromes and competing among each other for growth and food. Physics-based algorithms are based on the laws of physics to solve global optimization problems, e.g., Gravitational Search Algorithm (GSA) [4], Heat Transfer Search [5], Lévy flight distribution [6], Henry Gas Solubility Optimization (HGSO) [7], and Archimedes optimization algorithm [8]. Human-based algorithms simulate human intelligence for optimization, such as Tabu Search (TS) [9], inspired by human memory mechanisms, and Teaching Learning-Based Optimization (TLBO) [10], which mimics classroom learning dynamics. Swarm optimization algorithms replicate animal group behaviors, where prey represents the optimal solution and the swarm symbolizes candidate solutions. Examples include Cuckoo Search Algorithm [10], Particle Swarm Optimization (PSO) [11], Bat Algorithm (BAT) [12], Whale Optimization Algorithm (WOA) [13], Crow Search Algorithm (CSA) [14], Gray Wolf Optimization (GWO) [15], and Aquila Algorithm [16].

S. Thapar (✉) · A. Chhabra · A. Kaur
Department of Computer Engineering and Technology, Guru Nanak Dev University, Amritsar 143005, India
e-mail: shivanicet.rsh@gndu.ac.in

M. S. Kaiser et al. (eds.), *ICT: Smart Systems and Technologies*, Lecture Notes in Networks and Systems 878, https://doi.org/10.1007/978-981-99-9489-2_34

Modern meta-heuristic algorithms benefit from hybridization, as no single algorithm universally solves all optimization problems, in line with the No Free Lunch theorem [17]. This paper reviews the AOA meta-heuristic algorithm [18], its variants, and applications.

AOA utilizes fundamental arithmetic operators for optimizing both diversification and intensification phases [18]. The optimization procedure begins by randomly generating a set of candidate solutions (Y) as described in Eq. (1), and the best candidate solution evolves through iterations.

$$Y = \begin{bmatrix} y_{1,1} & \cdots\cdots & y_{1,j} & y_{1,n-1} & y_{1,n} \\ y_{2,1} & \cdots\cdots & y_{2,j} & \cdots & y_{2,n} \\ \cdots & \cdots\cdots & \cdots & \cdots & \cdots \\ \vdots & \vdots\;\vdots & \vdots & \vdots & \vdots \\ y_{N-1,1} & \cdots\cdots & y_{N-1,j} & \cdots & y_{N-1,n} \\ y_{N,1} & \cdots\cdots & y_{N,j} & y_{N,n-1} & y_{N,n} \end{bmatrix}. \tag{1}$$

AOA dynamically selects between its two search phases, diversification or intensification, by generating random values (r_1, r_2, r_3) within the range of 0 to 1. A key component in this decision is the Math Optimizer Accelerated function (MOA), computed using Eq. (2), and it plays a crucial role in guiding subsequent search phases.

$$\text{MOA(C\_Iter)} = \text{Min} + \text{C\_Iter} \times \left(\frac{\text{Max} - \text{Min}}{\text{M\_Iter}}\right). \tag{2}$$

In Eq. (2), MOA(C_Iter) represents the function value at the tth iteration, with C_Iter denoting the current iteration ranging from 1 to the maximum number of iterations. Min and Max represent the minimum and maximum values of the MOA function, respectively. When r_1 exceeds MOA, the algorithm proceeds to execute both diversification and intensification search phases. Division is employed when r_2 < 0.5, while multiplication is applied when r_2 > 5.

$$x_{i,j}(\text{C\_Iter} + 1) = \begin{cases} \text{best}(x_j) \div (MOP + \upsilon) \times ((\text{UB}_j - \text{LB}_j) \times \mu + \text{LB}_j), & r^2 < 0.5 \\ \text{best}(x_j) \times \text{MOP} \times ((\text{UB}_j - \text{LB}_j) \times \mu + \text{LB}_j), & \text{otherwise} \end{cases}. \tag{3}$$

In Eq. (3), x_i(C_Iter + 1) represents the ith solution in the next iteration, x_(i, j)(C_Iter) is the jth position of the ith solution in the current iteration, and "best" (x_ j) denotes the jth position in the best solution achieved thus far. ϵ is a small integer, UB_j and LB_j represent the upper and lower bounds for the jth position, and μ is a control parameter used to modulate the search process, typically set to 0.5 in the original AOA [18].

$$\text{MOP}(\text{C\_Iter}) = 1 - \frac{\text{C\_Iter}^{1/\alpha}}{\text{M\_Iter}^{1/\alpha}}. \tag{4}$$

In Eq. (4), Math Optimizer Probability (MOP) serves as a coefficient. MOP(C_Iter) represents the function value at iteration t, with C_Iter denoting the current iteration and (M_Iter) indicating the maximum number of iterations. The parameter α serves as a sensitivity constraint, signifying the level of exploitation accuracy achieved throughout the iterations. Subtraction and addition operators, known for their reduced dispersion, can produce closely packed solutions. This characteristic aligns them with the objective of the intensification search phase, which strives to approach the target solution. Therefore, these operators are now utilized during the intensification stage.

$$x_{i,j}(\text{C\_Iter} + 1) = \begin{cases} \text{best}\left(x_j\right) - \text{MOP} \times \left(\left(\text{UB}_j - \text{LB}_j\right) \times \mu + \text{LB}_j\right), \ r_3 < 0.5 \\ \text{best}\left(x_j\right) + \text{MOP} \times \left(\left(\text{UB}_j - \text{LB}_j\right) \times \mu + \text{LB}_j\right), \ \text{otherwise} \end{cases}. \tag{5}$$

The subtraction operator is chosen when $r_3 < 0.5$, while the addition operator is selected when $r_3 > 0.5$. These operators in AOA help prevent the algorithm from getting stuck in localized search areas, ensuring that it explores a wider range of potential solutions.

2 Research Methodology

In August 2023, we conducted an extensive review of English articles related to AOA published within the last two years (2021–2023). We performed this review by searching databases including Web of Science, IEEE Xplore, Science Direct, Scopus, and Springer Link. To enhance our article search strategy, we constructed a keyword-based search string, which encompassed terms such as "arithmetic optimization algorithm," "arithmetic optimization algorithm hybrid," "arithmetic optimization variant," "Bio-inspired meta-heuristic methods," "optimization problems," and "applications of arithmetic optimization algorithm."

Our article selection process adhered to specific criteria. We included articles related to AOA algorithms, meta-heuristic algorithms, their hybrids, variants, and applications, provided that they presented comprehensive content with results and validation. In contrast, we excluded commentaries, reviews, articles lacking full-text access, book chapters, and those without experiments or results. Following these criteria, we curated a selection of 40 journal articles that demonstrated effective outcomes across diverse application domains, as depicted in Fig. 1, showcasing the publishers of the chosen journal articles in this comprehensive review.

Fig. 1 Publishers of articles selected for the review

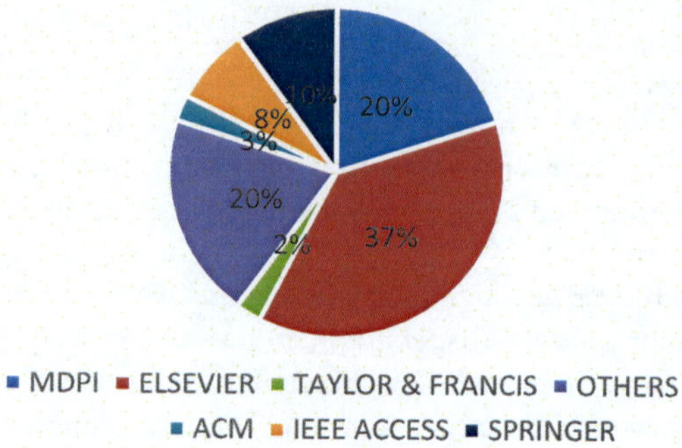

3 Applications and Variants of AOA

Within this section, we have organized the variants and hybrids of AOA based on their respective application domains.

3.1 *Application in Engineering Design Problems*

Zheng et al. [19] introduced Improved AOA (IAOA), a variant of AOA featuring a forced switching mechanism at specific iteration intervals to enhance exploratory search and the incorporation of a random math optimizer constraint to broaden the global solution search space. IAOA's performance was assessed across various scenarios, including classical benchmark functions, CEC2020 test functions, a multi-layer perceptron (MLP) model, and classical engineering design problems like the tree bar truss problem (TBTP), pressure vessel design (PVD), and tension–compression spring problem (TSCP).

Zheng et al. [20] proposed DESMAOA, which combines concepts from SMA and AOA. DESMAOA leverages SMA's RCS to avoid local stagnation and adds AOA-inspired subtraction and addition strategies for intensification. It was evaluated on classical test functions and engineering design problems like PVD, TCSP, and TBT.

Chauhan et al. [21] introduced OBAOA, integrating opposition-based learning into AOA and SMA. OBAOA improves diversification and intensification by utilizing both algorithms' solutions. It was tested on 23 benchmark functions and various engineering design problems, including WBD, PVDP, TCSP, and TBT.

In 2021, Agushaka et al. [25] improved AOA by incorporating dense mathematical values from natural logarithm and exponentials. They used the beta distribution method for initial solutions and tested it on 33 benchmark functions and three engineering design problems, including WBD, PVDP, and TCSP.

In 2022, Zhang et al. [30] fused AOA with the Aquila algorithm to boost convergence speed and optimize solutions. They introduced an energy parameter inspired by the Harris Hawk algorithm, balancing diversification and intensification processes. This approach was tested on various engineering problems like WBD, PVDP, TBT, SRD, and unimodal/multimodal functions.

Liu et al. [33] enhanced AOA with circular chaotic mapping, a composite cycloid in the MOA function, and an optimal mutation strategy. Evaluation included the Rastrigin function, 20 benchmark functions, CEC2019 test functions, and applications in AC Motor PID Control and PVD problems.

Hu et al. [34] improved standard AOA with optimal neighborhood learning and the crisscross strategy, enhancing its performance on 23 benchmark functions, IEEE CEC 2019/2020 functions, and eight engineering design problems.

3.2 Applications in IoT Scheduling

Elaziz et al. [24] suggested use of AOA in hybrid task scheduling for IoT devices in fog computing environments. The efficiency of model is tested using HP2CN workloads. Bahmanyar et al. [34] proposed multiobjective AOA which helps in energy-efficient scheduling of tasks among home appliances in a smart home environment.

3.3 Applications in Robot Path Planning

Wang et al. [22] used parallel messaging policy to encourage the communication and information exchange among group of solutions to delay convergence and avoid stagnation in local search space. A parameter adaptive equation is also introduced to monitor the sensitive constraint which helps to balance the diversification and intensification phases of AOA. This new Adaptive Parallel AOA (APAOA) is applied to 2D robot path planning. The results are verified using MATLAB environment and Robot workplace model. Gul et al. [35] proposed use of AO for multi-robot cluster for space exploration.

3.4 Applications in Feature Selection

In 2021, Ali Ibrahim et al. [23] introduced an enhanced strategy called Electric Fish Optimization AOA (EFAOA) to improve AOA's diversification ability. EFAOA was applied as a feature selection technique to remove insignificant features, enhancing the efficiency and accuracy of standard classification techniques. The algorithm's performance was assessed on eighteen datasets from the UC Irvine medical repository, comparing its efficacy with renowned feature selection methods.

Ewees et al. [36] proposed the AOAGA feature selection method in 2021, which is a hybrid of AOA and Genetic Algorithm (GA). This method was tested on UCI datasets. In 2022, Alweshah [37] introduced two wrapper feature selection strategies based on AOA and the Great Deluge Algorithm for medical diagnosis.

Abualigah et al. [26] suggested the use of the Differential Evolution technique in 2021 to enhance AOA's diversification capabilities. The improved algorithm was applied to image classification using eight test images from two groups (nature and CT-19 images) and to enhance multilevel thresholding of COVID-19 CT images. The results were verified using MATLAB.

Dahou et al. [38] improved the performance of a human activity recognition system in 2021 by fusing Convolutional Neural Network (CNN) and Binary AOA (BAOA). CNN was used to learn and extract features from input data, and the best features were selected from open datasets, including UCI-HAR, WISDM-HAR, and KU-HAR, using the modified AOA technique BAOA. The results demonstrated the competitive performance of the proposed model.

3.5 *Applications in Power Systems*

In 2022, Elkasem et al. [29] introduced an eagle-inspired enhancement to the AOA algorithm, aiming to prevent rapid convergence and ensure comprehensive exploration of candidate solutions within the search space. This improved algorithm was employed to fine-tune fractional-order proportional–integral–derivative (FOPID) parameters and PID controllers, contributing to the frequency stabilization of a hybrid power system utilizing both renewable and conventional energy sources. The algorithm's effectiveness was verified using 23 benchmark functions.

Hao et al. [39] proposed an enhanced AOA in 2022, incorporating elementary functional disturbance to address power demand while minimizing overall generation costs under various operational constraints. Experimental results demonstrated that the proposed AOA outperformed other methods in solving economic load dispatch problems. In 2021, Guo et al. [40] introduced an improved AOA for the design of combined cooling, heating, and power systems, with a focus on energy conservation.

3.6 *Miscellaneous*

In 2021, Sharma et al. [27] presented a novel opposition learning-based AOA for identifying undetermined parameters in proton-exchange membrane fuel cells. The results were evaluated using Ballard Mark V PEM fuel cells and standard benchmark functions through MATLAB. Aydemir [31] introduced a hybrid AOA algorithm in

2021, incorporating ten different chaotic maps to prevent premature convergence and explore a wide range of optimal solutions within the search space, avoiding local stagnation. The proposed CAOA was tested on eighteen benchmark problems and compared to the original base algorithm, with statistical significance established through the Wilcoxon sign-rank method.

The applications and variants of AOA are summarized in Table 1.

Table 1 Applications and variants of arithmetic optimization algorithm

Author	Approach	Highlights of the approach	Application	Tools used
Abualigah et al. [18]	AOA	Original AOA is motivated by the main arithmetic operators like addition, multiplication, subtraction, and division for diversification and intensification purposes	1. Twenty-nine benchmark functions 2. Engineering design problems	MATLAB
Zheng et al. [19]	IAOA	This AOA variant incorporates a forced switching mechanism to facilitate more effective exploratory search and utilizes random math optimization to enhance population diversity for global exploration	Three classical engineering design problems	MATLAB
Zheng et al. [20]	DESMAOA	The hybrid algorithm SMAOA combines the features of SMA and AOA to enhance the exploration capability of the original SMA	1. Twenty-three classical test functions 2. Engineering design problems	MATLAB
Chauhan et al. [21]	HAOASMA	Hybrid of AOA and SMA combined with lens opposition-based learning to diversify population matrix and stretch the search space	1. Twenty-three benchmark functions 2. Engineering design problems	MATLAB
Wang et al. [22]	APAOA	Parameter adaptive equation is introduced to control the sensitive constraint which helps in balancing the diversification and intensification phases of AOA	Two-dimensional robot path planning	MATLAB Robot workplace model
Ibrahim et al. [23]	EFOAOA	Electric fish optimization is fused with AOA with enhanced exploration ability which is used for feature selection	Feature selection problem	Eighteen standard datasets from UC Irvine medical repository

(continued)

Table 1 (continued)

Author	Approach	Highlights of the approach	Application	Tools used
Elaziz et al. [24]	AOAM	An effective workflow scheduling approach is proposed to boost AOA's efficiency in fog computing environment using IoT devices	Cloud scheduling problem using HP2CN workloads	MATLAB
Agushaka et al. [25]	nAOA	The natural logarithm ability to produce high density values is combined with exponential operators to enhance the diversification search phase of the AOA	1.Thirty-three benchmark functions 2. Mechanical engineering design problems	MATLAB
Sharma et al. [27]	OBAOA	Novel opposition-based arithmetic optimization algorithm (OBAOA) for identifying the unspecified parameters of proton-exchange membrane fuel cell (PEMFC)	Ballard Mark V PEM fuel cell (proton-exchange membrane) and standard benchmark functions	MATLAB
Abualigah et al. [29]	AOASC	Hybrid of AOA and sine–cosine algorithm and levy flight operators to enhance the exploitation ability of AOA in local search phase	Ten CEC 2019 benchmark functions and five common engineering optimization problems	MATLAB
Elkasem et al. [29]	ESAOA	Eagle strategy enhanced arithmetic optimization algorithm to increase the convergence rate	Twenty-three benchmark functions Power system design optimization problem	MATLAB
Zhang et al. [30]	AOAAO	AOA was fused with Aquila algorithm along with an energy parameter E inspired by Harris Hawk algorithm was also incorporated to switch between diversification and intensification and the randomness of the energy parameter was reduced by linear maps	Classical engineering design problems, nine unimodal, nine multimodal non-scalable, and nine multimodal scalable benchmark functions	MATLAB
Aydemir [31]	CAOA	The enhanced AOA algorithm incorporates ten different chaotic maps to accelerate convergence	Eighteen benchmark functions and Wilcoxon sign-rank method	MATLAB

(continued)

Table 1 (continued)

Author	Approach	Highlights of the approach	Application	Tools used
Liu et al. [33]	Multistrategy improved AOA	Four strategies—circular chaotic mapping, composite cycloid-enhanced MOA, and the sparrow elite mutation approach with Cauchy disturbances—are integrated into the standard AOA algorithm to enhance convergence speed and diversify the search space	Twenty benchmark test functions, the CEC S2019 test functions, engineering, PVD, AC motor PID control design problem	MATLAB
Hu et al. [34]	CSOAOA	The standard AOA algorithm is enriched with an optimal neighborhood learning strategy and a crisscross strategy to boost convergence speed and prevent local stagnation	Eight engineering design problems, 23 classical benchmark functions, CEC 2019 test suite, and CEC 2020 benchmark functions	MATLAB
Ewees et al. [36]	AOAGA	A hybrid of AOA and Genetic Algorithm which was tested on UCI datasets for medical diagnosis	Feature selection	MATLAB
Alweshah [37]	Standard AOA	Two wrapper feature functions based on AOA and great deluge algorithm for medical diagnosis	Feature selection	MATLAB
Dahou et al. [38]	BAOA	A fusion of convolutional neural network (CNN) and BAOA for human activity recognition	Feature selection	MATLAB
Hao et al. [39]	Improved AOA	AOA was improved with elementary functional disturbance to solve economic load dispatch problems	Power system design	MATLAB
Hao et al. [40]	Improved AOA	Improved AOA for enhancing combined cooling, heating, and power systems for saving energy	Power system design	MATLAB

4 Conclusion

In this paper, we have demonstrated diversity and effectiveness of AOA algorithm by studying wide range of problems, where it has been applied by researchers globally. We observed that AOA is motivated by basic mathematical functions to reach the best solutions traversing the randomly generated solution matrix. Current research paper also covered how AOA is further improved and optimized with the help of other meta-heuristics for solving real-world optimization problems like feature selection, engineering design problems, IoT scheduling, robot path planning, power system design, fuel cells, etc.

References

1. Wang C, Liu K, Shen P (2020) A novel genetic algorithm for global optimization. Acta Math Appl Sin Engl Ser 36:482–491. https://doi.org/10.1007/s10255-020-0930
2. Georgioudakis M, Plevris V (2020) A comparative study of differential evolution variants in constrained structural optimization. Front Built Environ 6:102. https://doi.org/10.3389/fbuil.2020.00102
3. Cheraghalipour A, Hajiaghaei-Keshteli M (2017) Tree growth algorithm (TGA): an effective metaheuristic algorithm inspired by trees' behavior
4. Rashedi E, Nezamabadi-pour H, Saryazdi S (2009) GSA: a gravitational search algorithm. Inf Sci 179:2232–2248. https://doi.org/10.1016/j.ins.2009.03.004
5. Patel V, Savsani V (2015) Heat transfer search (HTS): a novel optimization algorithm. Inf Sci 324:217–246. https://doi.org/10.1016/j.ins.2015.06.044
6. Houssein E, Saad M, Hashim FA, Shaban H, Hassaballah M (2020) Lévy flight distribution: a new metaheuristic algorithm for solving engineering optimization problems. Eng Appl Artif Intell 94. https://doi.org/10.1016/j.engappai.2020.103731
7. Hashim FA, Houssein E, Mabrouk M, Al-Atabany W, Mirjalili S (2019) Henry gas solubility optimization: a novel physics-based algorithm. Future Gener Comput Syst 101. https://doi.org/10.1016/j.future.2019.07.015
8. Hashim FA, Hussain K, Houssein E, Mabrouk M, Al-Atabany W (2021) Archimedes optimization algorithm: a new metaheuristic algorithm for solving optimization problems. Appl Intell 51:1–21. https://doi.org/10.1007/s10489-020-01893-z
9. Glover F (2022) Tabu search-Parts I and II
10. Rao RV, Savsani V, Vakharia D (2011) Teaching-learning-based optimization: a novel method for constrained mechanical design optimization problems. Comput-Aided Des 43:303–315. https://doi.org/10.1016/j.cad.2010.12.015
11. Poli R, Kennedy J, Blackwell T (2007) Particle swarm optimization. In: Swarm intelligence, vol 1, pp 33–57. https://doi.org/10.1007/s11721-007-0002-0
12. Yang X, Gandomi AH (2012) Bat algorithm: a novel approach for global engineering optimization. Eng Comput 29(5):464–483. https://doi.org/10.1108/02644401211235583
13. Mirjalili S, Lewis A (2016) The whale optimization algorithm. Adv Eng Softw 95(95):51–67. https://doi.org/10.1016/J.ADVENGSOFT.2016.01.008
14. Askarzadeh A (2016) A novel metaheuristic method for solving constrained engineering optimization problems: crow search algorithm. Comput Struct 169:1–12. https://doi.org/10.1016/J.COMPSTRUC.2016.03.001
15. Mirjalili S, Mirjalili SM, Lewis A (2014) Grey Wolf optimizer. Adv Eng Softw 69:46–61. https://doi.org/10.1016/J.ADVENGSOFT.2013.12.007
16. Abualigah L, Yousri D, Abd Elaziz ME, Ewees A, Al-qaness MAA, Gandomi A (2021) Aquila optimizer: a novel meta-heuristic optimization algorithm. Comput Ind Eng 157. https://doi.org/10.1016/j.cie.2021.107250
17. Mirjalili S (2015) Moth-flame optimization algorithm: a novel nature-inspired heuristic paradigm. Knowl-Based Syst 89. https://doi.org/10.1016/j.knosys.2015.07.006
18. Abualigah L, Diabat A, Mirjalili S, Abd Elaziz ME, Gandomi A (2021) The arithmetic optimization algorithm. Comput Methods Appl Mech Eng 376:113609. https://doi.org/10.1016/j.cma.2020.113609

19. Zheng R, Jia H, Abualigah L, Liu Q, Wang S (2021) An improved arithmetic optimization algorithm with forced switching mechanism for global optimization problems. Math Biosci Eng (MBE) 19:473–512. https://doi.org/10.3934/mbe.2022023
20. Zheng R, Jia H, Abualigah L, Liu Q, Wang S (2021) Deep ensemble of slime mold algorithm and arithmetic optimization algorithm for global optimization. Processes 9. https://doi.org/10.3390/pr9101774
21. Chauhan S, Vashishtha G, Kumar A (2021) A symbiosis of arithmetic optimizer with slime mould algorithm for improving global optimization and conventional design problem. J Supercomput. https://doi.org/10.1007/s11227-021-04105-8
22. Wang R, Wang W-F, Xu L, Pan J-S, Chu S-C (2021) An adaptive parallel arithmetic optimization algorithm for robot path planning. J Adv Transp 2021:1–22. https://doi.org/10.1155/2021/3606895
23. Ibrahim R, Abualigah L, Ewees A, Al-qaness MAA, Yousri D, Alshathri S, Abd Elaziz ME (2021) An electric fish-based arithmetic optimization algorithm for feature selection. Entropy 23:1189. https://doi.org/10.3390/e23091189
24. Abd Elaziz ME, Abualigah L, Ibrahim R, Attiya I (2021) IoT workflow scheduling using intelligent arithmetic optimization algorithm in fog computing. Comput Intell Neurosci 2021:1–14. https://doi.org/10.1155/2021/9114113
25. Agushaka O, Ezugwu A (2021) Advanced arithmetic optimization algorithm for solving mechanical engineering design problems. PLoS ONE 16:1–29. https://doi.org/10.1371/journal.pone.0255703
26. Abualigah L, Diabat A, Sumari P, Gandomi A (2021) A novel evolutionary arithmetic optimization algorithm for multilevel thresholding segmentation of COVID-19 CT images. Processes 9:1155. https://doi.org/10.3390/pr9071155
27. Sharma A, Khan R, Sharma A, Kashyap D, Rajput S (2021) A novel opposition-based arithmetic optimization algorithm for parameter extraction of PEM fuel cell. Electronics 10:2834. https://doi.org/10.3390/electronics10222834
28. Abualigah L, Ewees A, Al-qaness MAA, Abd Elaziz ME, Yousri D, Ibrahim R, Altalhi M (2022) Boosting arithmetic optimization algorithm by sine cosine algorithm and levy flight distribution for solving engineering optimization problems. Neural Comput Appl. https://doi.org/10.1007/s00521-022-06906-1
29. Elkasem A, Kamel S, Hosny M, Khamies M, Ahmed E (2022) An eagle strategy arithmetic optimization algorithm for frequency stability enhancement considering high renewable power penetration and time-varying load. Mathematics. https://doi.org/10.3390/math10060854
30. Zhang Y-J, Yan Y-X, Zhao J, Gao Z-M (2022) AOAAO: the hybrid algorithm of arithmetic optimization algorithm with aquila optimizer. IEEE Access 1–1. https://doi.org/10.1109/ACCESS.2022.3144431
31. Aydemir S (2022) A novel arithmetic optimization algorithm based on chaotic maps for global optimization. Evol Intel. https://doi.org/10.1007/s12065-022-00711-4
32. Liu Z, Li M, Pang G, Song H, Yu Q, Zhang H (2022) A multi-strategy improved arithmetic optimization algorithm. Symmetry 14:1011. https://doi.org/10.3390/sym14051011
33. Hu G, Zhong Y, Du B, Wei G (2022) An enhanced hybrid arithmetic optimization algorithm for engineering applications. Comput Methods Appl Mech Eng 394:114901. https://doi.org/10.1016/j.cma.2022.114901
34. Bahmanyar D et al (2022) Multi-objective scheduling of IoT-enabled smart homes for energy management based on arithmetic optimization algorithm: a node-RED and NodeMCU module-based technique. Knowl-Based Syst 247:108762. https://doi.org/10.1016/j.knosys.2022.108762
35. Gul F et al (2021) Multi-robot space exploration: an augmented arithmetic approach. IEEE Access 9:107738–50. https://doi.org/10.1109/ACCESS.2021.3101210
36. Ewees AA et al (2021) Boosting arithmetic optimization algorithm with genetic algorithm operators for feature selection: case study on cox proportional hazards model. Mathematics 9(18):2321. https://doi.org/10.3390/math9182321

37. Alweshah M (2022) Hybridization of arithmetic optimization with great deluge algorithms for feature selection problems in medical diagnoses. Jordanian J Comput Inform Technol 1. https://doi.org/10.5455/jjcit.71-1639410312
38. Dahou A et al (2022) Human activity recognition in IoHT applications using arithmetic optimization algorithm and deep learning. Measurement 199:111445. https://doi.org/10.1016/j.measurement.2022.111445
39. Hao W-K et al (2022) Arithmetic optimization algorithm based on elementary function disturbance for solving economic load dispatch problem in power system. Appl Intell 52(10):11846–72. https://doi.org/10.1007/s10489-021-03125-4
40. Guo H et al (2021) Optimal model of the combined cooling, heating, and power system by improved arithmetic optimization algorithm. In: Energy sources, Part A: recovery, utilization, and environmental effects, Aug 2021, pp 1–23. https://doi.org/10.1080/15567036.2021.1966138

Dark Channel Prior-Based Single-Image Dehazing Using Type-2 Fuzzy Sets for Edge Enhancement in Dehazed Images

Nisha Amin, B. Geeta, R. L. Raibagkar, and G. G. Rajput

1 Introduction

Over the years the increasing use of outdoor imaging applications has highlighted the importance of having clear and visually attractive images even when faced with difficult weather conditions. The presence of haze, fog, or other forms of airborne particulate matter can significantly degrade the quality of captured images, leading to diminished visibility, loss of detail, and reduced overall perceptual quality. Image dehazing, the process of enhancing hazy images to reveal obscured details and improve visual clarity, has emerged as a fundamental task in computer vision with widespread applications in domains such as surveillance, remote sensing, autonomous driving, and outdoor photography.

Traditional image dehazing methods often struggle to accurately restore the underlying scene content due to the complex interplay between light scattering and absorption by atmospheric particles. As a result, advanced computational techniques have been developed to address this challenge and provide more effective solutions. Among these, the dark channel prior (DCP) technique has gained prominence for its ability to estimate scene depth and haze density, forming a basis for effective haze removal.

However, while DCP proves powerful in capturing the coarse structure of the scene, it may still fall short in handling uncertainty and ambiguity present in real-world hazy images. A mathematical framework called fuzzy logic has been effectively used in a number of image processing applications to describe uncertainty and

N. Amin (✉) · R. L. Raibagkar
Department of Computer Science, Karnataka State Akkamahadevi Women's University, Vijayapura, India
e-mail: skanda.nsa@gmail.com

B. Geeta · G. G. Rajput
Department of Applied Electronics, Gulbarga University, Kalaburgi, India
e-mail: ggrajput@kswu.ac.in

M. S. Kaiser et al. (eds.), *ICT: Smart Systems and Technologies*, Lecture Notes in Networks and Systems 878, https://doi.org/10.1007/978-981-99-9489-2_35

imprecision. To this end, type-2 fuzzy set theory offers an extension of traditional fuzzy logic, enabling the representation of higher degrees of uncertainty. By integrating type-2 fuzzy set theory into the dehazing process, the system becomes more adaptive and capable of capturing the diverse range of atmospheric conditions and image characteristics encountered in practice.

Motivated by the need for more robust and adaptable image dehazing techniques, this research proposes a novel approach that combines the power of the DCP technique with the expressive capabilities of type-2 fuzzy sets. The goal is to enhance the dehazing process by accounting for uncertainties related to varying haze densities, lighting conditions, and scene complexities. The proposed method is evaluated using the I-Haze dataset, a comprehensive collection of hazy images captured in diverse outdoor environments. The efficiency of the approach is assessed through a range of quantitative and qualitative metrics, including PSNR, SSIM, LOE, and NIQE that are commonly used metrics to assess image quality in terms of signal to noise ratio, similarity index, lightness order error, and naturalness.

This research contributes to the advancement of image dehazing techniques by providing an efficient framework that leverages the strengths of both DCP and type-2 fuzzy set theory. The rest of the article is organized as follows. The techniques are reviewed in Sect. 2 in dehazing images. A description of the proposed methodology is presented in Sect. 3, experimental results are discussed in Sect. 4, and a conclusion is provided in Sect. 5. The main objective of this project is to improve the appearance of images taken under difficult weather conditions thus pushing the boundaries of outdoor photography applications even further.

2 Literature Review

Archana [1] proposed a comprehensive approach involving the utilization of a depth map to recover depth information, the implementation of an adaptive linear model with color attenuation prior characteristics, the precise removal of haze from images, and the introduction of a straightforward method for reconstructing scene radiance using a tropospheric dispersion model. The foundational image can be efficiently restored through the derived depth map. Despite efforts to address fog-induced issues such as image brightness and saturation to establish scene depth, challenges persist. This is due to the constant nature of the scattering coefficient under consistent atmospheric conditions, preventing effective recalibration. The study emphasizes the necessity for a versatile model, as current single-image dehazing techniques heavily rely on ongoing assumptions for their outcomes.

Yu et al. [2] introduced a novel strategy termed "viewbased cluster segmentation" tailored for image and video dehazing. Their approach aims to enhance the visibility of sky and white objects while circumventing distortions in sky regions through a viewbased cluster segmentation technique. In this approach, adjustments are made to the estimation of the sky region based on distance to mitigate distortion, while the initial clustering of depth employs a Gaussian mixture model (GMM). This method proposes the utilization of individual components such as GMM clustering, color attenuation prior, transmission estimation, and atmospheric light estimation for both the hazy image and depth map. To mitigate color distortion and enhance overall contrast, the approach is refined by incorporating viewbased cluster segmentation. Furthermore, the technique offers a means for video dehazing employing an online GMM cluster.

Park et al. [3] presented a rapid implementation methodology for dehazing outdoor videos using dark channel priors. It achieves the swift implementation of the dark channel priors approach aimed at the dehazing of outdoor footage. While maintaining the original method's dehazing quality, up to 49% less execution time is utilized generally. The signal processing method known as "dehazing" is used to remove haze. Every pixel in the picture has a different level of haze density. Therefore, finding the black pixel in a picture clears out any haze. Camera records the blurry image and locates the air light. A cost function made up of a term for the standard deviation and a term for the histogram uniformity is developed to assess the contrast. In the end the test results reveal that the proposed approach effectively enhances the clarity of the scenery by eliminating haze.

Yu et al. [4] outlined a technique for addressing single-image blurriness through a fractional-order dark channel prior. This study offers a method for deblurring photographs that uses a fractional-order dark channel prior (FODCP) and a fractional-order operator to improve hazy images. Additionally, the half-quadratic splitting approach is used to solve the non-convex problem, and many measures are put in place to gage the effectiveness of the deblurred photographs. The findings of quantitative and qualitative trials support the idea that when used on artificially and naturally blurred photographs, the suggested approach produces cutting-edge results.

Hari et al. [5] presented a paper titled "Enhancing Visibility in Day and Night Images through Combined Dark and Bright Channel Priors for Haze Removal." By combining the dark channel prior with the multi-scale retinex theory, they created an improved haze removal method for this investigation. They were able to accomplish dynamic optimization of the transmission map, leading to better visibility, by combining the combined dark channel prior (DCP) and bright channel prior (BCP) with the multi-scale retinex (MSR) method. The suggested approach efficiently decreases picture noise while utilizing retinex theory's basic ideas. The testing results show that the suggested strategy considerably improves image quality in situations when fog predominates at night as well as throughout the day. The outcomes demonstrate that the new method significantly improves picture quality, as seen by significant gains in measures like PSNR and SSIM.

Li et al. [6] introduced the research titled in this study, the authors introduce a novel method for single-image dehazing by utilizing improved bright channel prior (BCP) and dark channel prior (DCP) to yield more precise estimations of transmission maps and atmospheric light, ultimately leading to the restoration of clearer, haze-free images. The initial proposition addresses a shortcoming of the DCP method in handling sky regions. To overcome this, the authors propose Otsu segmentation using particle swarm optimization (PSO) to separate hazy images into sky and non-sky regions. Subsequently, parameter estimation is performed utilizing the improved

BCP and DCP methods. To further enhance parameter fusion between BCP and DCP, weighted fusion functions are proposed. These functions contribute to more accurate transmission map and atmospheric light estimations, respectively.

Huang et al. [7] have presented a novel approach titled "Enhancing Automated Driving Image Clarity through Enhanced Dark Channel Prior Based Defogging." The authors of this paper suggest a novel method to enhance the defogging process of dark channel images. This approach includes automatic color equalization, quick bilateral filtering for transmittance optimization, and adaptive domain dark channel computing. Experimental outcomes validate the effectiveness of this method. The suggested technique has outstanding features in terms of picture clarity and general brightness. Moreover, enhancements are observed in mitigating undesirable halo effects and block artifacts. The method also succeeds in maintaining favorable image contrast, color saturation, and realism. As a consequence, this method offers a robust solution that holds promise for improving image detection and processing within automated driving systems.

Zhu et al. [8] have introduced an innovative approach titled the phrase "Dehazing Algorithm Based on Atmospheric Light Estimation for Remote Sensing Images." The aim of this work is to enhance the visual quality of the haze-affected remote sensing photographs. The initial step for creating scene depth maps for remote sensing photographs is to optimize the parameters of a linear scene depth model using a differentiable function. The following scene depth map is used to approximate the atmospheric light present in each foggy remote sensing picture. The researchers then successfully remove the haze from the images using an atmospheric scattering model. The researchers put up a dataset of 100 remote sensing photographs taken under cloudy conditions to test their methodology. The effectiveness of the suggested dehazing method is supported by thorough comparison studies as well as theoretical research.

Ngo et al. [9] introduced a novel and straightforward approach for eliminating haze in single images, specifically designed for real-time vision-based systems. The authors presented an image enhancement-oriented method that effectively removes haze while producing satisfactory results. Their strategy is based on the finding that haze tends to mask picture details and boost overall brightness. The authors use a single blurry image as the source to create a series of under-exposed, detail-enhanced photographs to demonstrate their method. The fusion procedure is directed by weight maps that were generated using the dark channel prior (DCP) approach, a well-known haze indicator. The authors use adaptive tone remapping as a post-processing step to improve the findings even further. This process aids in increasing the final image's dynamic range, which enhances visual quality.

3 Methodology

The recommended approach is primarily built upon foundational principles of the dark channel prior method. The schematic representation of our suggested approach is illustrated in Fig. 1. This methodology takes hazy images as its input. Subsequently, it disentangles the color constituents and evaluates both the transmission map and atmospheric lighting based on these constituents. Notably, to enhance the precision of detection, we incorporate the type-2 fuzzy set Canny edge detection technique. The subsequent section provides a comprehensive elucidation of our proposed method.

In the following sections, we explain the proposed method's step-by-step workflow.

3.1 Dataset Preparation

Begin by selecting an appropriate dataset for training and evaluation. In this research, the I-Haze and O-Haze datasets, containing a diverse collection of hazy images captured under various atmospheric conditions, are chosen.

3.2 Dark Channel Prior

Dark channel prior (DCP) is a fundamental and widely used technique for image dehazing, designed to estimate and remove the haze present in outdoor images [21]. The DCP method exploits the statistical properties of the dark channel for the evaluation the haze transmission and recovers the scene radiance, resulting in improved visibility and clarity. Below, I will provide a more detailed explanation of the dark channel prior dehazing process:

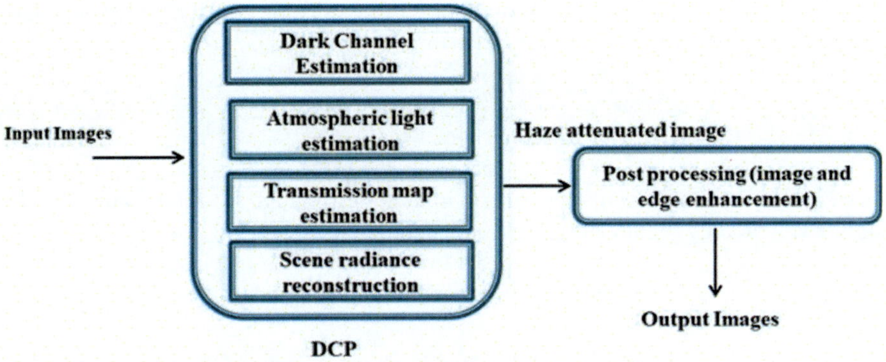

Fig. 1 Block diagram of the proposed method

3.2.1 Dark Channel Calculation

The dark channel of an image is essentially the minimum intensity value within a local window around each pixel. Mathematically, for a pixel at location (x, y), the dark channel is calculated as follows:

$$\text{Dark}(x, y) = \min(\min(R, G, B)), \text{ for all pixels within a local window around}(x, y)$$

Here, the image's red, green, and blue channel values are represented by *R*, *G*, and *B*.

3.2.2 Haze Transmission Estimation

A critical characteristic of outdoor photographs is their dark channel: a non-hazy region is likely to have low intensity values in at least one color channel. The quantity of haze present in the image is indicated by the haze transmission, which is estimated using this attribute. The haze transmission (t) is calculated using the following equation:

$$t(x, y) = 1 - w * \text{Dark}(x, y)$$

In this case, the amount of haze that must be eliminated is governed by the constant w. It is often set empirically or based on an estimate of the average atmospheric light.

3.2.3 Atmospheric Light Estimation

To obtain the scene radiance correctly, it is necessary to estimate atmospheric light (A). An atmospheric light represents a light that directly reaches the camera sensor without being scattered by haze. A common approach is to select the top intensities in the original image corresponding to outdoor regions (sky, bright areas) and calculate the maximum intensity across all color channels.

3.2.4 Scene Radiance Restoration

Once haze transmission and atmospheric light are estimated, the scene radiance (J) can be recovered by means of the following equation:

$$J(x, y) = (I(x, y) - A)/\max(t(x, y), t\_\min) + A$$

Here, $I(x, y)$ represents the hazy input image, and $t\_\min$ is a small value to avoid division by zero issues.

3.3 Type-2 Fuzzy Set

Fuzzy logic is employed for image processing to facilitate edge detection in images. When two regions with consistent intensity levels converge to form an edge, the neighboring pixel intensities play a role in edge identification. However, minor intensity variations between adjacent pixels do not always indicate an edge because defining uniform regions can be challenging. Alternatively, differences in intensity can indicate shading effects. Fuzzy logic, when applied to image processing, allows us to determine the extent to which a pixel belongs to an edge or a uniform area using membership functions.

To detect edges in grayscale images, the widely recognized and efficient Canny edge detection technique is utilized. In cases where inadequate lighting causes image boundaries to become unclear, creating uncertainty in the gradient image, we propose a method. This method automatically selects threshold values for segmenting the gradient within an image using the traditional Canny edge detection approach, incorporating the concept of type-2 fuzzy sets [23, 24]. Our results demonstrate the effectiveness of this algorithm, particularly in enhancing the quality of images affected by foggy conditions.

3.4 Evaluation Metrics

To evaluate the performance of the proposed technique, quantitative analysis is used. As a measure of quality and perceptual accuracy, we use a number of metrics, including Peak Signal to Noise Ratio (PSNR), Structural Similarity Index (SSIM), Lightness Order Error (LOE), and Naturalness Image Quality Evaluator (NIQE).

3.5 Comparative Analysis

The results obtained by proposed method are compared with the results of existing state-of-the-art dehazing techniques. Quantitative metrics are used to demonstrate the efficacy of the proposed method in terms of image quality improvement and haze removal.

4 Experimental Results

The proposed methodology was tested using I-Haze and O-Haze datasets. The outcomes of applying the suggested technique to foggy and hazy images are depicted in Fig. 2. The implementation of the proposed approach was executed using MATLAB R2020a (version 8.1.0.430) on a system equipped with an Intel(R) Core(TM) i5-1035G1 CPU running at 2.60 GHz (with a turbo frequency of 1.19 GHz) and 8 GB of RAM. The computed performance times for sample images are given in Table 1. The reconstructed images notably retain finer borders that were obscured in the hazy counterparts. It is worth noting that the proposed method effectively restores these intricate details that were obscured by the haze in the original images. However, the resulting output images exhibit reduced brightness compared to the input hazy images (Fig. 4).

The achieved results are subjected to a comparative analysis with the outcomes obtained by employing the methods presented by He [26] and Chen [21] from the existing literature. The visual outcomes for selected sample images are displayed in Fig. 3. To gage the efficacy of the different approaches, we employ both the Peak Signal to Noise Ratio (PSNR), Structural Similarity Index Measure (SSIM) as quality

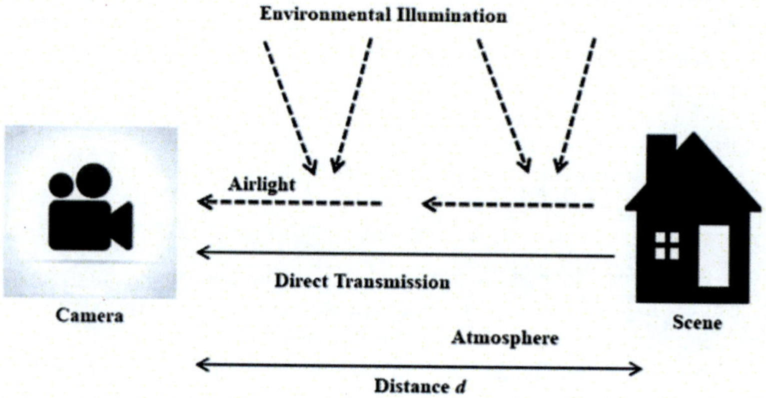

Fig. 2 Refraction of light from air particles

Table 1 Comparative analysis of the suggested approach with other methods

Method	Images	Execution time (s)	PSNR (dB)	SSIM
He [26]	1	9.5	13.8	0.5
	2	13.6	17.4	0.6
	3	6.9	12.3	0.5
	4	13.4	18.6	0.4
Chen [21]	1	9.8	14.1	0.5
	2	22.5	15.2	0.4
	3	5.8	11.6	0.5
	4	12.9	12.3	0.4
DCP	1	7.2	16.9	0.9
	2	8.2	18.8	0.8
	3	5.9	15.2	0.8
	4	10.8	21.2	0.8

(a) Original images

(b) Results of applying Type 2 Fuzzy set to enhance the edges in the images

(c) Dehazed images obtained by applying proposed method

Fig. 3 Results of proposed method on sample images from I-Haze and O-Haze datasets

metrics, Lightness Order Error LOE, and the Naturalness Image Quality Evaluator (NIQE).

PSNR, traditionally utilized in image compression, serves as an indicator of the value of the reconstructed image. A higher PSNR value corresponds to superior reconstruction quality. The PSNR is calculated using the equation:

$$\text{PSNR} = 10 \log 10(\text{peakval2})/\text{MSE},$$

The concept of "peakval" in this context refers to the image's highest intensity value, while MSE stands for mean square error.

On other hand, the SSIM delves into capturing how variations in structural information are perceived, particularly focusing on spatially confined or interdependent pixels. This metric is employed to measure the value of images and videos. Images that are similar to the originals and reconstructed are quantified by SSIM and calculated as:

$$\text{SSIM}(x, y) = [l(x, y)]^{\alpha}.[c(x, y)]^{\beta}.[s(x, y)]^{\gamma}$$

The luminance, the contrast, and the structure information are represented by l, c, and s, respectively.

(a) Haze transmission image on K.He method

(b) Haze transmission image on K Chen method

(c) Haze transmission image on proposed method

Fig. 4 Results on transmission image

Here l represents luminance, c is contrast, and s denotes structural information. The positive constants α, β, and γ play a role in the calculation.

A comprehensive comparison is presented in Table 1, contrasting the presentation of the proposed methodologies with those of alternative approaches available in the literature. This comparison is based on execution time, PSNR, and SSIM parameters. Visualizations of PSNR and SSIM values are depicted graphically in Fig. 5.

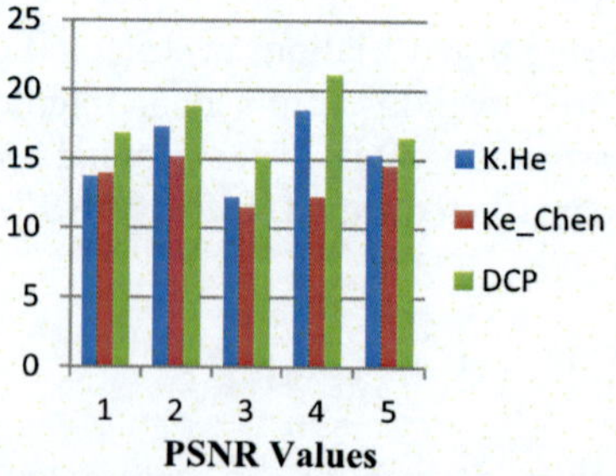

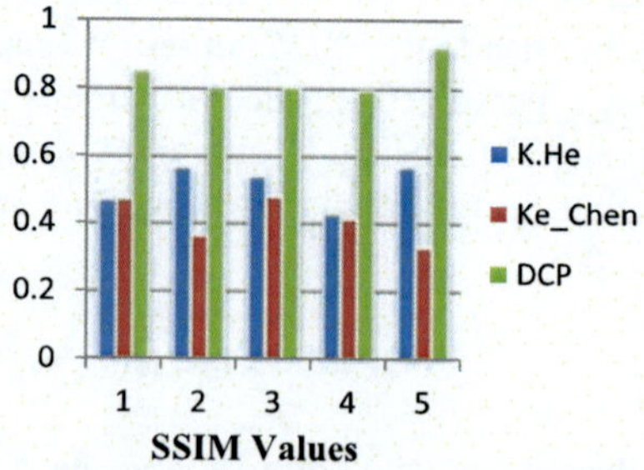

Fig. 5 Comparison of methods on the basis of PSNR and SSIM quality parameters

Based on the analysis conducted, it has been determined that the preservation of naturalness is contingent upon the relative order of lightness [27]. In order to provide an objective assessment of naturalness preservation, we propose the utilization of the LOE measure. The relative order of lightness is indicative of both the direction of the light source and the variation in lightness. As such, the naturalness of an enhanced image is dependent upon the relative order of lightness in various local areas. To this end, we have defined the LOE measure as a quantitative assessment of the Lightness Order Error between the original image, I, and its enhanced version, I_e. The lightness, $L(x, y)$, of an image is determined by the maximum value of its three color channels.

$$L(x, y) = \begin{matrix} \max I^c(x, y) \\ c \in \{r, g, b\} \end{matrix} \tag{1}$$

The relative order difference of the lightness between the original image I and its enhanced version I_e is defined for every pixel (x, y) in the following manner:

$$\mathrm{RD}(x, y) = \sum_{i=1}^{m} \sum_{j=1}^{n} (U(L(x, y), L(i, j)) \oplus U(L_e(x, y), L_e(i, j))) \tag{2}$$

$$U(x, y) = \begin{cases} 1, & x \geq y \\ 0, & \text{else} \end{cases} \tag{3}$$

In the context of this study, the variables "m" and "n" represent the height and width, respectively. The function "$U(x, y)$" denotes the unit step function, while the symbol "$\oplus$" signifies the exclusive OR operator.

Based on the definition of LOE, it is evident that a lower LOE value corresponds to a higher degree of preservation of the lightness order.

NIQE, which stands for Naturalness Image Quality Evaluator, is a metric used to assess the quality of natural images. It is designed to evaluate how similar an image is to typical, natural scenes. NIQE is based on the premise that natural images tend to have certain statistical properties that can be measured to assess their quality. It is often used as a no-reference image quality assessment metric, meaning it can evaluate image quality without the need for a reference (i.e., it does not require a high-quality version of the same image for comparison).

The NIQE metric is calculated using the following formula:

$$\mathrm{NIQE}\sqrt{\frac{1}{N}x^2 \sum_{k=1}^{N} \left(\frac{X_k - \mu_k}{\sigma_k} \right)^2}$$

where

- N is the number of local patches in the image.
- X_k is the feature vector of the k-th local patch.
- μ_k is the mean of the feature vector X_k across all patches.

Table 2 Comparison of NIQE and LOE for sample images of I-Haze and O-Haze datasets

NIQE		LOE
Original image	Dehazed image	
3.6	**2.9**	**0.026**
3.1	**2.8**	**0.177**
3.6	**3.0**	**0.002**
4.1	**2.9**	**0.017**

- σ_k is the standard deviation of the feature vector X_k across all patches.

The idea behind NIQE is to compute the statistical similarity between the local patches in an image and the statistics of natural scenes. Higher NIQE scores indicate lower image quality, while lower scores indicate higher image quality. It is a useful tool in various image processing and computer vision applications for assessing image quality in quantitative method (Table 2).

5 Conclusion

The realms of photography and image processing have undergone substantial expansion. Enhanced by potent lenses and electronic devices, photograph quality has vastly improved. Nevertheless, atmospheric fog or snow can still detrimentally impact visual quality in photographs and films. The system proposed in this paper employs a haze reduction technique aligned with the dark channel prior. In comparison with the prevailing techniques documented in the literature, our proposed approach yielded better results in terms of image quality by swiftly generating haze-free visuals.

References

1. Archana S, Abiraha A (2016) An effective haze removal algorithm using color attenuation prior model. In: International conference on emerging engineering trends and science (8), pp 387–392
2. Yu F, Qing C, Xu X, Cai B (2016) Image and video dehazing using view-based cluster segmentation
3. Park Y, Kim T-H (2018) Fast execution scheme for dark-channel-prior outdoor video dehazing. IEEE ACCESS 6:10003–10014
4. Yu X, Xie W, Yu J (2022) A single image deblurring approach based on a fractional order dark channel prior. Int J Appl Math Comput Sci 32(3):441–454. https://doi.org/10.34768/amcs-2022-0032
5. Hari U, Bevi AR (2022) Dark and bright channel priors for haze removal in day and night images. Intell Autom Soft Comput 34(2):957–967
6. Li C, Yuan C, Pan H, Yang Y, Wang Z, Zhou H, Xiong H (2023) Single-image dehazing based on improved bright channel prior and dark channel prior. Electronics 12:299. https://doi.org/10.3390/electronics12020299

7. Huang S, Qin H, Li Q, Yuan H (2022) A new method for automated driving image defogging based on improved dark channel prior. J Phys Conf Ser 2246. In: 2022 8th International symposium on sensors, mechatronics and automation system, 14/01/2022–16/01/2022
8. Zhu Z, Luo Y, Wei H, Li Y, Qi G, Mazur N, Li Y, Li P (2021) Atmospheric light estimation based remote sensing image dehazing. Rem Sens 13:2432. https://doi.org/10.3390/rs13132432
9. Ngo D, Lee S, Nguyen Q-H, Ngo TM, Lee G-D, Kang B (2020) Single image haze removal from image enhancement perspective for real-time vision-based systems. Sensors 20:5170. https://doi.org/10.3390/s20185170
10. Manjunath V, Phatate R (2016) A single image haze removal algorithm using color attenuation prior. Int J Sci Res Publ (6):291–297
11. Zhang J, Cao Y, Wang Z (2015) Nighttime haze removal with glow and multiple light colors. In: 2015 IEEE international conference on computer vision
12. Gopika M, Sirisha M (2017) Visibility enhancement of hazy image using depth estimation concept. IRJET 4(7):3300–3305
13. Pal NS, Lal S, Shinghal K (2018) Visibility enhancement of images degraded by hazy weather conditions using modified non-local approach. Optik 163:99–113. ISSN 0030-4026. https://doi.org/10.1016/j.ijleo.2018.02.067
14. Lee S, Yun S, Nam JH et al (2016) A review on dark channel prior based image dehazing algorithms. J Image Video Proc 2016:4. https://doi.org/10.1186/s13640-016-0104-y
15. Gupta N, Jha RK, Mohanty SK (2014) Enhancement of dark images using dynamic stochastic resonance in combined DWT and DCT domain
16. Shi J, Yang K (2016) An improved method of removing fog and haze effect from images. In: 4th International conference on machinery, materials and information technology applications (ICMMITA 2016) Copyright © 2017. Published by Atlantis Press
17. Guo F, Tang J, Cai Z-X (2014) Image dehazing based on haziness analysis. Int J Autom Comput 11(1):78–86. https://doi.org/10.1007/s11633014-0768-7
18. Shi Z, Zhu MM, Guo B, Zhao M, Zhang C (2018) Night time low illumination image enhancement with single image using bright/dark channel prior. EURASIP J Image Video Process 13:1–15
19. Kim W, Lee R, Park M, Lee S (2019) Low-light image enhancement based on maximal diffusion values. IEEE Access 7:129150–129163
20. Lee H, Sohn K, Min D (2020) Unsupervised low-light image enhancement using bright channel prior. IEEE Sig Process Lett 27:251–255
21. Ke N, Chen J (2013) Real-time visibility restoration from a single image. In: Proceedings on IEEE international conference on image processing, pp 923–927
22. Biswas R, Sil J (2012) An improved canny edge detection algorithm based on type-2 fuzzy sets. Procedia Technol 4:820–824. ISSN 2212-0173. https://doi.org/10.1016/j.protcy.2012.05.134
23. Eshwarappa L, Rajput GG (2022) Type-2 fuzzy sets-canny edge detection algorithm-based text extraction from complex video scene. Harbin Gongye Daxu Xuebao/J Harbin Inst Technol 54(4):76–83
24. Dong T et al (2019) Efficient traffic video dehazing using adaptive dark channel prior and spatial–temporal correlations. Sensors 19(7):1593
25. Yan Y et al (2017) Image deblurring via extreme channels prior. In: Proceedings of the IEEE conference on computer vision and pattern recognition
26. He K, Sun J, Tang X (2011) Single image haze removal using dark channel prior. IEEE Trans Pattern Anal Mach Intell 33(12):2341–2353. https://doi.org/10.1109/TPAMI.2010.168
27. Li B, Wang S, Geng Y (2011) Image enhancement based on Retinex and lightness decomposition. In: Proceedings of IEEE international conference on image processing, Sept 2011, pp 3417–3420

ECO-Guard: An Integrated AI Sensor System for Monitoring Wildlife and Sustainable Forest Management

Ch. Nikhilesh Krishna, Avishek Rauniyar, N. Kireeti Sai Bharadwaj, Sujay Bharath Raj, Vipina Valsan, Kavya Suresh, V. Ravikumar Pandi, and Soumya Sathyan

1 Introduction

Due to increased farming and wildfires, the world's forests, critical for biodiversity and climate control, are in danger of collapse. This is harmful to both the environment and wildlife. Wildlife populations have been declining since 1970, indicating the state of the ecosystem. These issues are highlighted in the Living Planet Index, which uses data from 30,000 wildlife communities. Italy's forest fires, which have burned 723,924 ha in 14 years, reflect a global trend. The problem was highlighted by the burning of 159,437 ha in 2021 alone [1]. Interactions with carnivores can also be dangerous, with casualties occurring over a seven-decade period. These issues differ according to region and activity. Deeper forest exploration, such as tourism, increases the risk of being lost. Innovative solutions to environmental degradation, wildlife threats, and human vulnerability are required.

This study adds to the important global issues covered by the Sustainable Development Goals (SDGs) of the United Nations. The SDGs 12 (Responsible Consumption and Production), 13 (Climate Action), 15 (Life on Land), 9 (Industry, Innovation, and Infrastructure), 11 (Sustainable Cities and Communities), and 17 (Partnerships for the Goals) are precisely aligned with and help to accomplish this goal. The important topics of environmental preservation, climate mitigation, and sustainable development are all covered by these interconnected SDGs. This study reflects a comprehensive approach to address the complex and interlinked obstacles facing

Ch. Nikhilesh Krishna · A. Rauniyar · N. Kireeti Sai Bharadwaj · S. B. Raj
Department of Computer Science and Engineering, Amrita School of Computing, Amrita Vishwa Vidyapeetham, Amritapuri, India

V. Valsan (✉) · K. Suresh · V. Ravikumar Pandi · S. Sathyan
Department of Electrical and Electronics Engineering, Amrita School of Engineering, Amrita Vishwa Vidyapeetham, Amritapuri, India
e-mail: deepthi@am.amrita.edu

M. S. Kaiser et al. (eds.), *ICT: Smart Systems and Technologies*, Lecture Notes in Networks and Systems 878, https://doi.org/10.1007/978-981-99-9489-2_36

our planet by focusing on the effective use of Internet of Things (IoT) technology in forest conservation, animal protection, and tourism. We aim to build a sustainable future that fits the demands of people with the preservation of Mother Nature via innovation, collaboration, and responsible resource management.

Using Internet of Things (IoT) technology, this study suggests an effective approach to improve monitoring and protection activities. Using IoT we can monitor movements in forest [2]. IoT uses interconnected devices and sensors to collect and exchange valuable information, enabling real-time forest monitoring and early hazard detection. This methodology highlights the vital role of network architecture, scalability, and dependability in ensuring effective data transfer like Wi-Fi [3] and LoRa [4]. Furthermore, it investigates various network topologies to enhance forest observation. The study recognizes the transformative potential of deep learning in computer science. With camera imaging, we can detect various objects like forest fire detection [5] and wildlife [6]. With computer vision, there can be easy forest surveillance [7]. Deep learning models can detect a variety of objects like fire [8] as well as animals [9]. Various machine learning algorithms can detect and send various unusual activities in forest for sensor readings [10]. YOLO is a popular machine learning algorithm that uses a single neural network to perform both classification and prediction of bounding boxes for detected objects. As such, it is heavily optimized for detection performance and can run much faster than running two separate neural networks to detect and classify objects separately. We can track the object after detecting them and get real-time updates of the object [11].

In conclusion, this article presents an integrated approach as shown in Fig. 1 that makes use of IoT technology, computer vision (YOLO), sensor networks, and modern tourist guidance system [12] to address the variety of difficulties facing our forests, wildlife, and tourists for the preservation and promotion of biodiversity. The effort provides an integrated plan for protecting the most priceless resources in our world, with a special focus on reducing environmental degradation, promoting safety, and maintaining biodiversity.

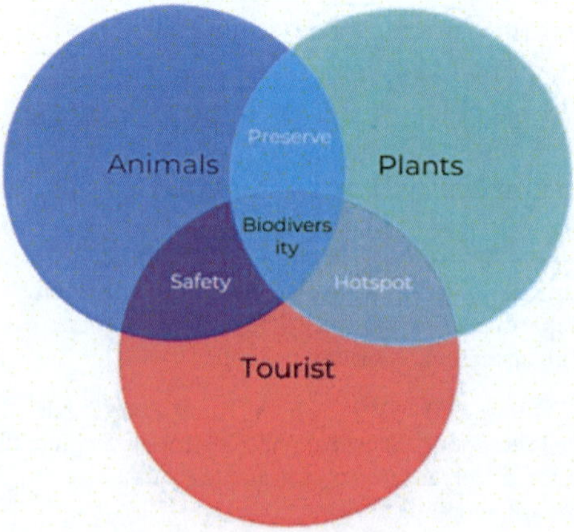

Fig. 1 Relationship between the targeted entities

2 Problem Statement

The forestry department faces many challenges due to unmanaged forests, including the declining forest resources driven by illegal activities and fires. These issues are exacerbated by unplanned reserves, resulting in various problems. These challenges encompass the increasing proximity of wild animals to humans, necessitating early detection to prevent potential harm. Poor navigation within nature reserves leads to unauthorized exploration, while combating illegal activities in vast forests requires real-time information for prompt response. Risk-taking despite warning signs highlights the need for monitoring and enforcement measures. Additionally, there is a growing need for studying animal behavior and habitats using technology, benefiting forest rangers in making informed decisions. Wild animal attacks require preventive measures, and addressing low connectivity in remote forests is crucial for communication and emergency response. Lastly, preventing unauthorized fires and halting the escalation of forest fires demand early detection and alert systems. These multifaceted challenges as shown in Fig. 2 describe the correlated aspects of ECO-Guard that underscore the importance of comprehensive and technologically advanced solutions for sustainable forest management.

3 Methodology

The methodology of the proposed ECO-Guard is shown graphically in Fig. 3. The details of the flow process are explained in this section from collection of data using intelligent sensors, detecting the beasts and communicating the same with navigation methods and setting the alarm system with premature stopping systems.

3.1 Collection of Biodata

By strategically positioning environmental sensors, the system captures and analyzes the precise routes and rhythms of wildlife motion. The setup of a Wi-Fi hotspot using satellite internet or other connectivity sources like cable wires allows multiple devices to connect to the internet within a limited range. These then allow tracked motion patterns, ranging from daily routines to the very seasonal migrations to be monitored. Augmented with a mesh-type architecture, any specific beast could be hand tracked and position could be under radar. The project capability extends benefits in informed policy design wherein any of policies are taken in interests of the exact habits and requirements of wild and wildlife. For instance, areas which witness high concentrations of seasonal bird migrations can be designated as protected zones during the critical migration periods and could be made sure no habitat degradation with regrowth of vegetation if exploited any during post migratory period. Researchers

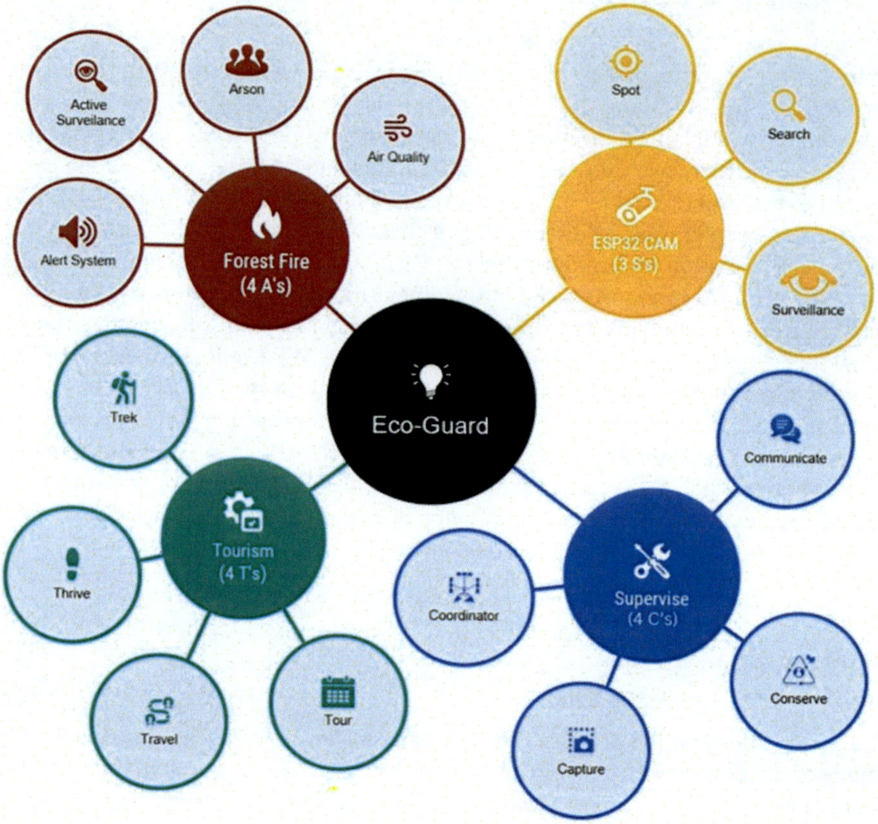

Fig. 2 Correlated aspects of ECO-Guard

possess the flexibility to tailor capture points according to their research objectives, catering to specific interests such as vegetation or wildlife, whether investigating plant life or observing wildlife behaviors.

3.2 *Beast Detection and Communication*

The device uses a camera module which captures real-time images using OpenCV library, and the machine learning model (YOLO) is applied to detect the presence of animals and various objects in the area. The machine learning-based CNN ML model classifies the objects, and then it shows us the object with the label [13] as shown in the above picture. Also, with this, the model is developed to detect any unusual activities in the forest. The warnings are shown according to various grouping of various objects such as:

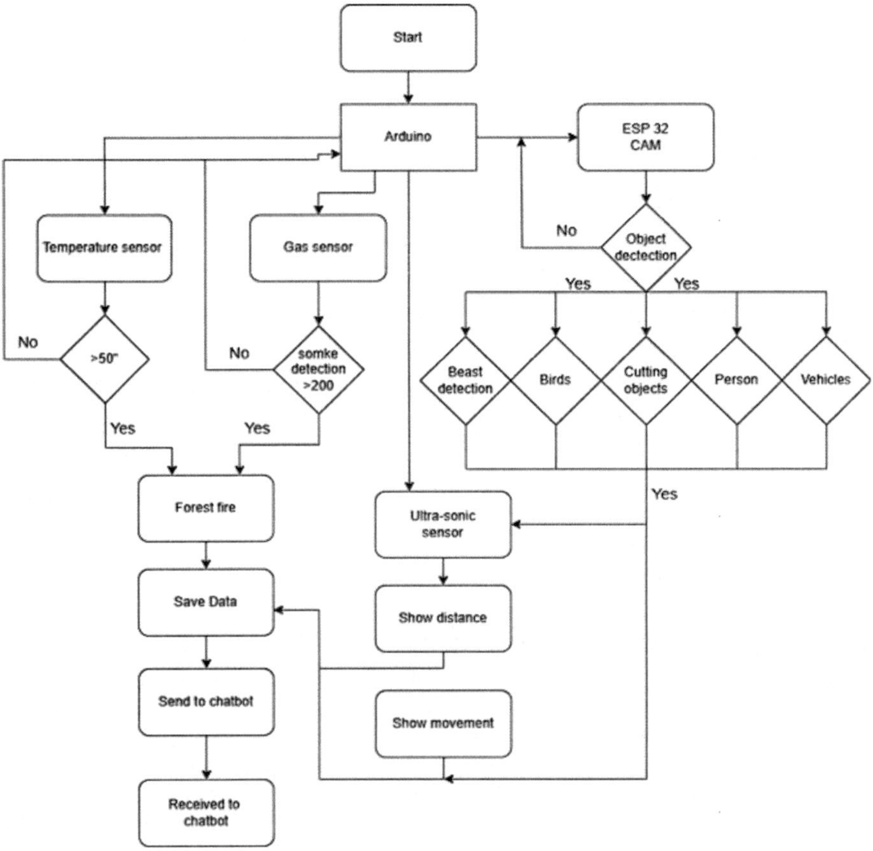

Fig. 3 Flowchart of working of ECO-Guard

['beasts', 'birds', 'person', 'vehicle', 'signs', 'cutting_objects', 'danger'.]

This means that whenever lion or cheetah is encountered in cam then the data is transmitted in the same ways as of giraffe or elephant is sent to set some warning of some priority order. Similarly, all the grouping characteristics work as a part of algorithm. The network of various devices is created and divided into certain areas for pinpointing of various devices. A unique id element is assigned to all devices, and a mesh-like transmission Bluetooth media (in a zone of no internet connection) is created to transfer the data from places to another, i.e., to a device that will transmit the data through web services to be used by all the entities. Whenever the device camera detects an unusual moment in forest or any wildlife or any harmful object groups then the data of the event with the device id and area data will be transmitted to a telegram bot. The area and id of device will be pre-fitted with the device which will be communicated through some mesh topology structure via Bluetooth technology or Wi-Fi to a device that can send the data online. The whole data is sent to the server or forest department to analyze the vulnerability of the event. The data that the department wants to be displayed to the server will be transmitted to a telegram channel by the bot itself to get the entities informed about the event by providing minimal data to the other entities. The action focuses on integration which mixes any landmass with surveillance to include, for example, anti-poaching effort with tourist community management.

3.3 Navigating Methods

The paper encompasses the study of wildlife behavioral intricacies through efficient tracking facilitated by a 180° servo motor, ensuring comprehensive monitoring and motif of movement. Utilizing sophisticated algorithms akin to flight radar systems, the system meticulously could capture every movement [14]. Through continuous data assessment, this approach could become more prominent in research data collection. The immense threat of losing track of the way is common to users especially in areas of wild or to put even in the villages. The working of our model could be extended in surveillance capabilities by the will of the user if they feel the risk of getting lost. The network of the devices could facilitate navigation which could provide the user with their area location by using the information of the nearest device implanted [15]. The device could also work as a helping hand if the user intends to be directed to a targeted location which our system taking in the ids of their current location and destination and by using in the smallest path algorithm could provide the destination route.

3.4 Forest Fire Alerting and Premature Stopping System

It is essential to keep an eye on forest fires. Smoke and gases from combustion are detected by the MQ2 gas sensor as shown in Fig. 4. When smoke levels are high, it sets off alarms. When combined with the LM35 temperature sensor, it helps in early fire detection and understanding of weather patterns [16]. The LM35 provides reliable temperature data. Overall, the MQ2 gas sensor and LM35 temperature sensor work together to monitor forest fires. They improve reliability and provide early detection by requiring both sensors to trigger alerts at the same time. This system is necessary for environmental protection and climate monitoring.

As the device starts, the Arduino starts computing the signal values received from various connected sensors and makes decisions for alerting to the authorities with the reading of different sensor values. The Arduino will identify high temperature value, then it sends the alert to the authorities that there can be forest fire. The gas sensor value is also taken to say whether forest fire has occurred or not. Generally, the smoke value lies within the range 50–80 varying with the air quality of the region. In case of fire, the temperature will increase and go to more than 50 and smoke value

Fig. 4 Hardware of ECO-Guard

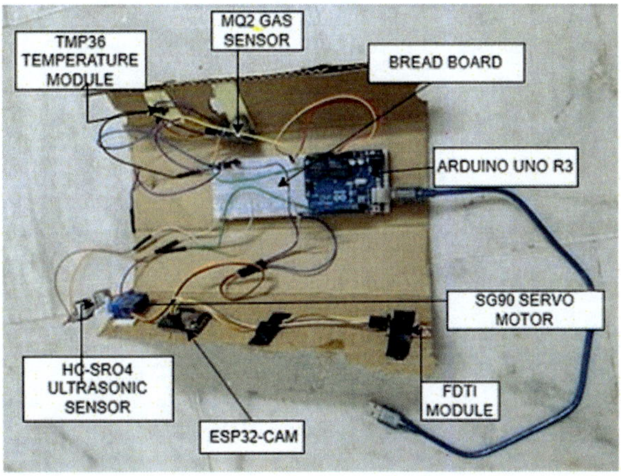

greater than 200 which confirms the occurrence of forest fire in the region. ESP32 cam used as an imaging device, tries to detect the objects, classifies it with the use of YOLO, and checks the ultrasonic sensor reading value which is the distance between the object and the device. Then all the abnormal data are stored but Arduino and sent to a telegram channel using telegram service. Now, the authorities can view the data, store data, and use it for surveillance.

4 Results and Discussions

4.1 Fire and Smoke Detection and Distance Measurement

We have used three sensors in our model, namely ultrasonic, gas, and temperature sensors as presented in Fig. 4. The sample readings are as shown in Table 1. The values were received in Arduino IDE when the device was tested. Whenever the ultrasonic sensor received some unusual value, we received warnings according to the sensor such as object detected as shown above. Similarly, when the gas value starts to increase temperature value also got increased which shows the alert message for fire detection in the forest. The device has detected some abnormal changes both in gas sensor and temperature sensor, so a message will be sent to the server indicating an ensured forest fire.

The device consists of an Arduino microcontroller that handles all the components used in the device. MQ2 gas sensor that gives gas values in ppm (parts per million) and helps in detecting smoke together with TMP36 temperature sensor is used to measure the temperature and say whether fire has occurred in a place or not. FDTA is a programmer kit used to program and control the ESP32 cam module which captures real-time images and sends them to the authorities by classifying vulnerability with

Table 1 Sample sensor readings of ECO-Guard in Arduino IDE

S. No.	Distance (cm)	Gas sensor value (ppm)	Temperature (°C)	Remarks
1	8	46	31.12	Object detected (close)
2	30	45	31.42	Object detected (close)
3	300	117	64.28	Object detected Smoke detected
4	2312	201	78.63	Possible case of fire
5	2391	252	98.23	Possible case of fire
6	2392	272	101.25	Possible case of fire
7	2393	290	125.16	Possible case of fire
8	2394	310	149.62	Possible case of fire
9	2393	326	149.6	Possible case of fire
10	2394	345	149.62	Possible case of fire

machine learning algorithms. When any object is detected in camera, the ultrasonic sensor sends the distance of the object to the authority, and with the use of ultrasonic sensor, the object is traced to get more valuable information.

4.2 *Person and Cutting Objects*

Case 1: Beast or Danger and Person: "Alert! Person and Beast Detected Together."

When person and beast are detected together, there can be a threat either to the person or the animal. In this case, when the camera detects both objects the message will be automatically generated and sent to the authorities as shown in Fig. 5.

Fig. 5 Case I. Beast or danger and person (alert message obtained)

Case 2: Signs and Person: "Alert! Person is spotted in proximity of Danger Zone."

When person and sign are detected together, there can be a threat as person is trespassing and going into danger area. In this case, the message will be automatically sent to the authorities as shown in Fig. 6.

Case 3: Cutting Objects and Person: "Alert! Possible Logging."

When person and cutting objects are detected together, there is a threat the person might be in for some illegal activities. In this case, the message will be automatically sent to the authorities as shown in Fig. 7. The device which can detect various pre-programmed cases provided a warning through a message automatically generated in telegram by the device as shown above.

Fig. 6 Case II. Beast or danger and person (alert message obtained)

Fig. 7 Case III. Person and cutting objects detected (alert message obtained)

5 Conclusion

In conclusion, this paper presents a comprehensive system that combines sensor technologies, wireless communication, and AI models to revolutionize tourism, wildlife conservation, and environmental monitoring. By integrating devices carried by tourists with environmental sensor nodes, real-time data on wildlife movements, population densities, tree cutting, and forest fires can be collected and analyzed. The use of AI models enhances data processing and enables informed decision-making. This work contributes to the SDGs 12, 13, and 15 thereby promoting responsible consumption and production, climate action, and the preservation of life on land. Through ECO-Guard's innovative approach of forest surveillance, it can transform the way we experience tourism while fostering a chance for greater understanding and conservation of our natural resources and endangered species. The technology pushes wildlife into rich governance with less physical interference and conceals habitat disruption.

Acknowledgement We would like to convey our gratitude to Amrita Vishwa Vidyapeetham and Chancellor Sri Mata Amritananda Mayi Devi for providing us a chance to showcase our efforts toward sustainable forest ecosystem.

References

1. Embargo 13–10–2022_LPR 2022 full 2 (panda.org)
2. Saad W, Alsayyari AS (2020) Real time IoT based camel-vehicle collision avoidance system for KSA. In: 2020 3rd international conference on computer applications and information security (ICCAIS), Riyadh, Saudi Arabia, pp 1–4. https://doi.org/10.1109/ICCAIS48893.2020.9096714
3. Wang Z, Zhang JA, Xu M, Guo YJ (2023) Single-target real-time passive WiFi tracking. IEEE Trans Mob Comput 22(6):3724–3742. https://doi.org/10.1109/TMC.2022.3141115
4. Bandari G, Devi PLN, Srividya P (2022) Wild animal detection using a machine learning approach and alerting using LoRa communication. In: 2022 International conference on smart generation computing, communication and networking (SMART GENCON), Bangalore, India, pp 1–5. https://doi.org/10.1109/SMARTGENCON56628.2022.10083577
5. Srishilesh PS, Parameswaran L, Tharagesh RSS, Kumar TS, Sridhar P (2019) Dynamic and chromatic analysis for fire detection and alarm raising using real-time video analysis. In: Proceedings of 3rd international conference on computational vision and bio inspired computing, Cham
6. Balaji SA, Geetha P, Soman KP (2016) Change detection of forest vegetation using remote sensing and GIS techniques in Kalakkad Mundanthurai Tiger Reserve—a case study. Indian J Sci Technol 9(30). https://doi.org/10.17485/ijst/2016/v9i30/99022
7. Yang T, Zhang H, Li Y, Ma Z, Huang R, Li S (2017) The visual simulation technology in formatting forest management plan at unit level based on WF. In: 2017 2nd International conference on image, vision and computing (ICIVC), Chengdu, China, pp 739–745. https://doi.org/10.1109/ICIVC.2017.7984654
8. Sanjana S, Premkumar A, Sneha CK, Sony J, Geetha M (2022) Deep learning models for fire detection using surveillance cameras in public places. In: 2022 13th International conference

on computing communication and networking technologies (ICCCNT), Kharagpur, India, pp 1–7. https://doi.org/10.1109/ICCCNT54827.2022.9984601

9. Kaur A, Jadli A, Sadhu A, Goyal S, Mehra A, Rahul (2021) Cloud based surveillance using ESP32 CAM. In: 2021 International conference on intelligent technology, system and service for internet of everything (ITSS-IoE), Sana'a, Yemen, pp 1–5. https://doi.org/10.1109/ITSS-IoE53029.2021.9615334
10. Phan A, Takejima K, Hirakawa T, Fukui H (2022) Forest-related SDG issues monitoring for data-scarce regions employing machine learning and remote sensing—a case study for Ena City, Japan. In: IGARSS 2022—2022 IEEE international geoscience and remote sensing symposium, Kuala Lumpur, Malaysia, pp 4343–4346. https://doi.org/10.1109/IGARSS46834.2022.9883037
11. Mondal K, Padmavathi S (2019) Wild animal detection and recognition from aerial videos using computer vision technique. Int J Emerg Trends Eng Res 7(5)
12. Gehlot A, Singh R (2022) Research guide for ML based smart tourist system. In: 2022 International interdisciplinary humanitarian conference for sustainability (IIHC), Bengaluru, India, pp 1427–1434. https://doi.org/10.1109/IIHC55949.2022.10059630
13. Megalingam RK, Babu DHTA, Sriram G, Yashwanthavvari VS (2021) Concurrent detection and identification of multiple objects using YOLO algorithm. In: 22nd Symposium on image, signal processing and artificial vision, STSIVA 2021. https://doi.org/10.1109/STSIVA53688.2021.9592012
14. Krishna NM, Reddy RY, Reddy MSC, Madhav KP, Sudham G (2021) Object detection and tracking using Yolo. In: 2021 Third international conference on inventive research in computing applications (ICIRCA), Coimbatore, India, pp 1–7. https://doi.org/10.1109/ICIRCA51532.2021.9544598
15. Sarkar JL, Majumder A, Ramasamy V, Panigrahi CR, Roy S (2022) MERIT: multiitinerary tourist recommendation engine for industrial internet of things. In: IEEE international conference on computer communications (IEEE INFOCOM), May 2022. https://doi.org/10.1109/INFOCOMWKSHPS54753.2022.9798002
16. Sineva IS, Molovtsev MD (2020) An integrated approach to the regression problem in forest fires detection. In: 2020 Systems of signals generating and processing in the field of on board communications, Moscow, Russia, pp 1–4. https://doi.org/10.1109/IEEECONF48371.2020.9078645

Machine-Learning-Based Diagnosis of Mental Health Issues

Sonali Chopra, Parul Agarwal, Jawed Ahmed, and Ahmed J. Obaid

1 Introduction

The present modifications to the social environment have played a role greatly in the rise in the prevalence of mental well-being issues and psychological illnesses [1]. The WHO defines "mental health" as an individual's capability for managing stresses throughout one's life while still working regularly as well as successfully, and contributing to the community [2]. Professional tension, poor economic status, marital issues, family troubles, and assault, as well as surrounding-related variables, are likely to have an impact on psychological well-being [3].

1.1 Need for Assessing Our Mental Well-Being

Stress in the mind has emerged as a problem of society in the twenty-first century. It has an impact on the functioning of ordinary labor as well as the economics of both

S. Chopra · P. Agarwal (✉) · J. Ahmed
Jamia Hamdard, New Delhi, India
e-mail: pagarwal@jamiahamdard.ac.in

S. Chopra
e-mail: Sonalichopra712@gmail.com

J. Ahmed
e-mail: jahmed2047@jamiahamdard.ac.in

A. J. Obaid
University of Kufa, Kufa, Iraq

Al-Ayen University, Thi-Qar, Iraq

A. J. Obaid
e-mail: ahmedj.aljanaby@uokufa.edu.iq

M. S. Kaiser et al. (eds.), *ICT: Smart Systems and Technologies*, Lecture Notes in Networks and Systems 878, https://doi.org/10.1007/978-981-99-9489-2_37

individuals and nations [4]. Job tension, meeting deadlines, testing, and other factors all contribute to enthusiastic tension [5]. These are many stress-initiating factors [6]. Scholars believe that the cerebrum of human beings is the principal source of tension [7, 8]. Inquiries are undertaken, and people are sometimes monitored, to research the impact of stress on various people [9]. Self-reported surveys are commonly employed to periodically assess an individual's stress levels [10–12].

At the current moment, efficient uses of machine learning in the detection of mood disorders were primarily driven by research employing neuroscience image information like fMRI scans, as demonstrated in [13–15], with just some machine learning investigations incorporating data gathered through self-report (e.g., in [16]). Current machine learning investigations of personalized bipolar disorder detection have primarily used neurological image information [17, 18].

1.2 Prevalence of Machine Learning

ML—Machine learning is an area of AI—Artificial Intelligence working with 3 issues, namely, regression, clustering, and classification. ML makes use of algorithms as well as data to replicate the way human beings learn when gradually enhancing effectiveness in a variety of activities [19]. Supervised learning constitutes one of the most extensively used ML algorithms for predicting mind-related diseases [20].

DL—Deep learning is one of the branches of ML [21]. Numerous studies and publications have explored the utilization of both deep learning (DL) and machine learning (ML) in enhancing the examination of diverse health concerns following the introduction of AI into the medical domain. This increased significance has extended AI's application in healthcare to the realm of psychological assessment [22].

A comprehensive psychiatric interview is often necessary to diagnose mental health problems. Finding psychiatric symptoms also benefits from the use of psychological testing and evaluation techniques [23].

1.3 Contribution of the Paper

- Numerous studies on the identification of mental illnesses have been conducted. We concentrated our search for this review by covering the major mental health diseases, namely, suicidal thoughts, stress, anxiety, depression, and bipolar disorder.
- To identify prospective research directions, it is essential to conduct a retrospective examination of prior studies, particularly due to the growing enthusiasm surrounding machine and deep learning technologies.
- The diagnoses of mind-related disorders, namely suicidal thoughts, stress, anxiety, depression, and bipolar disorder along with the minor aspects like mixed reality, emotion states (moods), mental health education, pharmacogenomics, precision

psychiatry, chronic disease contracting, body mass index, and wearable sensors were covered in this study by gathering 68 papers using the prominent keywords for searching.

1.4 Structure of the Paper

The structure of our paper is as follows: Sect. 2 describes the importance of mental health issue analysis. In Sect. 3, we have compiled relevant research within the field of mental health issue analysis that leverages machine learning or other advanced methodologies. Section 4 explains the diverse prospects of analyzing the mental health-related concerns. Section 5 discusses the limitations of small sample sizes in AI and ML research. The conclusion and future directions are discussed in Sect. 6.

2 Importance of Mental Health Issue Analysis

According to the World Health Organization (WHO), in 2017, an estimated 322 million individuals globally were afflicted by depressive disorders. Depression stands as the leading contributor to non-fatal health deterioration, and its impact continues to escalate annually [24]. According to [25] a diagnosis of major depressive disorder is established when an individual displays five or more symptoms of depression.

An adapted psycho-physical response to a stressor, sometimes known as a physical, social, or psychological stimulation, is known as stress [26]. Physiological, behavioral, emotional, or cognitive reactions to stress were all possible. Stress can increase the likelihood of developing various diseases, such as mental or cardiovascular conditions, depending on the kind, timing, and intensity of the stressor that was exposed to [27–31].

In order to boost the effectiveness and improve the quality of patient treatment, technology has been increasingly implemented in healthcare settings [32]. At first, it gained momentum slowly, but its adoption accelerated during the latest episode of the COVID-19 pandemic [33, 49]. Wearable sensors have the potential to revolutionize the manner in which we monitor and handle mental well-being in the near future [34, 35].

3 Previous Mental Health Issue Analysis Endeavors

In this section, we have gathered the related research in the genre of mental health issue analysis based on machine learning or other sophisticated techniques and presented under the following subsections.

3.1 Major Mental Health Issue Research

By utilizing PubMed/MEDLINE, PsychInfo, Web-of-Science, and EMBASE, [36] carried out a comprehensive assessment of ML studies assessing suicidal behavior. The PHQ-9's effectiveness in assessing suicidal thoughts was assessed in [37]. Then [38] concentrated on the most recent advancements in ML and AI techniques for precision psychiatric research. A study involving an examination of 28,755 entries retrieved from the Pregnancy Risk Assessment Monitoring System (PRAMS) for the years 2012 to 2013 revealed 3339 occurrences of postpartum depression alongside 25,416 instances of typical postpartum cases, was done by Shin et al. [39] in a retrospective cohort study. Now, [40] examined how the pandemic affected the stress levels of $N = 2053$ Italian people, identifying those who were more at risk based on sociodemographic characteristics and consistent psychological factors. The goal of [41] was to ascertain if voice could serve as a biomarker for detecting both moderate and severe depression. Next, we examined data from a group of 34 controls without post-stroke depression (PSD) who were age-matched, alongside 31 individuals diagnosed with PSD in [42]. The primary ML-based methods for detecting depression and other mental diseases using behavioral data were outlined in [43]. The most recent methods that use ML models for the first diagnosis of chronic illnesses (CDs) were reviewed in [44]. Now, [45] examined how affect-related psychological factors and BMI (i.e., body mass index) connect to one another. For improving SOC predictive designs, [46] offered machine learning techniques SVM, ANN, regression tree, RF, XGBoost (i.e., eXtreme Gradient Boosting), and traditional DNN (i.e., deep neural network).

3.2 Comparison of Some Notable Research

In addition to the above discussed works, we are presenting a few more of them and then compared in Table 1.

4 Diverse Prospects of Mental Health Issue Analysis

Having understood the diverse prospects of analyzing the mental health-related concerns, we are curating some pointers in the below section, which has been illustrated in Fig. 1.

Table 1 Comparison of some notable research focused on mental health issue analysis

Citation	Method adopted/contribution	Focus aspect of the research/study
[47]	LSTM—Long short-term memory and CNN—Convolutional neural network (deep learning)	Suicide Ideation
[48]	Comprehensive investigation of application context of machine learning in multi-variate biological data	Diverse types of data inputs collected by different sensor devices
[49]	Made use of three tools, namely PHQ—Patient Health Questionnaire, GAD—Generalized Anxiety Disorder Assessment, and GDS—Geriatric Depression Scale	Post-COVID-19 depression
[50]	Comprehensive Investigation of machine learning-based antidepressant selection with reference to symptoms	Depressive syndrome
[51]	NLP—Natural Language Processing, ML, and DL	Suicidal ideation
[52]	LSTM and RNN—Recurrent neural networks	Depression
[53]	Machine learning techniques along with correlation investigation	Depressive disorder along with emotion states
[54]	Comprehensive investigation of both the mixed reality and education aspect related to mental health issue	Context of mixed reality

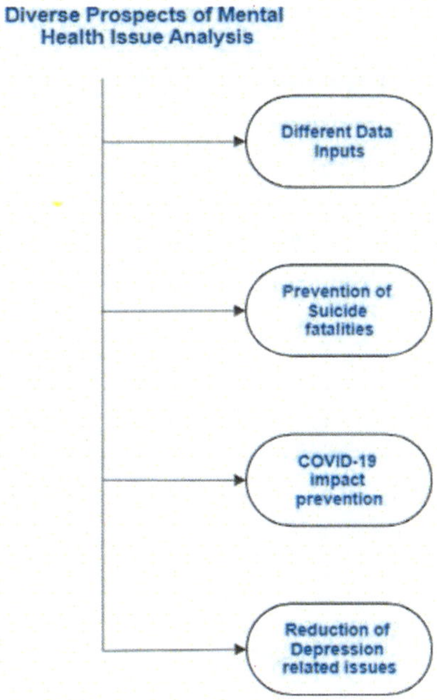

Fig. 1 Pictorial representation of prospects of mental health issue analysis

4.1 Different Data Inputs

A substantial quantum of investigation has focused on the use of ML approaches for obtaining, identifying, and categorizing psychological indicators in sensory datasets [55–60]. This biological sensor information is often a time sequence with many channels if not multiple modes [61]. For one to arrive at a proper detection, these signs usually need professional examination [62].

4.2 Prevention of Suicide Fatalities

Suicidal thoughts are defined as a desire to terminate one's own existence, which can range from sadness to planning an act of suicide to an extreme fixation toward self-harm [63]. Those who have suicidal thoughts (or those who plan to) and those who attempt suicide (or those who complete) are two types of individuals at risk [64]. The rate of mortality was earlier expected to rise to one for every 20s around the year 2020 [65].

5 Challenges

Due to the potential of overfitting during the training stage of AI and ML algorithms, one drawback of the small sample sizes used in these earlier research is that they do not allow for contributing to well-defined findings [66]. The use of global populations is also essential for generalizing these correlations to independent cohorts [67]. However, because large-scale data may not be accessible to facilitate further analysis, the majority of these research investigations did not employ replication data. It's possible that these findings won't be applicable to other situations. The fact that we would generalize to independent cohorts with multiple underrepresented ethnic groups, varied testing circumstances, many recruiting sites, numerous actual clinical situations, and larger global populations in the most appropriate ways is an open problem.

6 Conclusion

We hope that the findings from our study considering the most prevalent mental health ailments like suicidal thoughts, stress, anxiety, depression, and bipolar disorder along with the minor aspects like mixed reality, emotion states (moods), mental health education, pharmacogenomics, precision psychiatry, chronic disease contracting, body mass index, and wearable sensors will help prospective machine learning researchers to construct a readily available and exceptionally efficient detection paradigm utilizing machine learning as a successful communication medium involving mental wellness providers as well as people at risk.

The comprehensive survey that we have done will be insightful for the futuristic works carried out in the research genre of mental health issue analysis based on machine learning or other sophisticated techniques. The following are a few future directions:

- Having known the adverse mind-related consequences caused by the pandemic-like situations (for instance, COVID-19), it is also necessary to know and assess the mental disruptions caused due to the chronic and death resulting diseases.
- The researchers should look for more rich and diverse data inputs whenever conducting the mental illness analysis.
- It is of much importance to construct a practically viable online-based tools that could aid the physicians in the mental health centers using the state-of-the-art classification approaches for avoiding adverse impacts like suicidal deaths.

References

1. Abd Rahman R, Omar K, Noah SA, Danuri MS, Al-Garadi MA (2020) Application of machine learning methods in mental health detection: a systematic review. IEEE Access 8:183952–183964
2. World Health Organization (2004) Promoting mental health: concepts, emerging evidence, practice. Summary report, World Health Organization
3. Abd Rahman R, Omar K, Noah SA, Danuri MS (2018) A survey on mental health detection in Online Social Network. Int J Adv Sci Eng Inform Technol 8(4–2):1431
4. Gupta R, Alam MA, Agarwal P (2020) Modified support vector machine for detecting stress level using EEG signals. Comput Intell Neurosci 2020:1–4
5. Rabasa C, Dickson SL (2016) Impact of stress on metabolism and energy balance. Curr Opin Behav Sci 9:71–77
6. Khan M, Rizvi Z, Shaikh MZ, Kazmi W, Shaikh A (2014) Design and implementation of intelligent human stress monitoring system. Int J Innov Sci Res 10(1):179–190
7. Mantri S, Patil V, Mitkar R (2012) EEG based emotional distress analysis—a survey. Int J Eng Res Dev 4(6):24–28
8. Rigas G, Goletsis Y, Fotiadis DI (2011) Real-time driver's stress event detection. IEEE Trans Intell Transp Syst 13(1):221–234
9. Jun G, Smitha KG (2016) EEG based stress level identification. In: Proceedings of the 2016 IEEE international conference on systems, man, and cybernetics (SMC). IEEE, pp 003270–003274
10. Al-Shargie FM, Tang TB, Badruddin N, Kiguchi M (2016) Mental stress quantification using EEG signals. In: Proceedings of the international conference for innovation in biomedical engineering and life sciences. Springer Singapore, Putrajaya, Malaysia, pp 15–19
11. Liu TK, Chen YP, Hou ZY, Wang CC, Chou JH (2014) Noninvasive evaluation of mental stress using a refined rough set technique based on biomedical signals. Artif Intell Med 61(2):97–103
12. Houdmont J, Jachens L, Randall R, Hopson S, Nuttall S, Pamia S (2019) What does a single-item measure of job stressfulness assess? Int J Environ Res Public Health 16(9):1480
13. Pigoni A, Delvecchio G, Madonna D, Bressi C, Soares J, Brambilla P (2019) Can Machine Learning help us in dealing with treatment-resistant depression? A review. J Affect Disord 259:21–26

14. Claude LA, Houenou J, Duchesnay E, Favre P (2020) Will machine learning applied to neuroimaging in bipolar disorder help the clinician? A critical review and methodological suggestions. Bipolar Disord 22(4):334–355
15. Liu YS, Chokka S, Cao B, Chokka PR (2021) Screening for bipolar disorder in a tertiary mental health centre using EarlyDetect: a machine learning-based pilot study. J Affect Disord Rep 6:100215
16. Liu Y, Hankey J, Cao B, Chokka P (2021) Screening for major depressive disorder in a tertiary mental health centre using EarlyDetect: a machine learning-based pilot study. J Affect Disord Rep 3:100062
17. Sawalha J, Cao L, Chen J, Selvitella A, Liu Y, Yang C, Li X, Zhang X, Sun J, Zhang Y, Zhao L (2021) Individualized identification of first-episode bipolar disorder using machine learning and cognitive tests. J Affect Disord 282:662–668
18. Dwyer DB, Falkai P, Koutsouleris N (2018) Machine learning approaches for clinical psychology and psychiatry. Annu Rev Clin Psychol 14:91–118
19. Mitchell TM (1997) Machine learning. McGraw-Hill Science/Engineering/Math
20. Iyortsuun NK, Kim SH, Jhon M, Yang HJ, Pant S (2023) A review of machine learning and deep learning approaches on mental health diagnosis. Healthcare 11(3):285
21. LeCun Y, Bengio Y, Hinton G (2015) Deep learning. Nature 521(7553):436–444
22. Graham S, Depp C, Lee EE, Nebeker C, Tu X, Kim HC, Jeste DV (2019) Artificial intelligence for mental health and mental illnesses: an overview. Curr Psychiatry Rep 21:1–8
23. Jencks SF (1985) Recognition of mental distress and diagnosis of mental disorder in primary care. JAMA 253(13):1903–1907
24. World Health Organization (2017) Depression and other common mental disorders: global health estimates. World Health Organization, Geneva
25. Guha M (2014) Diagnostic and statistical manual of mental disorders: DSM-5. Ref Rev 28(3):36–37
26. McEwen BS (2017) Neurobiological and systemic effects of chronic stress. Chronic Stress 1:2470547017692328
27. Toussaint L, Shields GS, Dorn G, Slavich GM (2016) Effects of lifetime stress exposure on mental and physical health in young adulthood: how stress degrades and forgiveness protects health. J Health Psychol 21(6):1004–1014
28. Yang L, Zhao Y, Wang Y, Liu L, Zhang X, Li B, Cui R (2015) The effects of psychological stress on depression. Curr Neuropharmacol 13(4):494–504
29. Mayer SE, Lopez-Duran NL, Sen S, Abelson JL (2018) Chronic stress, hair cortisol and depression: a prospective and longitudinal study of medical internship. Psychoneuroendocrinology 92:57–65
30. Schultebraucks K, Rombold-Bruehl F, Wingenfeld K, Hellmann-Regen J, Otte C, Roepke S (2019) Heightened biological stress response during exposure to a trauma film predicts an increase in intrusive memories. J Abnorm Psychol 128(7):645
31. Colodro-Conde L, Couvy-Duchesne B, Zhu G, Coventry WL, Byrne EM, Gordon S, Wright MJ, Montgomery GW, Madden PA, Ripke S, Eaves LJ (2018) A direct test of the diathesis–stress model for depression. Mol Psychiatry 23(7):1590–1596
32. Wang Z, Xiong H, Zhang J, Yang S, Boukhechba M, Zhang D, Barnes LE, Dou D (2022) From personalized medicine to population health: a survey of mHealth sensing techniques. IEEE Internet Things J 9(17):15413–15434
33. Giansanti D (2021) The role of the mHealth in the fight against the Covid-19: successes and failures. Healthcare 9:58
34. Kang M, Chai K (2022) Wearable sensing systems for monitoring mental health. Sensors 22(3):994
35. Gedam S, Paul S (2021) A review on mental stress detection using wearable sensors and machine learning techniques. IEEE Access 9:84045–84066
36. Bernert RA, Hilberg AM, Melia R, Kim JP, Shah NH, Abnousi F (2020) Artificial intelligence and suicide prevention: a systematic review of machine learning investigations. Int J Environ Res Public Health 17(16):5929

37. Kim S, Lee HK, Lee K (2021) Which PHQ-9 items can effectively screen for suicide? Machine learning approaches. Int J Environ Res Public Health 18(7):3339
38. Lin E, Lin CH, Lane HY (2020) Precision psychiatry applications with pharmacogenomics: artificial intelligence and machine learning approaches. Int J Mol Sci 21(3):969
39. Shin D, Lee KJ, Adeluwa T, Hur J (2020) Machine learning-based predictive modeling of postpartum depression. J Clin Med 9(9):2899
40. Flesia L, Monaro M, Mazza C, Fietta V, Colicino E, Segatto B, Roma P (2020) Predicting perceived stress related to the Covid-19 outbreak through stable psychological traits and machine learning models. J Clin Med 9(10):3350
41. Shin D, Cho WI, Park CH, Rhee SJ, Kim MJ, Lee H, Kim NS, Ahn YM (2021) Detection of minor and major depression through voice as a biomarker using machine learning. J Clin Med 10(14):3046
42. Ryu YH, Kim SY, Kim TU, Lee SJ, Park SJ, Jung HY, Hyun JK (2022) Prediction of poststroke depression based on the outcomes of machine learning algorithms. J Clin Med 11(8):2264
43. Richter T, Fishbain B, Richter-Levin G, Okon-Singer H (2021) Machine learning-based behavioral diagnostic tools for depression: advances, challenges, and future directions. J Pers Med 11(10):957
44. Battineni G, Sagaro GG, Chinatalapudi N, Amenta F (2020) Applications of machine learning predictive models in the chronic disease diagnosis. J Pers Med 10(2):21
45. Delnevo G, Mancini G, Roccetti M, Salomoni P, Trombini E, Andrei F (2021) The prediction of body mass index from negative affectivity through machine learning: a confirmatory study. Sensors 21(7):2361
46. Emadi M, Taghizadeh-Mehrjardi R, Cherati A, Danesh M, Mosavi A, Scholten T (2020) Predicting and mapping of soil organic carbon using machine learning algorithms in Northern Iran. Rem Sens 12(14):2234
47. Tadesse MM, Lin H, Xu B, Yang L (2019) Detection of suicide ideation in social media forums using deep learning. Algorithms 13(1):7
48. Ehiabhi J, Wang H (2023) A systematic review of machine learning models in mental health analysis based on multi-channel multi-modal biometric signals. BioMedInformatics 3(1):193–219
49. Rao H, Gupta M, Agarwal P, Bhatia S, Bhardwaj R (2022) Mental health issues assessment using tools during COVID-19 pandemic. Innov Syst Softw Eng 12:1–12
50. Kim IB, Park SC (2021) Machine learning-based definition of symptom clusters and selection of antidepressants for depressive syndrome. Diagnostics 11(9):1631
51. Haque R, Islam N, Islam M, Ahsan MM (2022) A comparative analysis on suicidal ideation detection using NLP, machine, and deep learning. Technologies 10(3):57
52. Amanat A, Rizwan M, Javed AR, Abdelhaq M, Alsaqour R, Pandya S, Uddin M (2022) Deep learning for depression detection from textual data. Electronics 11(5):676
53. Kumar S, Chong I (2018) Correlation analysis to identify the effective data in machine learning: prediction of depressive disorder and emotion states. Int J Environ Res Public Health 15(12):2907
54. Rincon E, Rodriguez-Guidonet I, Andrade-Pino P, Monfort-Vinuesa C (2023) Mixed reality in undergraduate mental health education: a systematic review. Electronics 12(4):1019
55. Sabry F, Eltaras T, Labda W, Alzoubi K, Malluhi Q (2022) Machine learning for healthcare wearable devices: the big picture. J Healthcare Eng 2022
56. Goldstein M, Uchida S (2016) A comparative evaluation of unsupervised anomaly detection algorithms for multivariate data. PLoS ONE 11(4):e0152173
57. Manjunath S, Nathaniel A, Druce J, German S (2020) Improving the performance of fine-grain image classifiers via generative data augmentation. arXiv preprint arXiv:2008.05381
58. Li J, Izakian H, Pedrycz W, Jamal I (2021) Clustering-based anomaly detection in multivariate time series data. Appl Soft Comput 100:106919
59. Zhang C, Song D, Chen Y, Feng X, Lumezanu C, Cheng W, Ni J, Zong B, Chen H, Chawla NV (2019) A deep neural network for unsupervised anomaly detection and diagnosis in multivariate time series data. In: Proceedings of the AAAI conference on artificial intelligence 2019, vol 33. Springer, Heidelberg, pp 1409–1416

60. Lu W, Li J, Li Y, Sun A, Wang J (2020) A CNN-LSTM-based model to forecast stock prices. Complexity 2020:1–10
61. Mehdiyev N, Lahann J, Emrich A, Enke D, Fettke P, Loos P (2017) Time series classification using deep learning for process planning: a case from the process industry. In: Proceedings of the conference 2017, LNCS, vol 114. Springer, Heidelberg, pp 242–249
62. Li J, Wang Q (2022) Multi-modal bioelectrical signal fusion analysis based on different acquisition devices and scene settings: overview, challenges, and novel orientation. Inform Fusion 79:229–247
63. Beck AT, Kovacs M, Weissman A (1975) Hopelessness and suicidal behavior: an overview. JAMA 234(11):1146–1149
64. Silver MA, Bohnert M, Beck AT, Marcus D (1971) Relation of depression of attempted suicide and seriousness of intent. Arch Gen Psychiatry 25(6):573–576
65. World Health Organization (2018) National suicide prevention strategies: progress, examples, and indicators. World Health Organization, Geneva
66. Walter M, Alizadeh S, Jamalabadi H, Lueken U, Dannlowski U, Walter H, Olbrich S, Colic L, Kambeitz J, Koutsouleris N, Hahn T (2019) Translational machine learning for psychiatric neuroimaging. Prog Neuropsychopharmacol Biol Psychiatry 91:113–121
67. Bzdok D, Meyer-Lindenberg A (2018) Machine learning for precision psychiatry: opportunities and challenges. Biol Psychiatry Cogn Neurosci Neuroimaging 3(3):223–230

Hybrid CPU Scheduling Algorithm for Operating System to Improve User Experience

Ankit Saha, Tushar Mulwani, and Neelu Khare

1 Introduction

CPU scheduling is the process of deciding which process is executed next on the CPU. It is an important part of operating systems because it allows multiple processes to share the CPU and prevents the CPU from being idle. Additionally, it helps to ensure that the CPU is utilized efficiently, that is, all processes are given a fair chance to run, and that deadlocks are prevented. Herby it is a dynamic process that makes decisions quickly, fairly, and efficiently. The CPU scheduling algorithms are instrumental in achieving following objectives.

1. **CPU Utilization**: It refers to the amount of work performed by a CPU. Utilization varies based on the quantity and nature of managed computing tasks. Its optimized by efficiently assigning the CPU to processes, minimizing idle time, and maximizing the amount of work performed
2. **Throughput:** It is the number of processes finished in each amount of time. It is improved by prioritizing and scheduling processes effectively, resulting in a higher number of completed tasks within a given time frame.
3. **Turnaround Time (TT)**: It is the amount of time that passes between when a procedure is started and when it is completed. The overall amount of time spent waiting—on the CPU, in the ready queue, in memory, and during I/O—is known as turnaround time. It is reduced as the algorithms ensures that processes spend less time waiting in the ready queue.

A. Saha · T. Mulwani · N. Khare (✉)
Vellore Institute of Technology Vellore, Vellore, India
e-mail: Neelu.khare@vit.ac.in

A. Saha
e-mail: Ankit.saha2022@vitstudent.ac.in

T. Mulwani
e-mail: tushar.mulwani2022@vitstudent.ac.in

M. S. Kaiser et al. (eds.), *ICT: Smart Systems and Technologies*, Lecture Notes in Networks and Systems 878, https://doi.org/10.1007/978-981-99-9489-2_38

4. **Waiting Time (WT)**: The total amount of time spent in the ready queue is referred to as the waiting time. Therefore scheduling algorithms determine how long a process remains in the ready queue before it gets access to the CPU
5. **Response Time (RT)**: It is the time between submitting a request and receiving the first response. Scheduling algorithms ensures that high-priority processes are given prompt access to the CPU, help reduce response time and improve the overall user experience.

Various scheduling algorithms exist to accomplish the aforementioned goals. First, FCFS, in the order in which processes gain CPU access is based on their arrival time. Second, the SJF algorithm allocates CPU to the process on the basis of burst time, that is, the process with short BT is given access earlier. In a priority-based process, each process is assigned a numeric value, where a particular value may indicate a higher or lower priority, a process with a higher priority will receive CPU access sooner than one with a lower priority. In the Round Robin (RR) algorithm, a new process is added to the end of the ready queue, time quantum (TQ) which is the key component of this algorithm [1]. If burst time (BT) is more significant than TQ, the current process is replaced. If BT is less than TQ, the following process in the queue is granted CPU access at the completion time. In the Multilevel Queue and Multilevel Feedback Algorithm, the ready queue is divided into multiple queues, and each queue has its own scheduling algorithm.

Reducing turnaround, waiting, and reaction times is preferable while increasing CPU utilization and throughput. The average measure is typically what we optimize.

Nevertheless, these above-mentioned algorithms have their own drawbacks and limitations, FCFS suffers from the "convoy effect" where long processes in the front of the queue delay shorter processes. SJF suffers from the drawback of not being able to handle dynamic workloads efficiently. Priority scheduling may lead to starvation if a high-priority process continuously arrives, preventing lower-priority processes from executing. Round Robin (RR) has high context-switching overhead and may result in poor performance for long-running processes. Priority-based algorithms can neglect lower-priority processes, causing a lack of fairness. Finally, Multilevel Queue Scheduling suffers from the inability to dynamically adapt to changing workload priorities.

Therefore, to overcome the above issue, the proposed algorithm will minimize the variation in response time instead of simply optimizing for the average response time. This is because of the consistent and predictable response time which is more desirable than a system that boasts a high average response time but is highly variable in its performance.

Hybrid scheduling techniques combine multiple scheduling algorithms to address the limitations of single algorithms in modern operating systems. These techniques offer greater flexibility, allowing for adaptation to changes in workloads. Hybrid techniques often outperform single algorithms, particularly for specific workloads. By leveraging the strengths of different algorithms, hybrid scheduling techniques achieve superior results in managing diverse workloads. Additionally, hybrid scheduling techniques mitigate complexity compared to single algorithms, offering simpler implementation and management procedures.

This approach provides a more efficient and streamlined approach to managing diverse workloads. Adopting hybrid scheduling techniques in operating systems offers advantages in flexibility, performance optimization, and simplified management compared to relying solely on individual scheduling algorithms.

Earlier proposed hybrid scheduling algorithm uses RR and SJF [2]. In this algorithm, the waiting queue and the execution queue make up the ready queue. The waiting queue is empty. The CPU burst length is used to preserve the order of the execution queue, with the shortest burst length at the front. When a variable's total execution time exceeds the burst time for the most critical process, the essential process is given CPU time to run.

The reminder of this paper is laid out as follows: Sect. 2 presents a study of previously presented methods and approaches used to solve the above challenges, but still needs more improvements. Section 3 presents a thorough understanding of the structures and algorithms utilized to implement the proposed algorithms and its step by step demonstration. Section 4 simulates and tests the algorithm by analyzing the differences and concluding with achievements and future possibilities in Sect. 5. References in last section.

2 Literature Survey

Optimizing the scheduling objectives (TAT, WT) is the main objective of every scheduling algorithm. To ensure that the proposed algorithm goes in the same direction along with better results, past related works, are analyzed exhaustively and comprehensively, which are as follows:

Silberschatz et al. [3] provided the background information that is a prerequisite while designing a new scheduling algorithm, for example, the role of OS in job scheduling, scheduling objectives, and criteria.

The work gives a comprehensive analysis of the existing algorithm [1]. For example, it helps to identify the pros and cons of each existing algorithm. Though SJF solves issues of FCFS (high waiting time, less throughput), the primary issue is predicting CPU burst time, as there are several methods to predict the length of CPU burst.

The authors combine Round Robin (RR) with the Shortest Job First [2]. The authors concluded that compared to the other CPU scheduling policies, it produces the lowest average waiting time. The average turnaround time calculated for this algorithm is higher than the average turnaround time computed by "*Shortest time remaining first*" but lower than the average turnaround times computed for all other CPU scheduling policies. The book provides thorough learning about the Red–Black Tree [4]. It explains how it helps to reduce the searching time and how we can significantly reduce time complexity from O (n) to O (log n). Thombare et al. [5] give better CPU utilization by using a short burst time for the First Queue, making it synonymous with RR scheduling and, later, SJF before RR from the Second Queue. By using dynamic time quantum, they also prevented the starvation problem.

Adrian Colyer proposes Completely Fair Scheduling (CFS), which implements Weighted Fair Queueing. Here is a single-level or core system, CFS time-slices CPU among running threads, while in multi-core, it takes per-core run queues and a load balancing algorithm. It helps to reduce the duplicity in work. The author's proposed method modifies the time quantum in each cycle based on the remaining time of the process [6]. Compared to traditional RR, it provides a dynamic time quantum for each cycle and explicitly aids tasks with limited remaining burst time.

Using machine learning (ML) techniques such as K-NN, SVM, ANN, and DT, the authors of this study proposed a method for estimating the CPU burst length of processes that are waiting in a queue [7]. Then, on a grid workload called "GWA-T-4 Auver Grid," the most crucial characteristics were implemented. The results of this experiment demonstrate a robust linear relationship between the process characteristics and the burst CPU time, as well as the fact that K-NN outperforms alternative ML approaches in nearly all circumstances with respect to CC and RAE.

Farooq et al. [8] demonstrates in this paper that Adaptive Round Robin and Optimal Round Robin Scheduling Using Manhattan Distance are superior to Best Time Quantum Round Robin CPU Scheduling and Improved Round Robin Scheduling Algorithm. Using the advantages of MLQ, RBT, and Dynamic Time Quantum in RR scheduling, the proposed algorithm attempted to improve the user experience while also better optimizing scheduling objectives by utilizing the advantages of MLQ, RBT, and Dynamic Time Quantum, but failed to fully utilize CPU and provide faster turnaround times for high-priority tasks. Consequently, the proposed algorithm addresses these issues.

3 Proposed Methodology

The proposed algorithm uses a Red–Black Tree (RBT) and Multilevel Queue (MLQ). Following is the brief explanation:

3.1 Red–Black Tree

Invented by Rudolf Bayer, it is a partial self-balancing binary search tree it reduces the searching time to O(log 2 n) time [4]. It is utilized in a "completely fair scheduling" algorithm where it maintains jobs solely based on the time process that has already been executed. This approach treats all processes with the same priority, leading to a colossal turnaround time for high-priority jobs (Fig. 1).

The proposed algorithm uses it by storing processes based on burst time and selecting jobs with the lowest burst time.

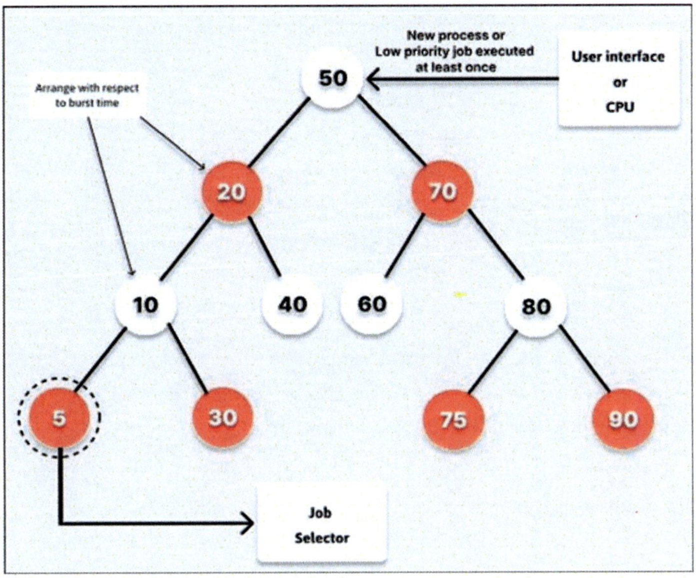

Fig. 1 Red–black tree of the proposed algorithm

3.2 *Multilevel Queue*

In operating systems, a ready queue data structure is used to store processes that are ready to be executed. In order to prioritize jobs and reduce turnaround time for high-priority jobs, the ready queue is frequently divided into multiple queues, with each queue containing processes with the same priority. This method facilitates the identification and execution of high-priority tasks, thereby enhancing the user experience. The CPU is then switched between different queues to select the next job to be executed, ensuring that jobs with a higher priority are executed before those with a lower priority. This method can significantly reduce turnaround time and enhance system's overall performance (Fig. 2).

3.3 *Incremental Time Quantum Round Robin (ITQRR)*

In this method, a novel Round Robin Algorithm is used, where the time quantum is calculated for each process based on burst time (or remaining burst time) so that each process can gain CPU for the appropriate amount of time without starving other processes. Every queue has a fixed block size that cannot be changed. Therefore, before a new TQI can be calculated, a certain number of processes must be carried out at least once (Fig. 3).

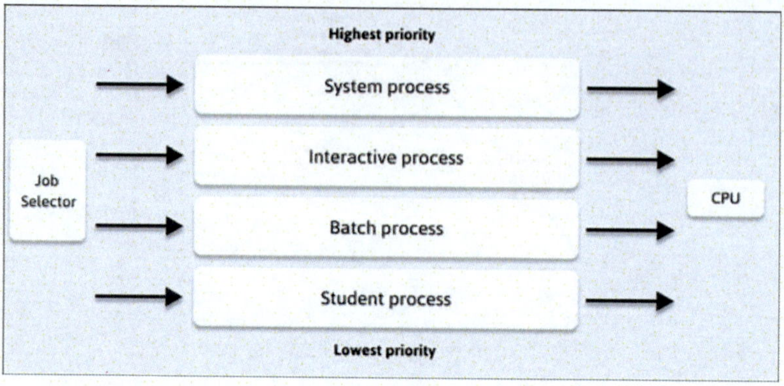

Fig. 2 Multi-queue architecture of the proposed algorithm

3.3.1 Steps Explanation

The algorithm will calculate burst distance (*bdi*), the distance between consecutive process burst times.

$$bd_i = \left(P_{(i+1)} - P_i\right)$$

The algorithm will calculate the Average Burst Distance (ABD), acting as a primary Time Quantum (TQ).

$$\text{ABD} = \frac{\sum bd_i}{\text{Block size}}$$

Time quantum will be added to ABD and stored in Time Quantum Incremental (TQI). Until the number of processes performed equals the block size or the queue is empty, the time quantum for each process will be increased using TQI.

$$\text{TQI} = \frac{(\text{Largest bust time} - 1)}{\text{Block size} - 1}$$

3.4 Proposed Algorithm

The proposed hybrid scheduling algorithm is primarily focused on better user experience by reducing the delay between command time and response time. The algorithm uses Red–Black Trees (RBT) and Multilevel Queues (MLQ). RBT will ensure O(log *n*) time complexity to get the shortest job and MLQ to prioritize the processes.

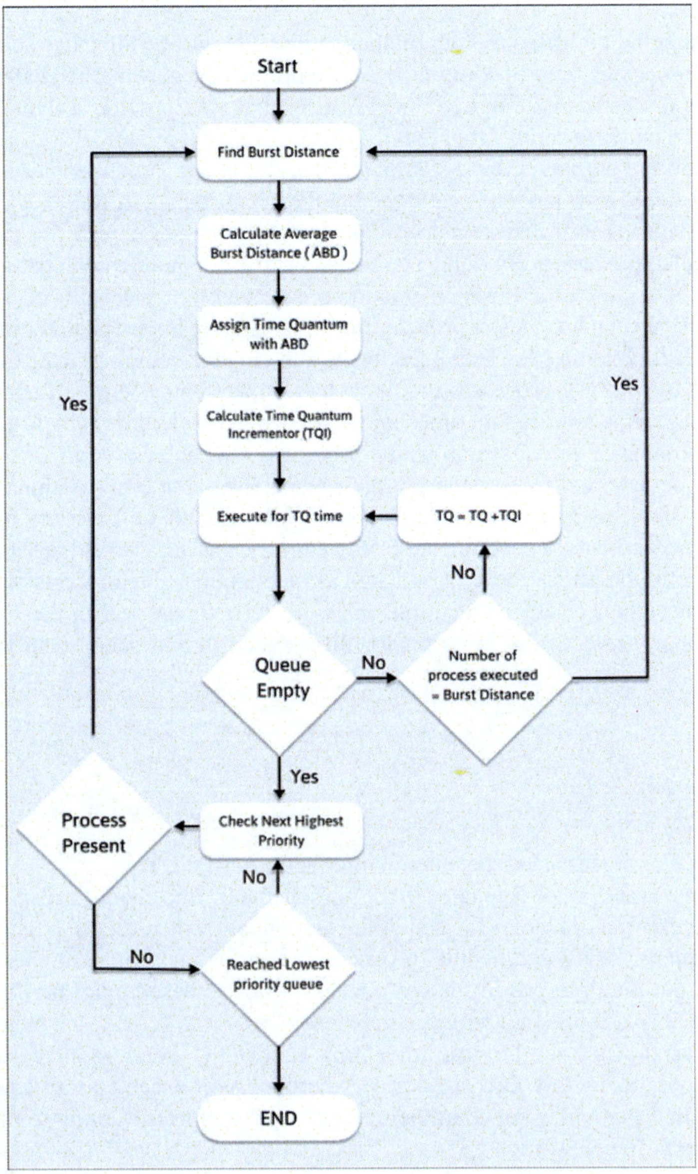

Fig. 3 Flow chart of Incremental Time Quantum Round Robin Algorithm

Consequently, by utilizing both methods, the CPU will be allocated to the most important processes first. In addition, we have proposed a new method called Incremental Time Quantum Round Robin (ITQRR) that will provide a dynamic time quantum and increase computer usage by approximately 50 percent, thereby reducing the likelihood of starvation by a significant margin. In the implementation section, we demonstrated how a queue is assigned to the proposed algorithm.

The job selector of this system prioritizes system processes, interactive processes, and then batch processes. By doing so, it ensures that the more crucial processes for the system's operation and user interaction are executed first, leading to improved computer performance. This approach prioritizes tasks that have a direct impact on the system's functioning and user experience, allowing the computer to operate more efficiently. In contrast, processes that have less impact on system performance are scheduled to run at appropriate times, optimizing overall system efficiency.

In the context of this research study, processes that possess equivalent priority levels will be assessed by considering their respective burst times within the same priority queue. This approach is adopted due to the fact that all priority queues in the system encompass a spectrum of priority levels, and as a result, any processes sharing a common priority level are treated as indistinguishable in terms of priority. Consequently, their relative execution times, or burst times, within the respective priority queue serve as the criterion for differentiation and scheduling decisions (Fig. 4).

3.5 Process Steps

1. At first, the process shall be entered into the Red–Black Tree.
2. Then, the task shall be delegated to the job scheduler, who takes the responsibility of selecting the processes located at the leaf nodes of the execution hierarchy.
3. Subsequently, the job scheduler proceeds to allocate tasks by assigning them to one of the multiple priority-based queues, which is based upon the individual priority levels associated with each process.
4. At the next stage, the Central Processing Unit (CPU) shall be allocated to the queue possessing the highest priority, wherein it will execute all tasks residing within that particular queue in their entirety, prior to transitioning to the queue of the next lower priority.
5. The Task execution shall adhere to the Iterative Task Queue Round Robin (ITQRR) scheduling policy before the completion of the Execution Cycle.
6. Upon completion of the designated execution cycle, tasks shall be reinserted into the Red–Black Tree, for the purpose of resource allocation to the subsequent tasks.

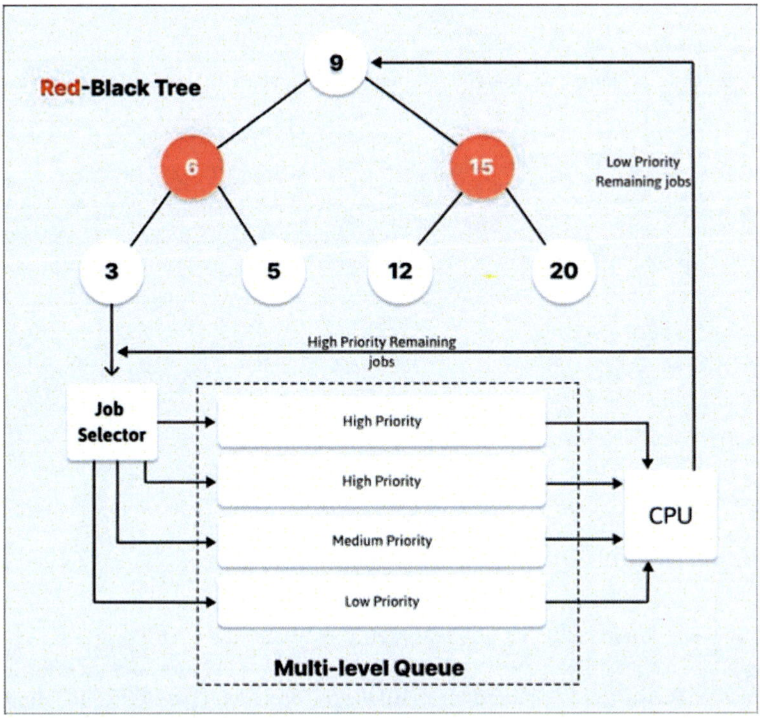

Fig. 4 Complete working of the proposed Incremental Time Quantum Round Robin algorithm

4 Experimentation

To evaluate and compare various scheduling approaches with ITQRR, a simulator was utilized. This simulator creates a realistic environment where processes can be entered with different criteria, including arrival time, priority, burst time, and input/output time. By simulating these processes, the simulator generates outcomes based on metrics such as throughput, waiting time, and response time. This allows for a comprehensive assessment of different scheduling approaches and facilitates the comparison with ITQRR in terms of their performance and effectiveness. The simulator provides a valuable tool for analyzing and optimizing scheduling algorithms in a simulated environment.

The previous research showed that the hybridization of algorithms are more efficient when compared to the other conventional scheduling algorithms, here the comparisons of ITQRR to Round Robin and Shortest Job First is presents in Table 1.

Table 1 The process detail table

	Arrival	Priority	1st Exec	1stI/O	2ndExec	2ndI/O	3rdExec
P1	0	3	2	–	–	–	–
P2	2	1	4	–	–	–	–
P3	3	2	3	1	3	1	1
P4	4	0	8	–	–	–	–

I/O, input/output duration; Exec, duration of execution; P1, system process; P2, batch process; P3, interactive process; P4, student process

4.1 Processes Details

See Table 1.

4.2 Experimentation Result and Analysis

The simulation experiments in this research were conducted utilizing a computing environment characterized by specific hardware and software specifications. The computational tasks were executed on an Intel(R) Core(TM) i7-9750H CPU, boasting a clock speed of 2.60 GHz with a base frequency of 2.59 GHz. The system was equipped with a total of 16.0 GB of RAM, with 15.9 GB being accessible for computational tasks, ensuring ample memory capacity for managing and manipulating. Additionally, the operating system employed was a 64-bit configuration with an ×64-based processor architecture. These meticulously chosen specifications contributed to the creation of a robust and capable computing environment, facilitating the accurate and reliable execution of the simulation experiments conducted as part of this research.

4.2.1 Response Time

See Figs. 5, 6, and 7.

4.2.2 Analyzing Response Time

Our proposed algorithm demonstrates notable improvements in response times for interactive processes (P1) and system processes (P3). This outcome signifies that the algorithm effectively prioritizes processes crucial for enhancing the user experience. In contrast, under traditional Round Robin (RR) and Shortest Job First (SJF) scheduling, P3 experiences a delay in CPU allocation. However, our algorithm addresses this issue by ensuring that necessary processes are promptly executed

Fig. 5 Propose algorithm

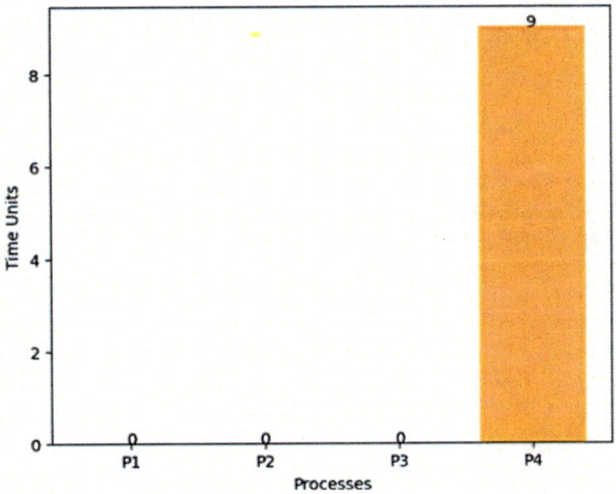

Fig. 6 Round Robin

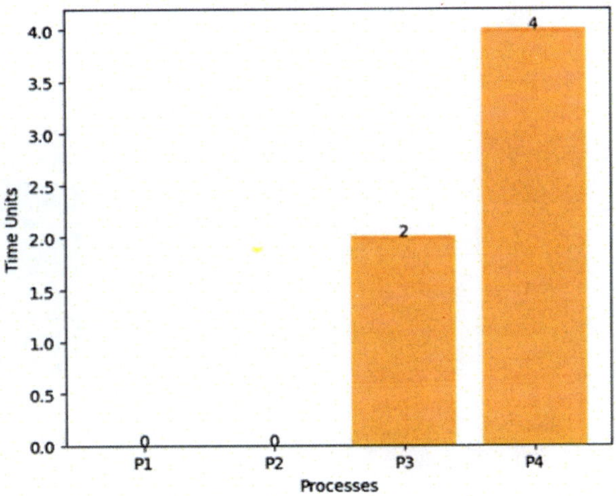

and ensuring users can interact with the system in a more dynamic and responsive manner.

4.2.3 Waiting Time

See Figs. 8, 9, and 10.

Fig. 7 Shortest Job First

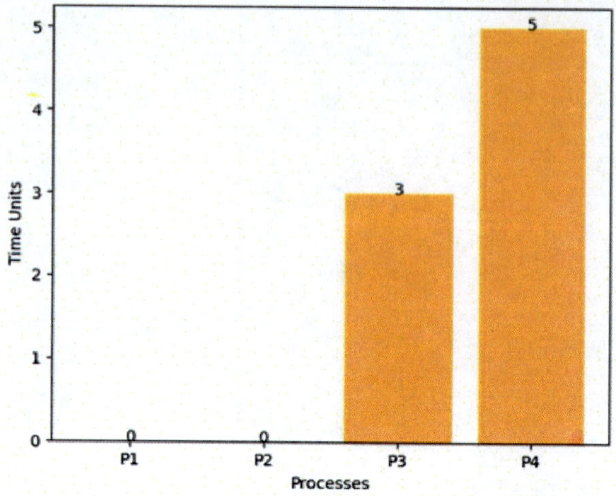

Fig. 8 Proposed algorithm

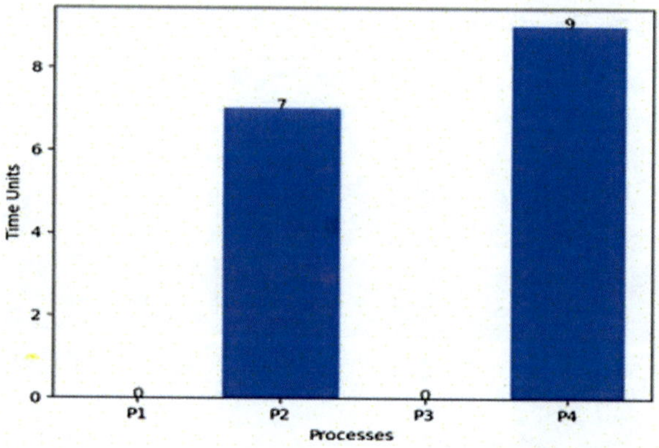

Fig. 9 Round Robin

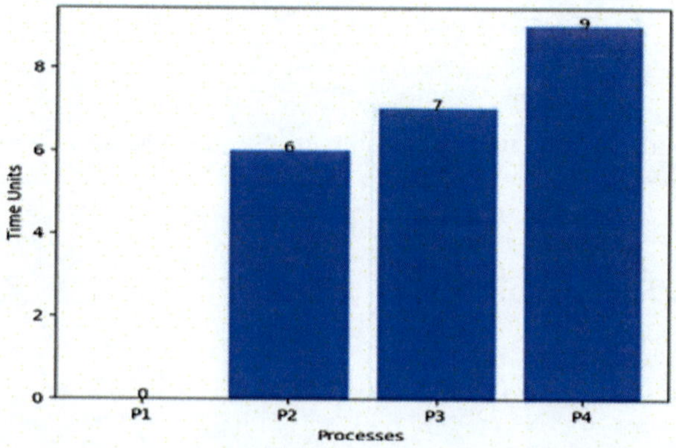

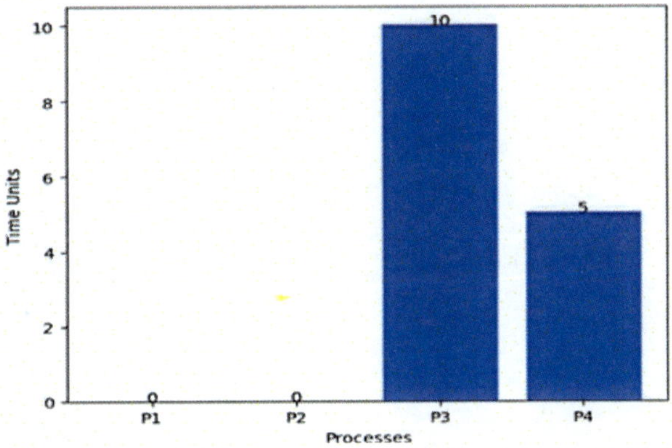

Fig. 10 Shortest Job First

4.2.4 Analyzing Waiting Time

Under our proposed algorithm, both interactive process P1 and system process P3 experience minimal or no waiting time for CPU allocation. This efficient scheduling ensures that these critical processes can run without delays, resulting in smooth and uninterrupted system operation. In contrast, in the Round Robin (RR) and Shortest Job First (SJF) algorithms, P3 incurs significant waiting time, leading to reduced system performance and abrupt execution. By addressing these limitations, tasks can be processed more quickly, now users can utilize the system more efficiently and also system can deliver responses faster with a seamless experience.

4.2.5 Turnaround Time

See Figs. 11, 12, and 13.

4.2.6 Analyzing Turn Around Time (TAT)

In comparison to the Round Robin (RR) and Shortest Job First (SJF) algorithms, our proposed algorithm exhibits a more balanced approach in terms of turnaround time (TAT) for different processes. While RR and SJF may result in low TAT for P2 processes, this can inadvertently increase the TAT for P3, which is a critical process. In contrast, our algorithm aims to maintain a closer proximity to the burst time for both P1 and P3, ensuring that these important processes are completed efficiently. As a result, there may be a slightly larger TAT for other processes, but the focus is on prioritizing and expediting the execution of crucial processes, leading to overall improved system performance and responsiveness.

However, this proposed algorithm needs to ensure minimum task delay in the execution of urgent tasks while lower-priority task is in execution. Additionally, the

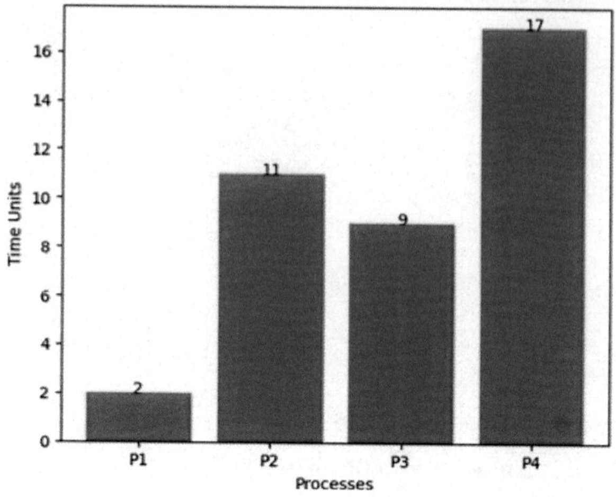

Fig. 11 Proposed algorithm

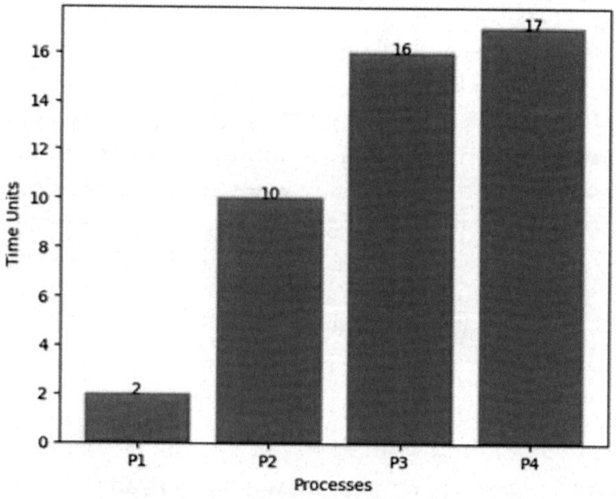

Fig. 12 Round Robin

proposed algorithm needs to ensure proper updation of each task in the Red–Black Tree so that it cannot lead to an unfair distribution of resources.

5 Potential Limitations

- The process of inserting and rebalancing nodes in the Red–Black Tree can be computationally expensive when the tree grows larger after a particular point.
- If there are always high-priority tasks in the queue, prioritizing tasks solely based on their priority may lead to lower-priority tasks being starved. This could result in unfairness or decreased performance for certain types of tasks.

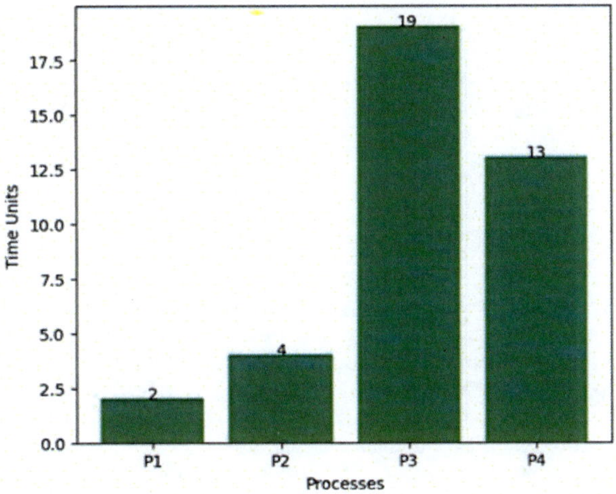

Fig. 13 Shortest Job First

- While ensuring the effectiveness of the system, maintaining multiple priority queues for different process priorities is a challenging task for the algorithm.

6 Conclusion and Future Work

In this study, we have developed an improved hybrid scheduling algorithm for the operating system that places a strong emphasis on enhancing the user experience compared to existing algorithms. Our approach encompasses a Red–Black Tree (RBT) combined with a Multilevel Queue (MLQ) scheduling algorithm to efficiently select jobs. In this, RBT will ensure efficient insertion and retrieval of tasks while maintaining the time complexity of O (log N) and MLQ based on process priority to ensure tasks are assigned appropriately according to their priority. Therefore, using RBT and MLQ will ensure a balance between efficient process management and priority-based task allocation. Additionally, we have introduced the concept of "Time Quantum Incremental" in Multilevel Queue (MLQ) scheduling, which dynamically increases the time quantum for subsequent processes within the same queue level. Through extensive experimentation, we have demonstrated that our proposed "Incremental Time Quantum Round Robin algorithm," (ITQRR) will help to increase the throughput of the system by allowing shorter tasks to be completed quickly additionally, it ensures deadlock be prevented, by allowing every task to make progress, even if periodically interrupted due to any reason.

At last, by using ITQRR in conjunction with RBT and MLQ we have observed that the proposed hybrid scheduling algorithm surpasses previous algorithms in terms of response time, average completion time, and resource utilization.

In subsequent studies, we can try to overcome the above-mentioned limitation (Sect. 5) by making the algorithm resource management, dynamic priority adjustment, and fault tolerance management more efficient. This can be achieved by making certain changes in proposed ITQRR as it has the potential to schedule real-time tasks effectively. Also in further studies, we can consider studying other factors like load balancing, process security, cost optimization, and machine learning integration, etc.

References

1. Pemasinghe S, Rajapaksha S (2022) Comparison of CPU scheduling algorithms: FCFS, SJF, SRTF, Round Robin, priority based, and multilevel queuing. In: 2022 IEEE 10th Region 10HumanitarianTechnology. Conference (R10-HTC), pp 318–323. https://doi.org/10.1109/R10-HTC54060.2022.9929533
2. Shah SNM, Mahmood AKB, Oxley A (2009) Hybrid scheduling and dual queue scheduling. In: 2009 2nd IEEE International Conference on Computer Science and Information Technology, pp 539–543. https://doi.org/10.1109/ICCSIT.2009.5234480
3. Silberschatz A, Galvin PB, Gagne G (2005) Operating system concepts, 7th edn. Wiley, New York
4. Weiss MA (2014) Data structures and algorithm analysis in C++, 4th edn. Pearson Education Limited, London
5. Thombare M, Sukhwani R, Shah P, Chaudhari S, Raundale P (2016) Efficient implementation of multilevel feedback queue scheduling. In: 2016 International Conference on Wireless Communications, Signal Processing and Networking (WiSPNET), pp 1950–1954. https://doi.org/10.1109/WiSPNET.2016.7566483
6. Mody S, Mirkar S (2019) Smart Round Robin CPU Scheduling algorithm for operating systems. In: 2019 4th International Conference on Electrical, Electronics, Communication, Computer Technologies, and Optimisation Techniques (ICEECCOT)
7. Helmy T, Al-Azani S, Bin-Obaidellah O (2015) A machine learning-based approach to estimate the CPU- burst time for processes in the computational grids. In: 2015 3rd International Conference on Artificial Intelligence, Modelling and Simulation (AIMS), pp 3–8. https://doi.org/10.1109/AIMS.2015.11
8. Farooq MU, Shakoor A, Siddique AB (2017) An efficient dynamic Round Robin Algorithm for CPU scheduling. In: 2017 International Conference on Communication, Computing and Digital Systems (C-CODE), pp 244–248. https://doi.org/10.1109/C-CODE.2017.7918936

AI-Enable Heart Sound Analysis: PASCAL Approach for Precision-Driven Cardiopulmonary Assessment

Ankit Kumar, Kamred Udham Singh, Gaurav Kumar, Tanupriya Choudhury, Teekam Singh, and Ketan Kotecha

1 Introduction

In the past, the ability to decipher the complicated symphony of heart sounds, which is essential for identifying a wide variety of cardiovascular disorders, depended on the acute hearing and analytical acuity of trained medical specialists. They painstakingly dissected each facet of these sounds in order to interpret the underlying cardiac tales, and they did this by distilling these sounds manually into waveforms. This traditional and hands-on method, although admirable, was burdened by the weight of its inherent constraints, the most significant of which were the tendency for human mistake and the sheer time-intensiveness of the process. Our study ushers in a revolutionary new

A. Kumar · G. Kumar
Department of Computer Engineering and Applications, GLA University, Mathura, Uttar Pradesh, India
e-mail: kumar.ankit@gla.ac.in

K. U. Singh (✉)
School of Computing, Graphic Era Hill University, Dehradun, India
e-mail: falsekamredudhamsingh@gmail.com

T. Choudhury (✉)
Department of Computer Science and Engineering, Graphic Era Deemed to Be University, Dehradun, India
e-mail: falsetanupriya1986@gmail.com

T. Singh
CSE Department, Symbiosis Institute of Technology, Symbiosis International University, Lavale Campus, Pune 411045, India
e-mail: gaurav.kumar_phd.ca21@gla.ac.in

K. Kotecha
Symbiosis Centre for Applied Artificial Intelligence, Symbiosis Institute of Technology, Symbiosis International University, Pune 411045, India
e-mail: director@sitpune.edu.in

M. S. Kaiser et al. (eds.), *ICT: Smart Systems and Technologies*, Lecture Notes in Networks and Systems 878, https://doi.org/10.1007/978-981-99-9489-2_39

age in the field of cardiovascular noise detection by presenting a cutting-edge technique that fuses automation with accuracy in a way that is completely seamless. This research not only brings the evaluation of heart sounds into the digital era but also favourably places it for integrations with machine learning, particularly in applications involving thoracentesis. In addition to this, it has a mutually beneficial partnership with the most cutting-edge clinical imaging technology, which ensures a comprehensive and up-to-date approach to cardiac sound analysis. The novel "accordion" technique, which we have designated as the PASCAL method, is essential to our methodology. PASCAL is distinctive in that it is able to collect reference data that is characterised by sudden shifts and combine that data with an exceptional classification model that is tuned for unsurpassed accuracy [1]. This capacity is shown in our Dataset for Characterising Heart Rate [2, 3]. The thoroughness of our assessments highlighted the remarkable performance metrics of our classification phase, demonstrating promising F1-scores and accuracy rates while also providing light on the dynamic nature of Electrocardiogram (ECG) data. Our in-depth investigations of primary (S_1) and secondary (S_2) sound durations, cardiac cycles, ventricular systolic durations, contraction timings, and systolic-to-diastolic ratios give a complete analytical toolset. These are only some of the aspects of heart sounds that have been investigated. The most important finding was that we were able to distinguish between the two principal peaks of the heart sounds known as S_1 and S_2, despite the fact that the expressions of these peaks might vary from one individual sample to the next. This variability was mostly related to variable sphygmomanometer locations.

1.1 Background

The natural diversity in heart sounds that exists across people is one of the aspects of identifying ECG data that presents one of the greatest challenges. The first heart sound, the second heart noise, and any murmurs that are present may all be linked to this variability. These differences may be attributed to a number of different physiological causes. For example, the amplitude, harmonic content, and length of heart signals may all be affected by the precise location of cardiac recording electrodes on the chest, the thickness of the skin, and the dynamics of blood circulation. Because of all of these difficulties, researchers have devoted a significant amount of time and effort to the development of reliable methods for analysing heart sound signals.

The envelope-based approach has been more well-liked among academics as a result of the ease with which it may keep the most important information from a data series intact while still maintaining the method's inherent simplicity [4]. During the process of segmenting heart sound waves, this methodology is used rather often. The normalised mean stochastic frequency, stochastic sensitivity Eigenvectors, and characteristic ECG signal waveforms are only few of the well-known envelope approaches that have been utilised in the past. A number of researchers have tried their hands at envelope-based approaches using a wide variety of sources of reference data.

These variables include maximal training, blank entry points, instantaneous amplitude, and regression coefficients. These references are helpful in segmenting and then afterwards distinguishing the individual properties of each heartbeat [5].

It is necessary to take a more sophisticated approach in order to deal with the inconsistencies as a result of the differences in the dependent variable that arise as a result of the varied installation locations for the sphygmomanometer. The inclusion of the accordion approach together with the insights gained from the maximal change junction detection technique and the abrupt modification juncture detection methodology is one option that has been offered [6]. This combination is an attempt to come up with an original solution to the problem of ECG signal segmentation and identification. The major purpose of this investigation is to put the technique that was suggested to use in order to partition the collected heartbeat audio signals that are included within the data series [7].

1.2 Related Work

A significant contribution to the field of conventional heart sound analysis. In [8] the work done by Smith laid the groundwork for understanding the fundamental properties of heart sounds, which in turn led to more study aimed at perfecting these approaches. This gave rise to the concept of digital stethoscopes in [9], which was made possible by the development of technology. The work done by Jones demonstrated the possibility of amplifying and recording heart sounds, which paved the door for further in-depth examination. The discipline of signal processing was revolutionised in [10], published a study in which he showed the use of the Fourier and Wavelet Transforms to extract essential information from heart sounds. This article is credited with starting the modern era of signal processing. The study that was published in [11] became an essential component in the application of algorithms to the prediction of cardiovascular abnormalities based on heart sound data. His technique, which included training models on large amounts of data, was revolutionary. In paper [12] has been very influential in terms of the integration of imaging technology with acoustic analysis. The study conducted used echocardiography as well as thoracentesis for more thorough examinations of the heart. The paper [13] made a substantial advance in her research by putting an emphasis on instruments that collect heart sounds without the need of invasive procedures. According to this paper's [14] core aim, the purpose of portable diagnostic equipment was to provide service to healthcare institutions all over the world, even those with a restricted number of available resources.

1.3 Research Gap

There is a large hole in the research that pertains to real-time analysis and forecasting. There has been very little research done on processing and diagnosing in real time, which is particularly problematic in a variety of scenarios, such as loud locations or with patients who have concomitant respiratory illnesses. The tools that are now available place the majority of their emphasis on the post-capture analysis of cardiovascular sounds. In addition, the incorporation of these techniques into wearable devices for the purpose of continuous monitoring is a path that has seen very little exploration. There is an urgent need to go beyond periodic diagnosis and towards continuous real-time monitoring of cardiovascular health. This will allow for prompt treatments, which will improve patient outcomes.

1.4 Research Objective

The aims of this investigation into the assessment of ECG signals may be broken down into the following categories:

- The Development of an Electrocardiogram Signal Evaluation Based on the Mailer Method, we will implement the mailer technique in order to do an ECG signal analysis. The mailer method provides an alternative methodology, which, in comparison to more traditional methods, may result in the discovery of unique insights.
- With the help of the PASCAL Datasets for Standard Cardiac Sounds, it is very necessary to evaluate the provided strategy using known datasets as a benchmark. We are able to assess how well the new method compares to established standards by using the PASCAL datasets, which include the standard cardiac sounds.
- Evaluation and Verification Utilising Various Other Databases, if the study were to rely entirely on the PASCAL datasets, it is possible that its reach would be restricted. The incorporation of data from several different databases will guarantee that the procedure may be used to any situation. The purpose of this more comprehensive comparison is to determine whether or not the approach is valid across a variety of datasets and settings.

2 Proposed Methodology

In the field of cardiovascular diagnostics, the computer-assisted spirometry system represents a cutting-edge innovation in diagnostic technology. This approach stands out from others since it does not include any intrusive procedures. It uses cutting-edge technology to detect cardiac beats across the surfaces of the lungs, making it a more up-to-date alternative to the conventional thoracentesis methods. This technique, as opposed to the traditional method, records the acoustic impulses that are emitted by the heart and then amplifies those signals using a complex digital device [15]. When these recordings of cardiac sounds have been obtained, the next step is to feed them into a computer network, where sophisticated statistical methods and signal analysis procedures will be used in order to conduct a thorough investigation.

Controlling quality is an essential part of the CAA network, particularly in the stages that come before beginning work on the cardiac acoustic processing applications. One of the primary goals is to simplify the data, which requires sorting through the many signals that are obtained from the cardiopulmonary system and dividing them up according to the cycles of individual heartbeats [16]. It is essential to be aware that the bulk of heart sound activities, depending on the particular condition, are normally contained within a predetermined cardiac cycle span. This is something that should be kept in mind. For the sake of clarity, a typical cardiac sound cycle consists of four basic components: the length of the first sound (S_1), the duration of the diastolic phase of the heart, the duration of the second sound (S_2), and the duration of the whole cardiac cycle. In contrast, an abnormal cardiovascular system will often produce distinctive noises during either the systolic or diastolic phase of the heartbeat. In circumstances when there are no murmurs present, a period of silence will often be noted between the S_1 and S_2 periods.

The mailer technique has been used to construct the fundamental framework for automated ECG signal analysis. This framework has been developed. This technique makes use of reference materials that place an emphasis on changes that are both substantial and sudden within the ECG signals. In this work, the PASCAL datasets were used, and typical heart sounds were utilised as instances of benchmarks. The difficulties manifested themselves throughout the procedures of classifying and recognising the properties of heart rhythm, particularly when illness conditions were taken into consideration. In these circumstances, the efficiency of the method diminished substantially, falling short of the realistic expectations that were set. It is fascinating to note that the conceptual method proven its efficiency in decomposing pulse signals [17]. These signals, which are famous for their constantly shifting maxima as a result of their deployment in diverse places, provided a substantial problem. However, the approach was quite good at dealing with the complexity that arose from the continual shifts. This is something that may be ascribed to the fact that the method had a single point of contact with the data.

The new insights that may be gleaned by contrasting our results with the information found in other databases come about as a result of this comparison. These first comparisons, which are shown in Figs. 1 and 2, provide the framework for further research, which may broaden the application of the developed technique and increase its level of detail.

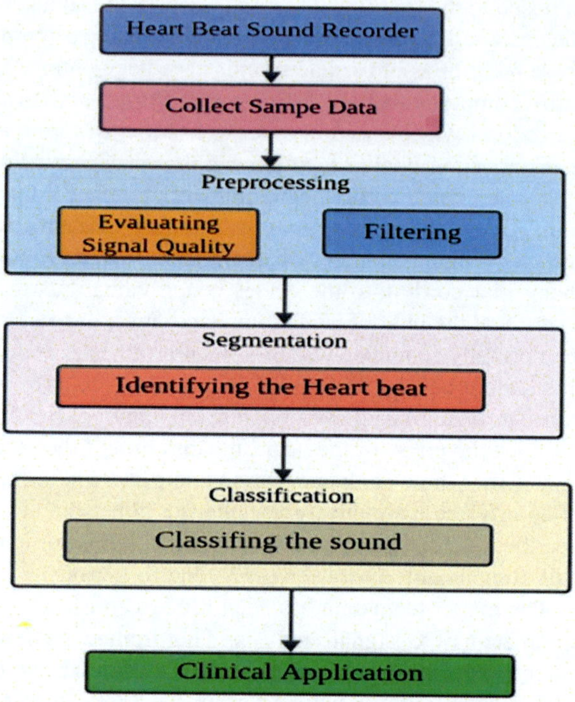

Fig. 1 The progression of the envisioned system

2.1 Structured Architecture

A structured circuit architecture is required for the creation of a computerised system that is capable of performing the segmentation and component identification of heart-beat sound signals. This system may be broken down into three significant steps, which are as follows:

- **Pre-processing:** The ECG signals are subjected to a variety of changes in order to get them ready for the succeeding stages. The levels of the signals are first changed in order to bring their waveforms into line with the 0y-axis in a manner that is consistent. After these changes have been made, the frequency bandwidth of the signals has been reduced to a level of just 2000 Hz. After that, a Fourier band-pass filter that has a frequency response that ranges from 25 to 250 Hz and has a 5th-degree order is applied [18]. This filtering is necessary because the principal components of the heartbeat, notably the S1 and S2 occurrences, occur within the spectral region that has been assigned for this purpose. By zeroing in on this frequency band, extraneous noise and data that is not related to the investigation may be successfully filtered away, and what is left are the most important aspects of the signal for examination.

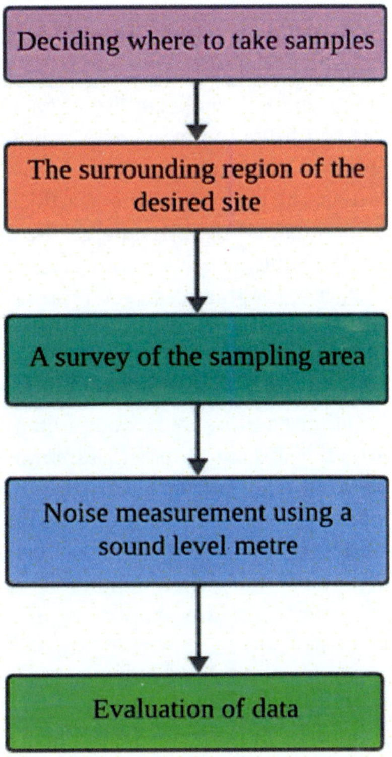

Fig. 2 Flow of the proposed system

- **Segmentation:** The signals that have been pre-processed are separated into their own independent segments. Each of these segments represents a different component or event that occurs throughout the heartbeat cycle [19]. This segmentation makes it possible for a more detailed examination as well as the detection of distinct characteristics included within the ECG signal.
- **Determination of Variables**: At this stage, the segmented signals are analysed in order to identify and recognise the particular variables that make up those signals. The major strategies that are used in this stage are the maximum discovery strategy and the abrupt alterations method, both of which aid in locating specific events or shifts within the segmented ECG signals.

The PASCAL database19 served as the primary resource for the analysis used in the assessment of this design. This database included a large dataset that was made up of two distinct collections of heartbeat sound signals. Each collection was recorded using a unique kind of electronic stethoscope. Because there were so many different kinds of data sources, it was possible to conduct a thorough analysis of the system under a wide range of different situations, which strengthened the system's potential for use in real-world settings.

The ECG signals were first normalised after the wrapping, flattening, and homing operations were finished, and then they were segmented after that. It is important to point out that the Savitzky-Golay filter was used in this procedure for both of its intended objectives. In the first place, as a method for smoothing, it assisted in improving the intelligibility of more modest cardiac sound events [20]. Second, it was helpful in removing brief murmuring episodes when used as a homing process. This was the case regardless of whether the murmuring occurred during the systolic or diastolic portions of the heartbeat.

Using the PASCAL dataset, the usefulness of the suggested accordion technique, which includes both maximum change and optimum abrupt threshold identification, was put through a series of rigorous tests to determine whether or not it is effective. This dataset has been divided up into two separate sets, which have been labelled Set A and Set B. Set A includes 45 recordings of heart sounds taken from healthy people and 48 recordings taken from people with pathological disorders. Set B, on the other hand, is more comprehensive than Set A and includes 203 recordings of normal heartbeat sounds in addition to 187 recordings of abnormal heartbeat sounds resulting from pathological cardiac events [21]. The capabilities of the accordion approach as well as any possible limits it may have may be properly analysed with the help of this varied dataset since it offers a complete background for doing so. The acquired signal, when combined with the wrapped signal generated by the abrupt alteration's detection approach, offers a transparent illustration of both normal and abnormal cardiac sound signals. In two separate cases, individual cardiac cycles are shown using the raw ECG data [22]. These cycles have been split and recorded with extreme precision for the purpose of the procedure, which offers insights into both physiological (healthy) and pathological situations. During the process of actually examining the heart, there was a discernible difference in the two significant spikes that were present for both healthy and sick cardiac acoustic signals. This highlights the need for more thorough approaches that can examine the complete range of heart sounds in a holistic manner, since this would ensure proper interpretation and diagnosis [23].

The average and standard deviation of the acoustic data associated with the heartbeat were used to determine the characteristics of the signals that were being sent. Nevertheless, a visual evaluation of the ECG signal data tables played an essential part as well, delivering an extra layer of scrutiny to evaluate the efficacy of the segmentation methods. This was an essential part of the process. The method that was suggested shown higher performance when it came to categorising and recognising healthy cardiac sound signals in comparison to those with pathological problems [24]. This variation in performance might be due to the variable transmission structures of sick cardiac sounds, which often rely on the precise location of the illness and the degree of its severity.

The consistency of the technique across a variety of datasets offers more illuminating perspectives. The suggested method has a reliability of 72.25% when applied to Database A, which has an emphasis on enhanced cardiac acoustic situations. While on the other hand, the computing efficiency of Datasheet B is estimated to be 93.52%. However, it is essential to highlight that the findings linked to pathological diseases across both datasets are fairly unimpressive. This is something that has to be taken into consideration. In spite of these obstacles, the technique shown a significant amount of promise by accurately predicting the classification and identification of heartbeat signals with an accuracy of 81.45% for Dataset A and an astonishing 94.38% for Dataset B. Despite these hurdles, the method has shown a promising potential. This demonstrates the potential flexibility and competence of the method across a wide range of circumstances; yet, there is still space for development, especially in pathological settings.

3 Conclusion

The investigation into the field of ECG signal analysis has been exhaustive and educational in equal measure. The results of our study shed light on the many intricacies that are involved in the categorization and recognition of heartbeat sounds throughout the healthy and pathological spectrums. In particular, a greater comprehension of the signal dynamics was achieved by combining visual assessment of ECG data with computational approaches. This layered analysis was reflected in the findings, which showed a noteworthy computational efficiency of 91% in Datasheet B and a reliability rate of 70.97% in Database A for enhanced cardiac acoustic situations. Both of these figures were derived from the improved cardiac acoustic scenarios. The difficulties emerged when the method was applied to diseased situations, despite the fact that it shown a promising trajectory when applied to the sounds of a heart that was healthy. The inherent complexity and variety of sick cardiac sounds, which are dependent on a number of factors like the resonance location and the degree of the illness, created challenges in terms of establishing consistent accuracy. This was made clear by the performance gaps that existed between Datasets A and B, with Dataset B achieving a recognition accuracy of 95.29% compared to Dataset A's 83.01%. Even if the findings from Dataset B are encouraging, it is clear from the findings as a whole that there is a pressing need for more methodological development, particularly in terms of identifying and categorising pathological diseases.

References

1. Pandya S et al (2022) Infused heart: a novel knowledge-infused learning framework for diagnosis of cardiovascular events. IEEE Trans Comput Social Syst. https://doi.org/10.1109/TCSS.2022.3151643
2. G. Dangi, T. Choudhury and P. Kumar, "A smart approach to diagnose Heart disease through machine learning and Springleaf Marketing Response," 2016 International Conference on Recent Advances and Innovations in Engineering (ICRAIE), Jaipur, India, 2016, pp. 1–6, doi: https://doi.org/10.1109/ICRAIE.2016.7939547.
3. Srivastava A, Jain S, Miranda R, Patil S, Pandya S, Kotecha K (2021) Deep learning based respiratory sound analysis for detection of chronic obstructive pulmonary disease. PeerJ Computer Science 7:e369. https://doi.org/10.7717/peerj-cs.369

4. Sathyanarayanan S, Murthy S, Chitnis S (2023) A comprehensive survey of analysis of heart sounds using machine learning techniques to detect heart diseases. J Popul Therap Clin Pharmacol 30(11):375–384
5. Ghosh SK, Nagarajan PR, Tripathy RK (2020) Heart sound data acquisition and preprocessing techniques: a review. In: Handbook of research on advancements of artificial intelligence in healthcare engineering, pp 244–264
6. Choi S, Jiang Z (2010) Cardiac sound murmurs classification with autoregressive spectral analysis and multi-support vector machine technique. Comput Biol Med 40(1):8–20
7. Sharma S, Aggarwal A, Choudhury T (2018) Breast cancer detection using machine learning algorithms. In: 2018 International Conference on Computational Techniques, Electronics and Mechanical Systems (CTEMS), Belgaum, pp 114–118. https://doi.org/10.1109/CTEMS.2018.8769187
8. Plog SC (1990) A carpenter's tools: an answer to Stephen LJ Smith's review of psychocentrism/allocentrism. J Travel Res 28(4):43–45
9. Jones A et al (1998) Acoustic performance of three stethoscope chest pieces. In: Proceedings of the 20th Annual International Conference of the IEEE Engineering in Medicine and Biology Society, Biomedical Engineering Towards the Year 2000 and Beyond (Cat. No. 98CH36286), vol 6. IEEE
10. Waddell AN et al (2003) Intranasal steroid sprays in the treatment of rhinitis: is one better than another? J Laryngol Otol 117(11):843–845
11. Sharma A, Choudhury T, Kumar P (2018) Health monitoring and management using IoT devices in a cloud based framework. In: 2018 International Conference on Advances in Computing and Communication Engineering (ICACCE), Paris, pp 219–224. https://doi.org/10.1109/ICACCE.2018.8441752
12. Dispenzieri A et al (2015) Treatment of immunoglobulin light chain amyloidosis: mayo stratification of myeloma and risk-adapted therapy (mSMART) consensus statement. Mayo Clin Proc 90(8):1054–1081
13. Shippee ND et al (2015) Patient and service user engagement in research: a systematic review and synthesized framework. Health Expect 18(5):1151–1166
14. Zhang J, Yi Lu (2018) Biocomputing for portable, resettable, and quantitative point-of-care diagnostics: making the glucose meter a logic-gate responsive device for measuring many clinically relevant targets. Angew Chem Int Ed 57(31):9702–9706
15. Bhoi AK, Sherpa KS, Khandelwal B (2015) Multidimensional analytical study of heart sounds: a review. Int J Bioautom 19(3):351
16. Khalili F, Taebi A (2021) Advances in computational fluid dynamics modeling of cardiac sounds as a non-invasive diagnosis method. In: ASME International Mechanical Engineering Congress and Exposition, vol 85598. American Society of Mechanical Engineers
17. Tavares C et al (2011) Computational tools for assessing cardiovascular variability. In: 1st Portuguese Biomedical Engineering Meeting. IEEE
18. Mandal D, Maity A, Misra IS (2019) Low-cost portable solution for real-time complete detection and analysis of heart sound components. Wireless Pers Commun 107:523–547
19. Thiyagaraja SR et al (2018) A novel heart-mobile interface for detection and classification of heart sounds. Biomed Signal Process Control 45:313–324
20. Walambe R, Marathe A, Kotecha K (2021) Multiscale object detection from drone imagery using ensemble transfer learning. Drones 5:66. https://doi.org/10.3390/drones5030066
21. Das R, Turkoglu I, Sengur A (2009) Diagnosis of valvular heart disease through neural networks ensembles. Comput Methods Programs Biomed 93(2):185–191
22. Choudrie J, Patil S, Kotecha K et al (2021) Applying and understanding an advanced, novel deep learning approach: a Covid 19, text based emotions analysis study. Inf Syst Front 23:1431–1465. https://doi.org/10.1007/s10796-021-10152-6
23. Gupta V (2023) Wavelet transform and vector machines as emerging tools for computational medicine. J Ambient Intell Humaniz Comput 14(4):4595–4605
24. Gahane A, Kotadi C (2022) An analytical review of heart failure detection based on IoT and machine learning. In: 2022 Second International Conference on Artificial Intelligence and Smart Energy (ICAIS). IEEE

Sentiment Analysis in Social Media Marketing: Leveraging Natural Language Processing for Customer Insights

Kamred Udham Singh, Ankit Kumar, Gaurav Kumar, Tanupriya Choudhury, Teekam Singh, and Ketan Kotecha

1 Introduction

Businesses are always looking for new and inventive methods to engage with their target demographic, cultivate customer loyalty, and remain one step ahead of their rivals in the fast-paced, digital landscape of current marketing. The social media landscape, with its pervasive presence and the large troves of user-generated material, has emerged as a crucial venue for the accomplishment of these goals [1]. Not only

K. U. Singh (✉)
School of Computing, Graphic Era Hill University, Dehradun, India
e-mail: falsekamredudhamsingh@gmail.com

A. Kumar · G. Kumar
Department of Computer Engineering and Applications, GLA University, Mathura, Uttar Pradesh, India
e-mail: kumar.ankit@gla.ac.in

G. Kumar
e-mail: gaurav.kumar_phd.ca21@gla.ac.in

T. Choudhury (✉)
CSE Department, Symbiosis Institute of Technology, Symbiosis International University, Lavale Campus, Pune 411045, India
e-mail: falsetanupriya1986@gmail.com

T. Singh
Department of Computer Science and Engineering, Graphic Era Deemed to Be University, Dehradun, India
e-mail: tsingh@ma.iitr.ac.in

K. Kotecha
Symbiosis Centre for Applied Artificial Intelligence, Symbiosis Institute of Technology, Symbiosis International University, Pune 411045, India
e-mail: director@sitpune.edu.in

M. S. Kaiser et al. (eds.), *ICT: Smart Systems and Technologies*, Lecture Notes in Networks and Systems 878, https://doi.org/10.1007/978-981-99-9489-2_40

do the thoughts, feelings, and sentiments that users share on social media platforms represent their own experiences, but they also offer a treasure mine of knowledge for astute marketers to glean from these interactions. In this context, sentiment analysis has emerged as a valuable tool for unlocking the dormant potential contained within social media data. This is made possible by the advancements that have been made in natural language processing (NLP). A area of natural language processing (NLP) that makes use of computer methods to identify, categories, and quantify sentiment in written material is called sentiment analysis. This subfield is also sometimes referred to as opinion mining. It enables marketers to go deeper than the surface level of interactions with their audience and grasp the underlying attitudes and feelings that lie behind the surface [2]. Since the development of complex natural language processing (NLP) algorithms and machine learning models, sentiment analysis has progressed from being only a curiosity to being a strategic need for companies that want to flourish in the digital age. In the context of social media marketing, this article will take you on a tour through the complex environment of sentiment analysis. This article examines the plethora of ways in which companies may tap into the power of NLP in order to obtain useful insights into the views and feelings of their customers. This paper gives a full review of the topic, from the principles that drive sentiment analysis to the actual applications in marketing that can be found throughout the study. When it comes to social media marketing, the significance of sentiment analysis simply cannot be understated [3]. It provides organisations with the means to interpret the feelings of their customers, monitor the public's impression of their brands, evaluate the effectiveness of marketing initiatives, and spot developing tendencies and problems in real time.

In sentiment analysis has the potential to play a pivotal role in the management of reputations, the improvement of customer service, and the creation of new products. Nevertheless, it is not devoid of any of its complexity. The data from social media platforms tend to be loud and unstructured, and they also include subtleties such as sarcasm and context-dependent sentiment, all of which provide substantial hurdles for sentiment analysis. In addition, there are moral aspects to take into account, such as concerns over privacy and biases introduced by algorithms, which highlight the need of responsible and conscientious implementation. This article [4] not only examines the theoretical underpinnings of sentiment analysis, but it also includes practical case studies that showcase the transformational influence that sentiment analysis can have on marketing tactics. This will help readers traverse the ever-changing marketplace. It encourages marketers to investigate the potential of natural language processing (NLP) and sentiment analysis (SA), highlighting the importance of these techniques in promoting data-driven decision-making and maintaining a competitive edge in the rapidly developing field of digital marketing. Businesses who are able to read the attitudes and emotions of their audience on social media have a significant advantage over their competitors in this age of abundant information [5]. The purpose of this study is to shed light on the potential, obstacles, and ethical issues that are inherent in sentiment analysis, and eventually provide a road map for using the power of natural language processing to reveal vital consumer insights inside the arena of social media marketing [6].

1.1 Understanding Sentiment Analysis with NLP

Sentiment analysis, also known as natural language processing (NLP), is a rapidly developing and ever-changing topic that has significant ramifications for comprehending the thoughts and feelings expressed by customers on social media platforms [7]. This kind of data analysis, which is also known as opinion mining, combines the capabilities of natural language processing (NLP) with machine learning in order to extract and evaluate sentiment from textual data provided by users on platforms such as Twitter, Facebook, Instagram, and online review sites. In the course of this in-depth conversation, we are going to investigate how employing NLP for sentiment analysis may provide useful insights into the feelings of customers, as well as how it operates, the applications it has, and the obstacles it encounters.

The Process Behind Sentiment Analysis

A mix of linguistic, statistical, and machine learning approaches are used in the process of sentiment analysis in order to identify the attitude conveyed by a piece of written text [8]. The following stages are involved in NLP-based sentiment analysis at its most fundamental level:

- Text Preprocessing: The raw text data is cleaned and processed to eliminate noise, which may contain special characters, emojis, and other information that is not important to the task at hand.
- Tokenization: It is the process of dissecting a text into its component words or phrases, which enables a more detailed level of analysis.
- Extraction of Features: Natural language processing methods are used in this step to turn text into numerical features that can then be processed by machine learning models. Methods such as word embedding and Term Frequency-Inverse Document Frequency (TF-IDF) are examples of these approaches.

In machine learning models, such as Support Vector Machines, Naive Bayes, or neural networks like Long Short-Term Memory (LSTM), are trained on labelled datasets to classify the sentiment of each piece of text as either positive, negative, or neutral [9]. This allows the models to determine if a piece of text contains positive, negative, or neutral information. The data are then analysed, and often visualised, in order to get insights into the general sentiment patterns that have emerged [10].

1.2 Problems that Can Arise When Performing Sentiment Analysis

Although NLP-based sentiment analysis has a tremendous deal of potential, it also has to contend with a number of obstacles. Ambiguity and context play a huge role in determining the meaning of words and phrases, since the context in which they are employed may drastically alter that meaning [12]. The ability for sentiment analysis algorithms to comprehend context is a tough task. Because sarcasm and irony are often communicated via oblique verbal clues, sentiment analysis has a difficult time identifying them in written communication. Multilingual Analysis: Due to the fact that social media is a worldwide platform, analysis of user sentiment must take into consideration a variety of languages and dialects. Emojis and Emoticons: The usage of emojis and emoticons may substantially alter the emotion of a communication, yet it can be difficult to precisely discern what is being said via their use [13]. In many datasets, one sentiment class (for example, neutral) may dominate, which may lead to unbalanced training data and bias in the model. Neutrality may be the most common sentiment class.

2 Related Work

Analysis of sentiment, often called opinion mining, is a technique that use computer approaches to recognise and extract sentiments from written material. This mostly assesses how individuals feel about certain goods, services, or subjects. Over the course of time, a significant amount of research has been concentrated on enhancing its accuracy via the use of natural language processing (NLP). For illustration purposes [14] used data from Twitter to classify tweets as either good, negative, or neutral based on certain language characteristics.

As the use of social media platforms increased, it became clearer that these sites are rich resources for gaining a knowledge of how customers feel. Twitter was analysed by [15] in order to get a better understanding of public sentiment and its impact on financial markets. Additionally, companies began to realise the value of sentiment research when it came to the process of developing marketing strategies. Fernández-Gavilanes et al. [16] shown that analysing comments left by customers online may provide organisations with a competitive advantage.

The most recent developments in NLP, most notably deep learning, have significantly improved the capabilities of sentiment analysis. It was the ground-breaking work of [17] using recursive deep models that marked major advancement in this field. In the end, the use of natural language processing (NLP) to gather consumer insights from social media rests on a foundation of considerable study and technological advancement, with the goal of capturing the plurality of voices online for the purpose of gaining strategic insights.

3 Methodology

The research methodology for the study named incorporates a methodical strategy to collect, analyse, and evaluate data pertaining to sentiment analysis in the context of social media marketing. The study was conducted using Leveraging Natural

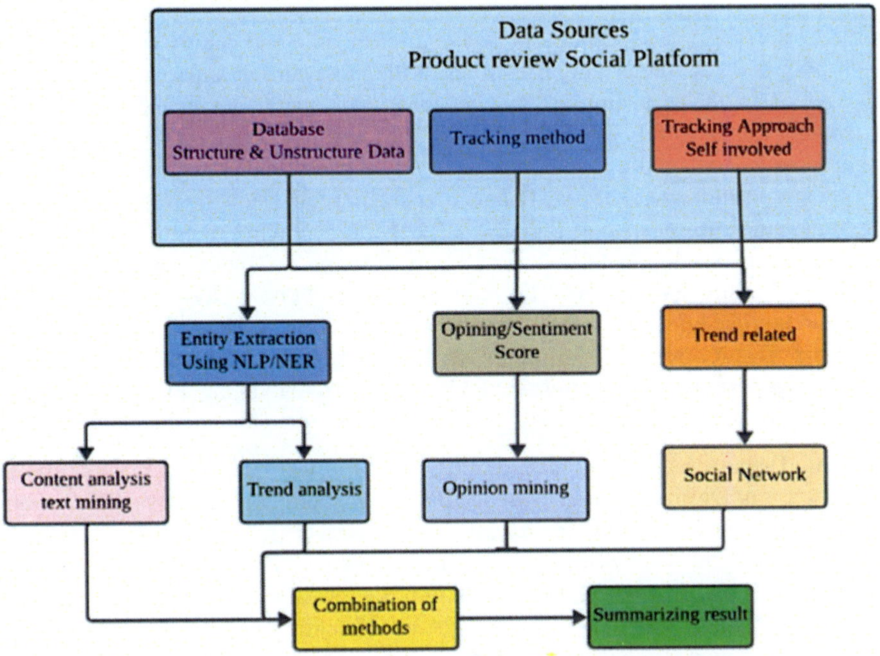

Fig. 1 The use of natural language processing in social media to determine product features

Language Processing (NLP) for Customer Insights. NLP in social media in the form of product features in Fig. 1.

The following is a comprehensive summary of the research methodology:

3.1 Research Design

Exploratory research is the first stage of the research process and lays the groundwork for learning more about complex topics like sentiment analysis, NLP, and social media marketing.

In start a thorough literature evaluation of sentiment analysis, NLP, and social media advertising. Scholarly articles, industrial reports, academic magazines, and books in the field are all carefully examined. By doing so, they are able to incorporate the knowledge of several authorities into their own work and ensure that their studies are founded on well-established theories, methods, and empirical results. The exploratory phase provides researchers with a solid theoretical basis by delving into the vast body of prior research to identify knowledge gaps and promising topics for future study.

Research into the theoretical underpinnings of sentiment analysis, as well as NLP techniques and their evolving function in the realm of social media marketing, is part of the exploratory phase. Scientists devote a lot of time and energy to deciphering everything from the underlying principles to the most current and ground-breaking innovations in these disciplines. In the context of social media marketing, this conceptual inquiry offers the framework for setting research questions, establishing hypotheses, and designing a methodological approach that matches perfectly with the intricacies of sentiment analysis. To conceptual comprehension, the researchers also investigate the current state of the relevant technology environment. They examine the most recent developments in terms of tools, systems, and innovations in the fields of NLP and sentiment analysis. This technical savvy not only guides their selection of research procedures, but it also places the investigation in the vanguard of current academic advancement [18]. It assures that the study will make use of the latest cutting-edge tools in order to mine valuable insights from the immense ocean of data gleaned from social media.

A. **Data Procurement**

- **Sources of Data:** In order to conduct this study, a complete data gathering procedure was carried out by tapping into a wide variety of social media networks. This all-encompassing strategy takes into account well-known sites like Twitter, Facebook, and Instagram, in addition to professional networks like LinkedIn.
- **Types of Data:** The data that were obtained for this study display a variety of features, which is reflective of the many different kinds of information that can be found in academic journals and social media platforms. The vast majority of it is made up of textual information taken from posts, comments, and reviews on various social media platforms. Because these textual examples serve as a rich bank of sentiments, expressions, and viewpoints, it is feasible to carry out an in-depth analysis of the manner in which users feel about a certain topic [19].

B. **Data Preprocessing:**

- **Data Cleansing:** A thorough data cleaning procedure is an essential first stage in the research process. This phase takes a lot of time and effort. During this stage, a concerted effort is undertaken to get rid of any unnecessary noise that might potentially skew the results of the upcoming study.
- **Tokenization:** Following the completion of the data purification phase, the textual data passes through the tokenization procedure. This requires severing the continuous flow of text into distinct components such as individual words, phrases, or even sentences [20].
- **Feature Engineering:** Word embeddings and Term Frequency-Inverse Document Frequency (TF-IDF) are two examples of the sophisticated NLP approaches that are used in order to make it possible to conduct an efficient analysis of the textual data.

C. **Sentiment Analysis**

It investigates a wide variety of methodologies, including as lexicon-based approaches, machine learning methods, and deep learning algorithms, among others. Lexicon-based techniques depend on established sentiment dictionaries to give sentiment scores to words and phrases, while machine learning methods use statistical

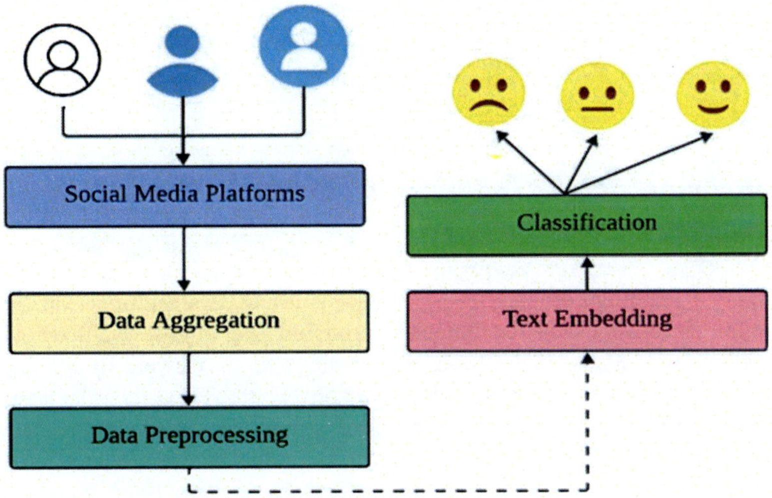

Fig. 2 Steps for analysing the mood of social media posts

models to categorise sentiments based on training data. Lexicon-based approaches are more accurate than machine learning methods.

The models are given access to these datasets, which include instances of text along with accompanying sentiment labels. This enables the models to understand the correlations that exist between linguistic variables and sentiments. Training includes activities such as feature extraction, model parameter adjustment, and optimisation, and its purpose is to guarantee that the models are capable of accurately categorising emotions based on the content of textual documents. The goal of this iterative approach is to acquire a high level of accuracy and reliability in the categorization of sentiments [21].

The effectiveness of algorithms for sentiment analysis is evaluated in a methodical manner utilising a set of evaluation criteria that has been clearly specified. A quantifiable measurement of the performance of the model may be obtained by using these measures, which include accuracy, precision, recall, and $F1$-score. Precision evaluates the capability of the model to avoid producing false positives, while accuracy estimates the percentage of attitudes that are properly identified (Fig. 2).

4 Case Study

Case studies provide tangible proof of the real-world impact that sentiment analysis in social media marketing has, proving the transforming influence that it has on marketing tactics. Case studies are going to be the main topic of our conversation. In this piece of writing, we will look into numerous famous cases that illustrate the power of natural language processing (NLP) in gleaning client insights in order to strengthen marketing tactics [22].

4.1 Streaming Service Netflix Is Improving Its Content Recommendations

The leading streaming company in the world, Netflix, uses sentiment analysis to improve the effectiveness of its content recommendation engine. Netflix obtains a profound comprehension of the tastes and feelings of its viewers by reading and analysing the comments, reviews, and online conversations that its users post [23]. This strategy, which is powered by data, makes it possible to provide personalised content recommendations that are specific to individual viewers, which increases engagement and customer loyalty. Sentiment analysis not only helps in the process of suggesting material, but it can also be used to guide the conception of new television shows and films that connect with the feelings and preferences of viewers.

4.2 Airbnb: Further Development of the User Experience

Airbnb makes use of sentiment analysis in order to provide a user experience that is unrivalled and to increase levels of customer happiness. Airbnb is able to get priceless information about how visitors feel about their lodgings and the interactions they have with their hosts by carefully analysing guest reviews and other online interactions. This method, which is centred on user input, gives Airbnb the ability to zero in on areas for improvement, hone down on property listings, and provide hosts with advice on how to provide memorable experiences for their guests.

The aforementioned case studies highlight the discernible influence that sentiment research has on marketing strategy. Companies like Netflix and Airbnb are able to not only comprehend the feelings of their consumers but also adjust their service and content offerings to correspond with these sentiments by utilising the power of natural language processing (NLP). This results in increased user engagement, contentment, and ultimately, the success of the companies themselves.

5 Result and Analysis

The exploration of the field of sentiment analysis in social media marketing, which was made possible by the capabilities of natural language processing (NLP), has resulted in the discovery of significant results and useful conclusions that highlight the revolutionary potential of the field. In this section, we condense these realisations into two essential features, namely the inherent worth of sentiment analysis and the influence that it has on marketing plans.

5.1 Sentiment Analysis' Inherent Usefulness

The results of this study provide more evidence that sentiment analysis is not only a technical advancement but rather a strategic need for modern-day companies doing business in the digital age. Organisations may get priceless insights into the emotional reactions, preferences, and views of their audience by identifying and measuring the feelings expressed by customers across social media platforms.

5.2 Effect on Advertising Methods

This study highlights the transformational influence that sentiment analysis can have on marketing tactics. The strategy, which is powered by technology, reshapes the fundamental core of marketing by providing firms with the ability to communicate with their audience in a way that is more personalised and relevant. To begin, sentiment analysis may be used to direct the generation of content and the message used in marketing so that it is in line with the predominant sentiment of the audience that is being targeted. It helps to identify the most successful communication channels and pinpoint the best time, which both increase the likelihood of the material hitting a chord with the population that is being targeted. When it comes to monitoring how people feel about a business in real time, sentiment analysis is an extremely important tool.

6 Conclusion

In today's dynamic social media landscape, understanding the emotions of online users is pivotal. Sentiment analysis, powered by natural language processing (NLP), offers unparalleled insights into customer sentiments. This isn't just a technical leap but the core of modern marketing, providing a real-time tool for companies to resonate with customer emotions. This method decodes vast human emotions across social platforms, offering businesses a real-time guide to understand and quantify their audience's feelings and preferences. It's a transformative strategy where marketing decisions are anchored not just on gut feeling but actual data from digital conversations. This approach shines a light in the digital maze, helping businesses navigate the Internet waters to reach a more customer-focused approach. It revolutionises marketing by facilitating genuine connections with audiences, letting businesses craft messages aligned with prevailing sentiments. Real-time insights from sentiment analysis shield a company's reputation and inform product development, enhancing customer satisfaction and loyalty. As the digital age progresses, sentiment analysis, backed by NLP, becomes essential for businesses in competitive social media marketing. It's a blend of emotional understanding and tech prowess, forming empathetic marketing techniques. Essentially, sentiment analysis is the essence of modern marketing, positioning businesses at the forefront in the digital age.

References

1. Gabrilovich E (2011) Ad retrieval systems in vitro and in vivo: knowledge-based approaches to computational advertising. In: Lecture Notes in Computer Science (including subseries Lecture Notes in Artificial Intelligence and Lecture Notes in Bioinformatics)
2. Singh L, Gautam K (2021) An attempt on Twitter 'likes' grading strategy using pure linguistic feature engineering: a novel approach. In: Lecture Notes in Electrical Engineering
3. Vikas K et al (2022) User gender classification based on Twitter Profile Using machine learning. In: Proceedings—International Conference on Applied Artificial Intelligence and Computing, ICAAIC 2022
4. Bhanage DA, Pawar AV, Kotecha K, Abrahim A (2023) Failure detection using semantic analysis and attention-based classifier model for IT Infrastructure log data. IEEE Access. https://doi.org/10.1109/ACCESS.2023.3319438
5. Suresh S, Rajan GTS, Gopinath V (2018) VOC-DL: revisiting voice of customer using deep learning. In: 32nd AAAI Conference on Artificial Intelligence, AAAI 2018
6. Sumathi N, Sheela T (2017) An empirical study on analyzing the distortion detection on OSN using NLP and SA in banking institution. In: Proceedings of the 2017 2nd International Conference on Computing and Communications Technologies, ICCCT 2017
7. Ramaswamy S, DeClerck N (2018) Customer perception analysis using deep learning and NLP. Proc Comput Sci 140:170–178
8. Kusal S, Patil S, Choudrie J et al (2023) A systematic review of applications of natural language processing and future challenges with special emphasis in text-based emotion detection. Artif Intell Rev 56:15129–15215. https://doi.org/10.1007/s10462-023-10509-0
9. Mamgain S, Balabantaray RC, Das AK (2019) Author profiling: prediction of gender and language variety from document. In: Proceedings—2019 International Conference on Information Technology, ICIT 2019
10. Mackey TK et al (2020) Big data, natural language processing, and deep learning to detect and characterize illicit COVID-19 product sales: infoveillance study on Twitter and Instagram. JMIR Public Health Surveill 6(3):e20794
11. Nahili W, Rezeg K (2018) Digital marketing with social media: what Twitter Says! In: Proceedings—PAIS 2018: international conference on pattern analysis and intelligent systems
12. Nezhad ZB, Deihimi MA (2019) A combined deep learning model for Persian sentiment analysis. IIUM Eng J 20(1):129–139
13. Kushwaha AK, Kar AK (2020) Language model-driven Chatbot for business to address marketing and selection of products. In: IFIP Advances in Information and Communication Technology
14. Agarwal A, Gupta S, Choudhury T (2018) Continuous and integrated software development using DevOps. In: 2018 International Conference on Advances in Computing and Communication Engineering (ICACCE), Paris, pp 290–293. https://doi.org/10.1109/ICACCE.2018.8458052
15. Khursheed B et al (2020) Microtargeting or Microphishing? Phishing unveiled. In: Lecture Notes in Computer Science (including subseries Lecture Notes in Artificial Intelligence and Lecture Notes in Bioinformatics)

16. Fernández-Gavilanes M et al (2019) Differentiating users by language and location estimation in sentiment analisys of informal text during major public events. Expert Syst Appl 117:15–28
17. Djebbi MA, Ouersighni R (2022) TunTap: a Tunisian dataset for topic and polarity extraction in social media. In: Lecture Notes in Computer Science (including subseries Lecture Notes in Artificial Intelligence and Lecture Notes in Bioinformatics)
18. Hameed RA, Abed WJ, Sadiq AT (2023) Evaluation of hotel performance with sentiment analysis by deep learning techniques. Int J Interact Mobile Technol 17(9):70–87
19. Farzindar A, Inkpen D (2018) Natural language processing for social media, Second Edition. In: Synthesis Lectures on Human Language Technologies, vol 10(2), p 1–197
20. Chandel K, Kunwar V, Sabitha S et al (2016) A comparative study on thyroid disease detection using K-nearest neighbor and Naive Bayes classification techniques. CSIT 4:313–319. https://doi.org/10.1007/s40012-016-0100-5
21. Kansal T, Bahuguna S, Singh V, Choudhury T (2018) Customer segmentation using K-means clustering. In: 2018 International Conference on Computational Techniques, Electronics and Mechanical Systems (CTEMS), Belgaum, India, pp 135–139. https://doi.org/10.1109/CTEMS.2018.8769171
22. Chathumali EJAPC, Thelijjagoda S (2020) Detecting human emotions on Facebook comments. In: Proceedings—International Research Conference on Smart Computing and Systems Engineering, SCSE 2020
23. Siraskar R, Kumar S, Patil S et al (2023) Reinforcement learning for predictive maintenance: a systematic technical review. Artif Intell Rev 56:12885–12947. https://doi.org/10.1007/s10462-023-10468-6

Author Index

M. S. Kaiser et al. (eds.), *ICT: Smart Systems and Technologies*, Lecture Notes in
Networks and Systems 878, https://doi.org/10.1007/978-981-99-9489-2